Climate Change 2014
Impacts, Adaptation, and Vulnerability

Summaries, Frequently Asked Questions, and Cross-Chapter Boxes

A Working Group II Contribution to the
Fifth Assessment Report of the
Intergovernmental Panel on Climate Change

Edited by

Christopher B. Field
Working Group II Co-Chair
Department of Global Ecology
Carnegie Institution for Science

Vicente R. Barros
Working Group II Co-Chair
Centro de Investigaciones del Mar y la Atmósfera
Universidad de Buenos Aires

David Jon Dokken
Executive Director

Katharine J. Mach
Co-Director of Science

Michael D. Mastrandrea
Co-Director of Science

T. Eren Bilir **Monalisa Chatterjee** **Kristie L. Ebi** **Yuka Otsuki Estrada** **Robert C. Genova** **Betelhem Girma**

Eric S. Kissel **Andrew N. Levy** **Sandy MacCracken** **Patricia R. Mastrandrea** **Leslie L. White**

Working Group II Technical Support Unit

ISBN 978-92-9169-141-8

Use the following reference to cite this publication:

IPCC, 2014: *Climate Change 2014: Impacts, Adaptation, and Vulnerability. Summaries, Frequently Asked Questions, and Cross-Chapter Boxes. A Contribution of Working Group II to the Fifth Assessment Report of the Intergovernmental Panel on Climate Change* [Field, C.B., V.R. Barros, D.J. Dokken, K.J. Mach, M.D. Mastrandrea, T.E. Bilir, M. Chatterjee, K.L. Ebi, Y.O. Estrada, R.C. Genova, B. Girma, E.S. Kissel, A.N. Levy, S. MacCracken,P.R. Mastrandrea, and L.L. White (eds.)]. World Meteorological Organization, Geneva, Switzerland, 190 pp. (in Arabic, Chinese, English, French, Russian, and Spanish)

Cover Photo:

Planting of mangrove seedlings in Funafala, Funafuti Atoll, Tuvalu. © David J. Wilson

Contents

Foreword, Preface, and Dedication

Foreword

Climate Change 2014: Impacts, Adaptation, and Vulnerability is the second volume of the Fifth Assessment Report (AR5) of the Intergovernmental Panel on Climate Change (IPCC) — *Climate Change 2013/2014* — and was prepared by its Working Group II. The volume focuses on why climate change matters and is organized into two parts, devoted respectively to human and natural systems and regional aspects, incorporating results from the reports of Working Groups I and III. The volume addresses impacts that have already occurred and risks of future impacts, especially the way those risks change with the amount of climate change that occurs and with investments in adaptation to climate changes that cannot be avoided. For both past and future impacts, a core focus of the assessment is characterizing knowledge about vulnerability, the characteristics and interactions that make some events devastating, while others pass with little notice.

Three elements are new in this assessment. Each contributes to a richer, more nuanced understanding of climate change in its real-world context. The first new element is a major expansion of the topics covered in the assessment. In moving from 20 chapters in the AR4 to 30 in the AR5, the Working Group II assessment makes it clear that expanding knowledge about climate change and its impacts mandates attention to more sectors, including sectors related to human security, livelihoods, and the oceans. The second new element is a pervasive focus on risk, where risk captures the combination of uncertain outcomes and something of value at stake. A framing based on risk provides a framework for utilizing information on the full range of possible outcomes, including not only most likely outcomes but also low probability but high consequence events. The third new element is solid grounding in the evidence that impacts of climate change typically involve a number of interacting factors, with climate change adding new dimensions and complications. The implication is that understanding the impacts of climate change requires a very broad perspective.

The IPCC was established by the World Meteorological Organization (WMO) and the United Nations Environment Programme (UNEP) in 1988, with the mandate to provide the world community with the most up-to-date and comprehensive scientific, technical, and socio-economic information about climate change. The IPCC assessments have since then played a major role in motivating governments to adopt and implement policies in responding to climate change, including the United Nations Framework Convention on Climate Change and the Kyoto Protocol. IPCC's AR5 provides an important foundation of information for the world's policymakers, to help them respond to the challenge of climate change.

The *Impacts, Adaptation, and Vulnerability* report was made possible thanks to the commitment and voluntary labor of a large number of leading scientists. We would like to express our gratitude to all Coordinating Lead Authors, Lead Authors, Contributing Authors, Review Editors, and Reviewers. We would also like to thank the staff of the Working Group II Technical Support Unit and the IPCC Secretariat for their dedication in organizing the production of a very successful IPCC report. Furthermore, we would like to express our thanks to Dr. Rajendra K. Pachauri, Chairman of the IPCC, for his patient and constant guidance through the process, and to Drs. Vicente Barros and Chris Field, Co-Chairs of Working Group II, for their skillful leadership. We also wish to acknowledge and thank those governments and institutions that contributed to the IPCC Trust Fund and supported the participation of their resident scientists in the IPCC process. We would like to mention in particular the Government of the United States of America, which funded the Technical Support Unit; the Government of Japan, which hosted the plenary session for the approval of the report; and the Governments of Japan, United States of America, Argentina, and Slovenia, which hosted the drafting sessions to prepare the report.

M. Jarraud
Secretary-General
World Meteorological Organization

A. Steiner
Executive Director
United Nations Environment Programme

Preface

The Working Group II contribution to the Fifth Assessment Report of the Intergovernmental Panel on Climate Change (IPCC WGII AR5) considers climate change impacts, adaptation, and vulnerability. It provides a comprehensive, up-to-date picture of the current state of knowledge and level of certainty, based on the available scientific, technical, and socio-economic literature. As with all IPCC products, the report is the result of an assessment process designed to highlight both big-picture messages and key details, to integrate knowledge from diverse disciplines, to evaluate the strength of evidence underlying findings, and to identify topics where understanding is incomplete. The focus of the assessment is providing information to support good decisions by stakeholders at all levels. The assessment is a unique source of background for decision support, while scrupulously avoiding advocacy for particular policy options.

Scope of the Report

Climate change impacts, adaptation, and vulnerability span a vast range of topics. With the deepening of knowledge about climate change, we see connections in expanding and diverse areas, activities, and assets at risk. Early research focused on direct impacts of temperature and rainfall on humans, crops, and wild plants and animals. New evidence points to the importance of understanding not only these direct impacts but also potential indirect impacts, including impacts that can be transmitted around the world through trade, travel, and security. As a consequence, few aspects of the human endeavor or of natural ecosystem processes are isolated from possible impacts in a changing climate. The interconnectedness of the Earth system makes it impossible to draw a confined boundary around climate change impacts, adaptation, and vulnerability. This report does not attempt to bound the issue. Instead, it focuses on core elements and identifies connecting points where the issue of climate change overlaps with or merges into other issues.

The integrative nature of the climate change issue underlies three major new elements of the WGII contribution to the AR5. The first is explicit coverage of a larger range of topics, with new chapters. Increasing knowledge, expressed in a rapidly growing corpus of published literature, enables deeper assessment in a number of areas. Some of these are geographic, especially the addition of two chapters on oceans. Other new chapters further develop topics covered in earlier assessments, reflecting the increased sophistication of the available research. Expanded coverage of human settlements, security, and livelihoods builds on new research concerning human dimensions of climate change. A large increase in the published literature on adaptation motivates assessment in a suite of chapters.

A second new emphasis is the focus on climate change as a challenge in managing and reducing risk, as well as capitalizing on opportunities. There are several advantages to understanding the risk of impacts from climate change as resulting from the overlap of hazards from the physical climate and the vulnerability and exposure of people, ecosystems, and assets. Some of the advantages accrue from the opportunity to evaluate factors that regulate each component of risk. Others relate to the way that a focus on risk can clarify bridges to solutions. A focus on risk can link historical experience with future projections. It helps integrate the role of extremes. And it highlights the importance of considering the full range of possible outcomes, while opening the door to a range of tools relevant to decision making under uncertainty.

A third new emphasis ties together the interconnectedness of climate change with a focus on risk. Risks of climate change unfold in environments with many interacting processes and stressors. Often, climate change acts mainly through adding new dimensions and complications to sometimes longstanding challenges. Appreciating the multi-stressor context of the risks of climate change can open doors to new insights and approaches for solutions.

Increased knowledge of the risks of climate change can be a starting point for understanding the opportunities for and implications of possible solutions. Some of the solution space is in the domain of mitigation, extensively covered by the Working Group III contribution to the AR5. The WGII AR5 delves deep into adaptation. But many opportunities exist in linking climate change adaptation, mitigation, and sustainable development. In contrast to past literature that tended to characterize adaptation, mitigation, and sustainable development as competing agendas, new literature identifies complementarities. It shines light on options for leveraging investments in managing and reducing the risks of climate change to enable vibrant communities, robust economies, and healthy ecosystems, in all parts of the world.

Structure of the Report

The Working Group II contribution to the IPCC Fifth Assessment Report consists of a brief summary for policymakers, a longer technical summary, and 30 thematic chapters, plus supporting annexes. A series of cross-chapter boxes and a collection of Frequently Asked Questions provide an integrated perspective on selected key issues. Electronic versions of all the printed contents, plus supplemental online material, are available at no charge at www.ipcc.ch.

The report is published in two parts. Part A covers global-scale topics for a wide range of sectors, covering physical, biological, and human systems. Part B considers the same topics, but from a regional perspective, exploring the issues that arise from the juxtaposition of climate change, environment, and available resources. Conceptually, there is some overlap between the material in Parts A and B, but the contrast in framing makes each part uniquely relevant to a particular group of stakeholders. For setting context and meeting the needs of users focused on regional-scale issues, Part B extracts selected materials from the Working Group I and Working Group III contributions to the Fifth Assessment Report. To acknowledge the different purposes for the two parts and the balanced contributions of the co-chairs, the listing order of the editors differs between the two parts, with Chris Field listed first on Part A and Vicente Barros listed first on Part B.

The 20 chapters in Part A are arranged in six thematic groups.

Context for the AR5

The two chapters in this group, (1) Point of departure and (2) Foundations for decision making, briefly summarize the conclusions of the Fourth Assessment Report and the Working Group I contribution to the AR5. They explain the motivation for the focus on climate change as a challenge in managing and reducing risks and assess the relevance of diverse approaches to decision making in the context of climate change.

Natural and Managed Resources and Systems, and Their Uses

The five chapters in this group, (3) Freshwater resources, (4) Terrestrial and inland water systems, (5) Coastal systems and low-lying areas, (6) Ocean systems, and (7) Food security and food production systems, cover diverse sectors, with a new emphasis on resource security. The ocean systems chapter, focused on the processes at work in ocean ecosystems, is a major element of the increased coverage of oceans in the WGII AR5.

Human Settlements, Industry, and Infrastructure

The three chapters in this group, (8) Urban areas, (9) Rural areas, and (10) Key economic sectors and services, provide expanded coverage of settlements and economic activity. With so many people living in and moving to cities, urban areas are increasingly important in understanding the climate change issue.

Human Health, Well-Being, and Security

The three chapters in this group, (11) Human health: impacts, adaptation, and co-benefits, (12) Human security, and (13) Livelihoods and poverty, increase the focus on people. These chapters address a wide range of processes, from vector-borne disease through conflict and migration. They assess the relevance of local and traditional knowledge.

Adaptation

An expanded treatment of adaptation is one of the signature changes in the WGII AR5. Chapters treat (14) Adaptation needs and options, (15) Adaption planning and implementation, (16) Adaptation opportunities, constraints, and limits, and (17) Economics of adaptation. This coverage reflects a large increase in literature and the emergence of climate-change adaptation plans in many countries and concrete action in some.

Multi-Sector Impacts, Risks, Vulnerabilities, and Opportunities

The three chapters in this group, (18) Detection and attribution of observed impacts, (19) Emergent risks and key vulnerabilities, and (20) Climate-resilient pathways: adaptation, mitigation, and sustainable development, collect material from the chapters in both Parts A and B to provide a sharp focus on aspects of climate change that emerge only by examining many examples across the regions of the Earth and the entirety of the human endeavor. These chapters provide an integrative view of three central questions related to understanding risks in a changing climate – what are the impacts to date (and how certain is the link to climate change), what are the most important risks looking forward, and what are the opportunities for linking responses to climate change with other societal goals.

The 10 chapters in Part B start with a chapter, (21) Regional context, structured to help readers understand and capitalize on regional information. It is followed by chapters on 9 world regions: (22) Africa, (23) Europe, (24) Asia, (25) Australasia, (26) North America, (27) Central and South America, (28) Polar regions, (29) Small islands, and (30) The ocean (taking a regional cut through ocean issues, including human utilization of ocean resources). Each chapter in this part is an all-in-one resource for regional stakeholders, while also contributing to and building from the global assessment. Regional climate-change maps, which complement the Working Group I Atlas of Global and Regional Climate Projections, and quantified key regional risks are highlights of these chapters. Each chapter explores the issues and themes that are most relevant in the region.

Process

The Working Group II contribution to the IPCC Fifth Assessment Report was prepared in accordance with the procedures of the IPCC. Chapter outlines were discussed and defined at a scoping meeting in Venice in July 2009, and outlines for the three Working Group contributions were approved at the 31st session of the Panel in November 2009, in Bali, Indonesia. Governments and IPCC observer organizations nominated experts for the author team. The team of 64 Coordinating Lead Authors, 179 Lead Authors, and 66 Review Editors was selected by the WGII Bureau and accepted by the IPCC Bureau in May 2010. More than 400 Contributing Authors, selected by the chapter author teams, contributed text.

Drafts prepared by the author teams were submitted for two rounds of formal review by experts, of which one was also a review by governments. Author teams revised the draft chapters after each round of review, with Review Editors working to assure that every review comment was fully considered, and where appropriate, chapters were adjusted to reflect points raised in the reviews. In addition, governments participated in a final round of review of the draft Summary for Policymakers. All of the chapter drafts, review comments, and author responses are available online via www.ipcc.ch. Across all of the drafts, the WGII contribution to the AR5 received 50,492 comments from 1,729 individual expert reviewers from 84 countries. The Summary for Policymakers was approved line-by-line by the Panel, and the underlying chapters were accepted at the 10th Session of IPCC Working Group II and the 38th Session of the IPCC Panel, meeting in Yokohama, Japan, from March 25-30, 2014.

Acknowledgments

For the AR5, Working Group II had an amazing author team. In many ways, the author team encompasses the entire scientific community, including scientists who conducted the research and wrote the research papers on which the assessment is based, and the reviewers who contributed their wisdom in more than 50,000 review comments. But the process really ran on the sophistication, wisdom, and dedication of the 309 individuals from 70 countries who comprise the WGII team of Coordinating Lead Authors, Lead Authors, and Review Editors. These individuals, with the support of a talented group of volunteer chapter scientists and the assistance of scores of contributing authors, demonstrated an inspirational commitment to scientific quality and public service. Tragically, three of our most experienced authors passed away while the report was being written. We greatly miss JoAnn Carmin, Abby Sallenger, and Steve Schneider.

We benefitted greatly from the advice and guidance of the Working Group II Bureau: Amjad Abdulla (Maldives), Eduardo Calvo Buendía (Peru), José M. Moreno (Spain), Nirivololona Raholijao (Madagascar), Sergey Semenov (Russian Federation), and Neville Smith (Australia). Their understanding of regional resources and concerns has been invaluable.

Throughout the AR5, we benefitted greatly from the wisdom and insight of our colleagues in the IPCC leadership, especially the IPCC chair, R.K. Pachauri. All of the members of the IPCC Executive Committee worked effectively and selflessly on issues related to the reports from all three working groups. We extend a heartfelt thanks to all of the members of the ExCom: R.K. Pachauri, Ottmar Edenhofer, Ismail El Gizouli, Taka Hiraishi, Thelma Krug, Hoesung Lee, Ramón Pichs Madruga, Qin Dahe, Youba Sokona, Thomas Stocker, and Jean-Pascal van Ypersele.

We are very appreciative of the enthusiastic cooperation of the nations that hosted our excellent working meetings, including four lead author meetings and the 10th Session of Working Group II. We gratefully acknowledge the support of the governments of Japan, the United States, Argentina, and Slovenia for hosting the lead author meetings, and the government of Japan for hosting the approval session. The government of the United States provided essential financial support for the Working Group II Technical Support Unit. Special thanks to the principals of the United States Global Change Research Program for orchestrating the funding across many research agencies.

We want very much to thank the Secretary of the IPCC, Renate Christ, and the staff of the IPCC Secretariat: Gaetano Leone, Carlos Martin-Novella, Jonathan Lynn, Brenda Abrar-Milani, Jesbin Baidya, Laura Biagioni, Mary Jean Burer, Annie Courtin, Judith Ewa, Joelle Fernandez, Nina Peeva, Sophie Schlingemann, Amy Smith, and Werani Zabula. Thanks to Francis Hayes who served as conference officer for the approval session. Thanks to the individuals who coordinated the organization for each of the lead authors meetings. This was Mizue Yuzurihara and Claire Summers for LAM1, Sandy MacCracken for LAM2, Ramiro Saurral for LAM3, and Mojca Deželak for LAM4. Students from Japan, the United States, Argentina, and Slovenia helped with the lead author meetings.

The WGII Technical Support Unit was fabulous. They combined scientific sophistication, technical excellence, artistic vision, deep resilience, and profound dedication, not to mention a marked ability to compensate for oversights by and deficiencies of the co-chairs. Dave Dokken, Mike Mastrandrea, Katie Mach, Kris Ebi, Monalisa Chatterjee, Sandy MacCracken, Eric Kissel, Yuka Estrada, Leslie White, Eren Bilir, Rob Genova, Beti Girma, Andrew Levy, and Patricia Mastrandrea have all made wonderful contributions to the report. In addition, the work of David Ropeik (frequently asked questions), Marcos Senet (assistant to Vicente Barros), Terry Kornak (technical edits), Marilyn Anderson (index), Liu Yingjie (Chinese author support), and Janak Pathak (UNEP communications) made a big difference. Kyle Terran, Gete Bond, and Sandi Fikes facilitated travel. Volunteer contributions from John Kelley and Ambarish Malpani greatly enhanced reference management. Catherine Lemmi, Ian Sparkman, and Danielle Olivera were super interns.

We extend a deep, personal thanks to our families and to the families of every author and reviewer. We know you tolerated many late nights and weekends with partners, parents, or children sitting at the computer or mumbling about one more assignment from us.

Vicente Barros
IPCC WGII Co-Chair

Chris Field
IPCC WGII Co-Chair

Dedication

Credit: Odd-Steinar Tøllefsen

Yuri Antonievich Izrael
(15 May 1930 to 23 January 2014)

The Working Group II contribution to the IPCC Fifth Assessment Report is dedicated to the memory of Professor Yuri Antonievich Izrael, first Chair of Working Group II from 1988 to 1992 and IPCC Vice Chair from 1992 to 2008. Professor Izrael was a pioneer, opening doors that have allowed thousands of scientists to contribute to the work of the IPCC.

Through a long and distinguished career, Professor Izrael was a strong proponent of environmental sciences, meteorology, climatology, and international organizations, especially the IPCC and the World Meteorological Organization. A creative researcher and tireless institution builder, Dr. Izrael founded and for more than two decades led the Institute of Global Climate and Ecology.

In the IPCC, Professor Izrael played a central role in creating the balance of IPCC efforts on careful observations, mechanisms, and systematic projections using scenarios. An outspoken advocate for the robust integration of scientific excellence and broad participation in IPCC reports, Dr. Izrael pioneered many of the features that assure the comprehensiveness and integrity of IPCC reports.

Summary for Policymakers

Summary for Policymakers

Drafting Authors:
Christopher B. Field (USA), Vicente R. Barros (Argentina), Michael D. Mastrandrea (USA), Katharine J. Mach (USA), Mohamed A.-K. Abdrabo (Egypt), W. Neil Adger (UK), Yury A. Anokhin (Russian Federation), Oleg A. Anisimov (Russian Federation), Douglas J. Arent (USA), Jonathon Barnett (Australia), Virginia R. Burkett (USA), Rongshuo Cai (China), Monalisa Chatterjee (USA/India), Stewart J. Cohen (Canada), Wolfgang Cramer (Germany/France), Purnamita Dasgupta (India), Debra J. Davidson (Canada), Fatima Denton (Gambia), Petra Döll (Germany), Kirstin Dow (USA), Yasuaki Hijioka (Japan), Ove Hoegh-Guldberg (Australia), Richard G. Jones (UK), Roger N. Jones (Australia), Roger L. Kitching (Australia), R. Sari Kovats (UK), Joan Nymand Larsen (Iceland), Erda Lin (China), David B. Lobell (USA), Iñigo J. Losada (Spain), Graciela O. Magrin (Argentina), José A. Marengo (Brazil), Anil Markandya (Spain), Bruce A. McCarl (USA), Roger F. McLean (Australia), Linda O. Mearns (USA), Guy F. Midgley (South Africa), Nobuo Mimura (Japan), John F. Morton (UK), Isabelle Niang (Senegal), Ian R. Noble (Australia), Leonard A. Nurse (Barbados), Karen L. O'Brien (Norway), Taikan Oki (Japan), Lennart Olsson (Sweden), Michael Oppenheimer (USA), Jonathan T. Overpeck (USA), Joy J. Pereira (Malaysia), Elvira S. Poloczanska (Australia), John R. Porter (Denmark), Hans-O. Pörtner (Germany), Michael J. Prather (USA), Roger S. Pulwarty (USA), Andy Reisinger (New Zealand), Aromar Revi (India), Patricia Romero-Lankao (Mexico), Oliver C. Ruppel (Namibia), David E. Satterthwaite (UK), Daniela N. Schmidt (UK), Josef Settele (Germany), Kirk R. Smith (USA), Dáithí A. Stone (Canada/South Africa/USA), Avelino G. Suarez (Cuba), Petra Tschakert (USA), Riccardo Valentini (Italy), Alicia Villamizar (Venezuela), Rachel Warren (UK), Thomas J. Wilbanks (USA), Poh Poh Wong (Singapore), Alistair Woodward (New Zealand), Gary W. Yohe (USA)

This Summary for Policymakers should be cited as:

IPCC, 2014: Summary for policymakers. In: *Climate Change 2014: Impacts, Adaptation, and Vulnerability. Summaries, Frequently Asked Questions, and Cross-Chapter Boxes. A Contribution of Working Group II to the Fifth Assessment Report of the Intergovernmental Panel on Climate Change* [Field, C.B., V.R. Barros, D.J. Dokken, K.J. Mach, M.D. Mastrandrea, T.E. Bilir, M. Chatterjee, K.L. Ebi, Y.O. Estrada, R.C. Genova, B. Girma, E.S. Kissel, A.N. Levy, S. MacCracken, P.R. Mastrandrea, and L.L. White (eds.)]. World Meteorological Organization, Geneva, Switzerland, pp. 1-32. (in Arabic, Chinese, English, French, Russian, and Spanish)

Contents

ASSESSING AND MANAGING THE RISKS OF CLIMATE CHANGE

Human interference with the climate system is occurring,[1] and climate change poses risks for human and natural systems (Figure SPM.1). The assessment of impacts, adaptation, and vulnerability in the Working Group II contribution to the IPCC's Fifth Assessment Report (WGII AR5) evaluates how patterns of risks and potential benefits are shifting due to climate change. It considers how impacts and risks related to climate change can be reduced and managed through adaptation and mitigation. The report assesses needs, options, opportunities, constraints, resilience, limits, and other aspects associated with adaptation.

Climate change involves complex interactions and changing likelihoods of diverse impacts. A focus on risk, which is new in this report, supports decision making in the context of climate change and complements other elements of the report. People and societies may perceive or rank risks and potential benefits differently, given diverse values and goals.

Compared to past WGII reports, the WGII AR5 assesses a substantially larger knowledge base of relevant scientific, technical, and socioeconomic literature. Increased literature has facilitated comprehensive assessment across a broader set of topics and sectors, with expanded coverage of human systems, adaptation, and the ocean. See Background Box SPM.1.[2]

Section A of this summary characterizes observed impacts, vulnerability and exposure, and adaptive responses to date. Section B examines future risks and potential benefits. Section C considers principles for effective adaptation and the broader interactions among adaptation, mitigation,

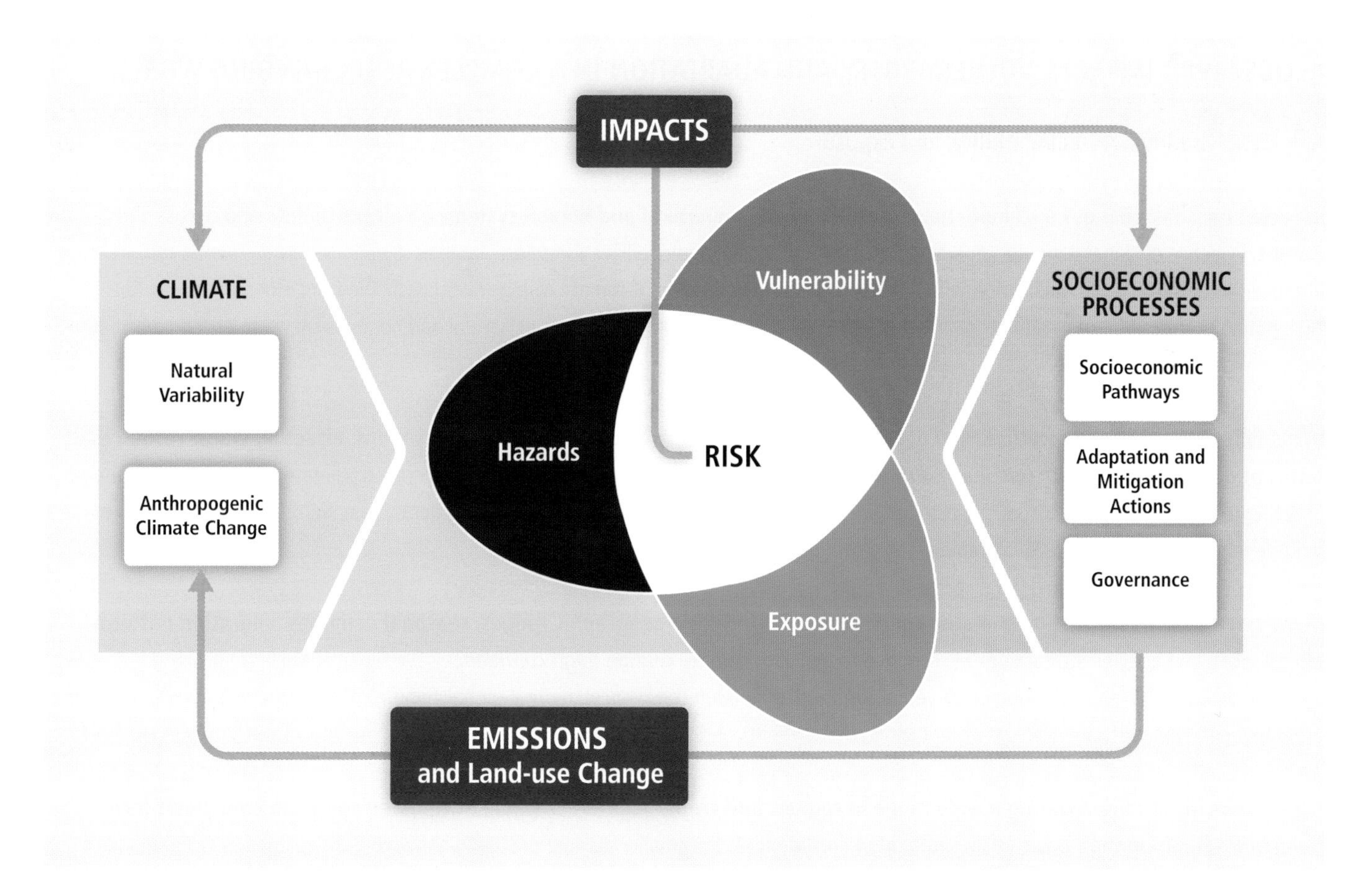

Figure SPM.1 | Illustration of the core concepts of the WGII AR5. Risk of climate-related impacts results from the interaction of climate-related hazards (including hazardous events and trends) with the vulnerability and exposure of human and natural systems. Changes in both the climate system (left) and socioeconomic processes including adaptation and mitigation (right) are drivers of hazards, exposure, and vulnerability. [19.2, Figure 19-1]

[1] A key finding of the WGI AR5 is, "It is *extremely likely* that human influence has been the dominant cause of the observed warming since the mid-20th century." [WGI AR5 SPM Section D.3, 2.2, 6.3, 10.3-6, 10.9]

[2] 1.1, Figure 1-1

Background Box SPM.1 | Context for the Assessment

For the past 2 decades, IPCC's Working Group II has developed assessments of climate-change impacts, adaptation, and vulnerability. The WGII AR5 builds from the WGII contribution to the IPCC's Fourth Assessment Report (WGII AR4), published in 2007, and the *Special Report on Managing the Risks of Extreme Events and Disasters to Advance Climate Change Adaptation* (SREX), published in 2012. It follows the Working Group I contribution to the AR5 (WGI AR5).[3]

The number of scientific publications available for assessing climate-change impacts, adaptation, and vulnerability more than doubled between 2005 and 2010, with especially rapid increases in publications related to adaptation. Authorship of climate-change publications from developing countries has increased, although it still represents a small fraction of the total.[4]

The WGII AR5 is presented in two parts (Part A: Global and Sectoral Aspects, and Part B: Regional Aspects), reflecting the expanded literature basis and multidisciplinary approach, increased focus on societal impacts and responses, and continued regionally comprehensive coverage.

and sustainable development. Background Box SPM.2 defines central concepts, and Background Box SPM.3 introduces terms used to convey the degree of certainty in key findings. Chapter references in brackets and in footnotes indicate support for findings, figures, and tables.

A: OBSERVED IMPACTS, VULNERABILITY, AND ADAPTATION IN A COMPLEX AND CHANGING WORLD

A-1. Observed Impacts, Vulnerability, and Exposure

In recent decades, changes in climate have caused impacts on natural and human systems on all continents and across the oceans. Evidence of climate-change impacts is strongest and most comprehensive for natural systems. Some impacts on human systems have also been attributed[5] to climate change, with a major or minor contribution of climate change distinguishable from other influences. See Figure SPM.2. Attribution of observed impacts in the WGII AR5 generally links responses of natural and human systems to observed climate change, regardless of its cause.[6]

In many regions, changing precipitation or melting snow and ice are altering hydrological systems, affecting water resources in terms of quantity and quality (*medium confidence*). Glaciers continue to shrink almost worldwide due to climate change (*high confidence*), affecting runoff and water resources downstream (*medium confidence*). Climate change is causing permafrost warming and thawing in high-latitude regions and in high-elevation regions (*high confidence*).[7]

Many terrestrial, freshwater, and marine species have shifted their geographic ranges, seasonal activities, migration patterns, abundances, and species interactions in response to ongoing climate change (*high confidence*). See Figure SPM.2B. While only a few recent species extinctions have been attributed as yet to climate change (*high confidence*), natural global climate change at rates slower than current anthropogenic climate change caused significant ecosystem shifts and species extinctions during the past millions of years (*high confidence*).[8]

Based on many studies covering a wide range of regions and crops, negative impacts of climate change on crop yields have been more common than positive impacts (*high confidence*). The smaller number of studies showing positive impacts relate mainly to

[3] 1.2-3
[4] 1.1, Figure 1-1
[5] The term *attribution* is used differently in WGI and WGII. Attribution in WGII considers the links between impacts on natural and human systems and observed climate change, regardless of its cause. By comparison, attribution in WGI quantifies the links between observed climate change and human activity, as well as other external climate drivers.
[6] 18.1, 18.3-6
[7] 3.2, 4.3, 18.3, 18.5, 24.4, 26.2, 28.2, Tables 3-1 and 25-1, Figures 18-2 and 26-1
[8] 4.2-4, 5.3-4, 6.1, 6.3-4, 18.3, 18.5, 22.3, 24.4, 25.6, 28.2, 30.4-5, Boxes 4-2, 4-3, 25-3, CC-CR, and CC-MB

Background Box SPM.2 | Terms Central for Understanding the Summary[9]

Climate change: Climate change refers to a change in the state of the climate that can be identified (e.g., by using statistical tests) by changes in the mean and/or the variability of its properties, and that persists for an extended period, typically decades or longer. Climate change may be due to natural internal processes or external forcings such as modulations of the solar cycles, volcanic eruptions, and persistent anthropogenic changes in the composition of the atmosphere or in land use. Note that the Framework Convention on Climate Change (UNFCCC), in its Article 1, defines climate change as: "a change of climate which is attributed directly or indirectly to human activity that alters the composition of the global atmosphere and which is in addition to natural climate variability observed over comparable time periods." The UNFCCC thus makes a distinction between climate change attributable to human activities altering the atmospheric composition, and climate variability attributable to natural causes.

Hazard: The potential occurrence of a natural or human-induced physical event or trend or physical impact that may cause loss of life, injury, or other health impacts, as well as damage and loss to property, infrastructure, livelihoods, service provision, ecosystems, and environmental resources. In this report, the term *hazard* usually refers to climate-related physical events or trends or their physical impacts.

Exposure: The presence of people, livelihoods, species or ecosystems, environmental functions, services, and resources, infrastructure, or economic, social, or cultural assets in places and settings that could be adversely affected.

Vulnerability: The propensity or predisposition to be adversely affected. Vulnerability encompasses a variety of concepts and elements including sensitivity or susceptibility to harm and lack of capacity to cope and adapt.

Impacts: Effects on natural and human systems. In this report, the term *impacts* is used primarily to refer to the effects on natural and human systems of extreme weather and climate events and of climate change. Impacts generally refer to effects on lives, livelihoods, health, ecosystems, economies, societies, cultures, services, and infrastructure due to the interaction of climate changes or hazardous climate events occurring within a specific time period and the vulnerability of an exposed society or system. Impacts are also referred to as *consequences* and *outcomes*. The impacts of climate change on geophysical systems, including floods, droughts, and sea level rise, are a subset of impacts called physical impacts.

Risk: The potential for consequences where something of value is at stake and where the outcome is uncertain, recognizing the diversity of values. Risk is often represented as probability of occurrence of hazardous events or trends multiplied by the impacts if these events or trends occur. Risk results from the interaction of vulnerability, exposure, and hazard (see Figure SPM.1). In this report, the term *risk* is used primarily to refer to the risks of climate-change impacts.

Adaptation: The process of adjustment to actual or expected climate and its effects. In human systems, adaptation seeks to moderate or avoid harm or exploit beneficial opportunities. In some natural systems, human intervention may facilitate adjustment to expected climate and its effects.

Transformation: A change in the fundamental attributes of natural and human systems. Within this summary, transformation could reflect strengthened, altered, or aligned paradigms, goals, or values towards promoting adaptation for sustainable development, including poverty reduction.

Resilience: The capacity of social, economic, and environmental systems to cope with a hazardous event or trend or disturbance, responding or reorganizing in ways that maintain their essential function, identity, and structure, while also maintaining the capacity for adaptation, learning, and transformation.

high-latitude regions, though it is not yet clear whether the balance of impacts has been negative or positive in these regions (*high confidence*). Climate change has negatively affected wheat and maize yields for many regions and in the global aggregate (*medium confidence*). Effects on rice and soybean yield have been smaller in major production regions and globally, with a median change of zero across all available data, which are fewer for soy compared to the other crops. Observed impacts relate mainly to production aspects of food security rather than access

[9] The WGII AR5 glossary defines many terms used across chapters of the report. Reflecting progress in science, some definitions differ in breadth and focus from the definitions used in the AR4 and other IPCC reports.

SPM

Background Box SPM.3 | Communication of the Degree of Certainty in Assessment Findings[10]

The degree of certainty in each key finding of the assessment is based on the type, amount, quality, and consistency of evidence (e.g., data, mechanistic understanding, theory, models, expert judgment) and the degree of agreement. The summary terms to describe evidence are: *limited*, *medium*, or *robust*; and agreement: *low*, *medium*, or *high*.

Confidence in the validity of a finding synthesizes the evaluation of evidence and agreement. Levels of confidence include five qualifiers: *very low*, *low*, *medium*, *high*, and *very high*.

The likelihood, or probability, of some well-defined outcome having occurred or occurring in the future can be described quantitatively through the following terms: *virtually certain*, 99–100% probability; *extremely likely*, 95–100%; *very likely*, 90–100%; *likely*, 66–100%; *more likely than not*, >50–100%; *about as likely as not*, 33–66%; *unlikely*, 0–33%; *very unlikely*, 0–10%; *extremely unlikely*, 0–5%; and *exceptionally unlikely*, 0–1%. Unless otherwise indicated, findings assigned a likelihood term are associated with *high* or *very high confidence*. Where appropriate, findings are also formulated as statements of fact without using uncertainty qualifiers.

Within paragraphs of this summary, the confidence, evidence, and agreement terms given for a bold key finding apply to subsequent statements in the paragraph, unless additional terms are provided.

or other components of food security. See Figure SPM.2C. Since AR4, several periods of rapid food and cereal price increases following climate extremes in key producing regions indicate a sensitivity of current markets to climate extremes among other factors (*medium confidence*).[11]

At present the worldwide burden of human ill-health from climate change is relatively small compared with effects of other stressors and is not well quantified. However, there has been increased heat-related mortality and decreased cold-related mortality in some regions as a result of warming (*medium confidence*). Local changes in temperature and rainfall have altered the distribution of some water-borne illnesses and disease vectors (*medium confidence*).[12]

Differences in vulnerability and exposure arise from non-climatic factors and from multidimensional inequalities often produced by uneven development processes (*very high confidence*). These differences shape differential risks from climate change. See Figure SPM.1. People who are socially, economically, culturally, politically, institutionally, or otherwise marginalized are especially vulnerable to climate change and also to some adaptation and mitigation responses (*medium evidence*, *high agreement*). This heightened vulnerability is rarely due to a single cause. Rather, it is the product of intersecting social processes that result in inequalities in socioeconomic status and income, as well as in exposure. Such social processes include, for example, discrimination on the basis of gender, class, ethnicity, age, and (dis)ability.[13]

Impacts from recent climate-related extremes, such as heat waves, droughts, floods, cyclones, and wildfires, reveal significant vulnerability and exposure of some ecosystems and many human systems to current climate variability (*very high confidence*). Impacts of such climate-related extremes include alteration of ecosystems, disruption of food production and water supply, damage to infrastructure and settlements, morbidity and mortality, and consequences for mental health and human well-being. For countries at all levels of development, these impacts are consistent with a significant lack of preparedness for current climate variability in some sectors.[14]

Climate-related hazards exacerbate other stressors, often with negative outcomes for livelihoods, especially for people living in poverty (*high confidence*). Climate-related hazards affect poor people's lives directly through impacts on livelihoods, reductions in crop

[10] 1.1, Box 1-1
[11] 7.2, 18.4, 22.3, 26.5, Figures 7-2, 7-3, and 7-7
[12] 11.4-6, 18.4, 25.8
[13] 8.1-2, 9.3-4, 10.9, 11.1, 11.3-5, 12.2-5, 13.1-3, 14.1-3, 18.4, 19.6, 23.5, 25.8, 26.6, 26.8, 28.4, Box CC-GC
[14] 3.2, 4.2-3, 8.1, 9.3, 10.7, 11.3, 11.7, 13.2, 14.1, 18.6, 22.3, 25.6-8, 26.6-7, 30.5, Tables 18-3 and 23-1, Figure 26-2, Boxes 4-3, 4-4, 25-5, 25-6, 25-8, and CC-CR

SPM

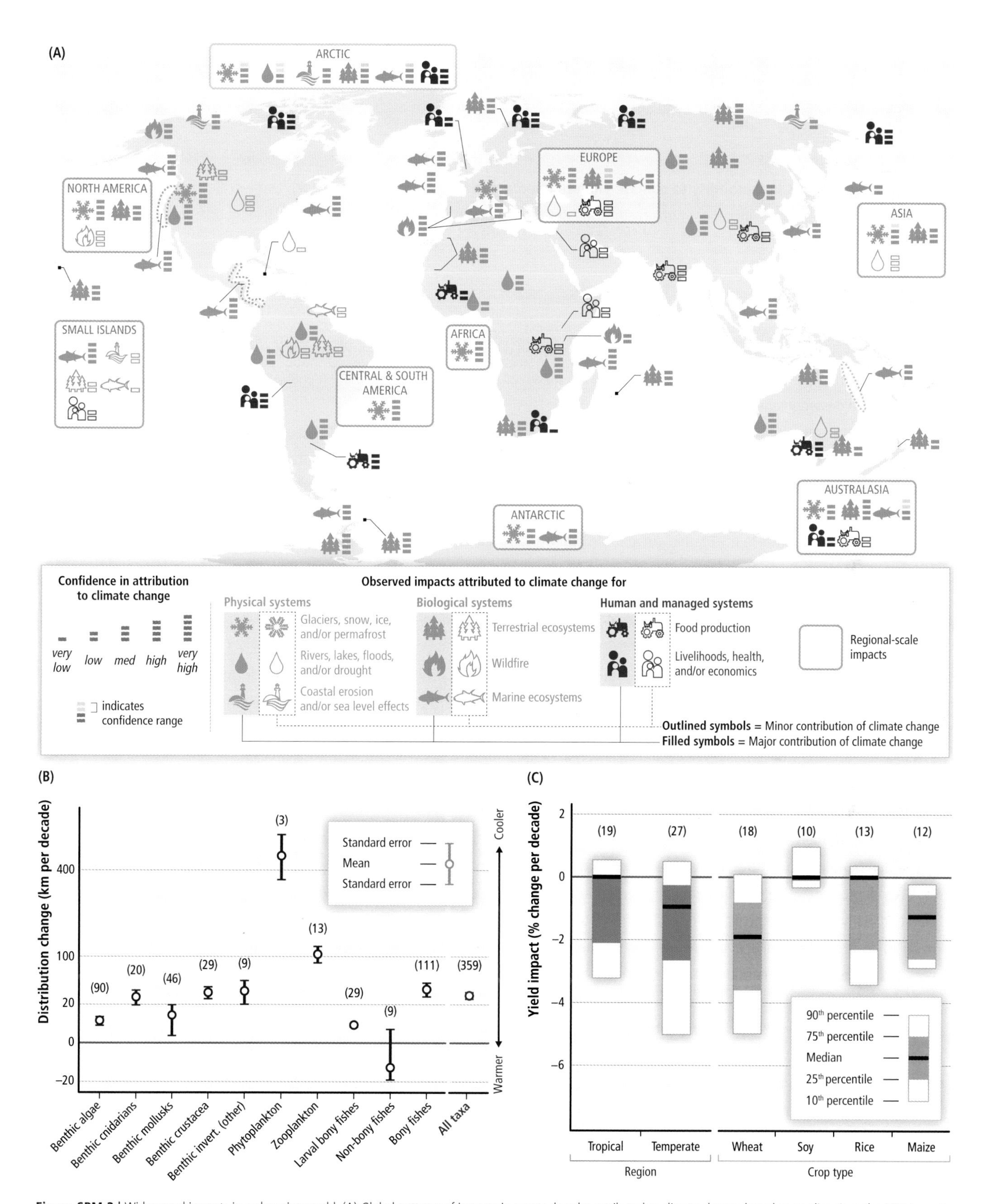

Figure SPM.2 | Widespread impacts in a changing world. (A) Global patterns of impacts in recent decades attributed to climate change, based on studies since the AR4. Impacts are shown at a range of geographic scales. Symbols indicate categories of attributed impacts, the relative contribution of climate change (major or minor) to the observed impact, and confidence in attribution. See supplementary Table SPM.A1 for descriptions of the impacts. (B) Average rates of change in distribution (km per decade) for marine taxonomic groups based on observations over 1900–2010. Positive distribution changes are consistent with warming (moving into previously cooler waters, generally poleward). The number of responses analyzed is given within parentheses for each category. (C) Summary of estimated impacts of observed climate changes on yields over 1960–2013 for four major crops in temperate and tropical regions, with the number of data points analyzed given within parentheses for each category. [Figures 7-2, 18-3, and MB-2]

yields, or destruction of homes and indirectly through, for example, increased food prices and food insecurity. Observed positive effects for poor and marginalized people, which are limited and often indirect, include examples such as diversification of social networks and of agricultural practices.[15]

Violent conflict increases vulnerability to climate change (*medium evidence, high agreement*). Large-scale violent conflict harms assets that facilitate adaptation, including infrastructure, institutions, natural resources, social capital, and livelihood opportunities.[16]

A-2. Adaptation Experience

Throughout history, people and societies have adjusted to and coped with climate, climate variability, and extremes, with varying degrees of success. This section focuses on adaptive human responses to observed and projected climate-change impacts, which can also address broader risk-reduction and development objectives.

Adaptation is becoming embedded in some planning processes, with more limited implementation of responses (*high confidence*). Engineered and technological options are commonly implemented adaptive responses, often integrated within existing programs such as disaster risk management and water management. There is increasing recognition of the value of social, institutional, and ecosystem-based measures and of the extent of constraints to adaptation. Adaptation options adopted to date continue to emphasize incremental adjustments and co-benefits and are starting to emphasize flexibility and learning (*medium evidence*, *medium agreement*). Most assessments of adaptation have been restricted to impacts, vulnerability, and adaptation planning, with very few assessing the processes of implementation or the effects of adaptation actions (*medium evidence*, *high agreement*).[17]

Adaptation experience is accumulating across regions in the public and private sector and within communities (*high confidence*). Governments at various levels are starting to develop adaptation plans and policies and to integrate climate-change considerations into broader development plans. Examples of adaptation across regions include the following:

- In Africa, most national governments are initiating governance systems for adaptation. Disaster risk management, adjustments in technologies and infrastructure, ecosystem-based approaches, basic public health measures, and livelihood diversification are reducing vulnerability, although efforts to date tend to be isolated.[18]
- In Europe, adaptation policy has been developed across all levels of government, with some adaptation planning integrated into coastal and water management, into environmental protection and land planning, and into disaster risk management.[19]
- In Asia, adaptation is being facilitated in some areas through mainstreaming climate adaptation action into subnational development planning, early warning systems, integrated water resources management, agroforestry, and coastal reforestation of mangroves.[20]
- In Australasia, planning for sea level rise, and in southern Australia for reduced water availability, is becoming adopted widely. Planning for sea level rise has evolved considerably over the past 2 decades and shows a diversity of approaches, although its implementation remains piecemeal.[21]
- In North America, governments are engaging in incremental adaptation assessment and planning, particularly at the municipal level. Some proactive adaptation is occurring to protect longer-term investments in energy and public infrastructure.[22]
- In Central and South America, ecosystem-based adaptation including protected areas, conservation agreements, and community management of natural areas is occurring. Resilient crop varieties, climate forecasts, and integrated water resources management are being adopted within the agricultural sector in some areas.[23]

---

[15] 8.2-3, 9.3, 11.3, 13.1-3, 22.3, 24.4, 26.8
[16] 12.5, 19.2, 19.6
[17] 4.4, 5.5, 6.4, 8.3, 9.4, 11.7, 14.1, 14.3-4, 15.2-5, 17.2-3, 21.3, 21.5, 22.4, 23.7, 25.4, 26.8-9, 30.6, Boxes 25-1, 25-2, 25-9, and CC-EA
[18] 22.4
[19] 23.7, Boxes 5-1 and 23-3
[20] 24.4-6, 24.9 Box CC-TC
[21] 25.4, 25.10, Table 25-2, Boxes 25-1, 25-2, and 25-9
[22] 26.7-9
[23] 27.3

- In the Arctic, some communities have begun to deploy adaptive co-management strategies and communications infrastructure, combining traditional and scientific knowledge.[24]
- In small islands, which have diverse physical and human attributes, community-based adaptation has been shown to generate larger benefits when delivered in conjunction with other development activities.[25]
- In the ocean, international cooperation and marine spatial planning are starting to facilitate adaptation to climate change, with constraints from challenges of spatial scale and governance issues.[26]

A-3. The Decision-making Context

Climate variability and extremes have long been important in many decision-making contexts. Climate-related risks are now evolving over time due to both climate change and development. This section builds from existing experience with decision making and risk management. It creates a foundation for understanding the report's assessment of future climate-related risks and potential responses.

Responding to climate-related risks involves decision making in a changing world, with continuing uncertainty about the severity and timing of climate-change impacts and with limits to the effectiveness of adaptation (*high confidence*). Iterative risk management is a useful framework for decision making in complex situations characterized by large potential consequences, persistent uncertainties, long timeframes, potential for learning, and multiple climatic and non-climatic influences changing over time. See Figure SPM.3. Assessment of the widest possible range of potential impacts, including low-probability outcomes with large consequences, is central to understanding the benefits and trade-offs of alternative risk management actions. The complexity of adaptation actions across scales and contexts means that monitoring and learning are important components of effective adaptation.[27]

Adaptation and mitigation choices in the near term will affect the risks of climate change throughout the 21st century (*high confidence*). Figure SPM.4 illustrates projected warming under a low-emission mitigation scenario and a high-emission scenario [Representative Concentration Pathways (RCPs) 2.6 and 8.5], along with observed temperature changes. The benefits of adaptation and mitigation occur over different but overlapping timeframes. Projected global temperature increase over the next few decades is similar across emission scenarios (Figure SPM.4B).[28] During this near-term period, risks will evolve as socioeconomic trends interact with the changing climate. Societal

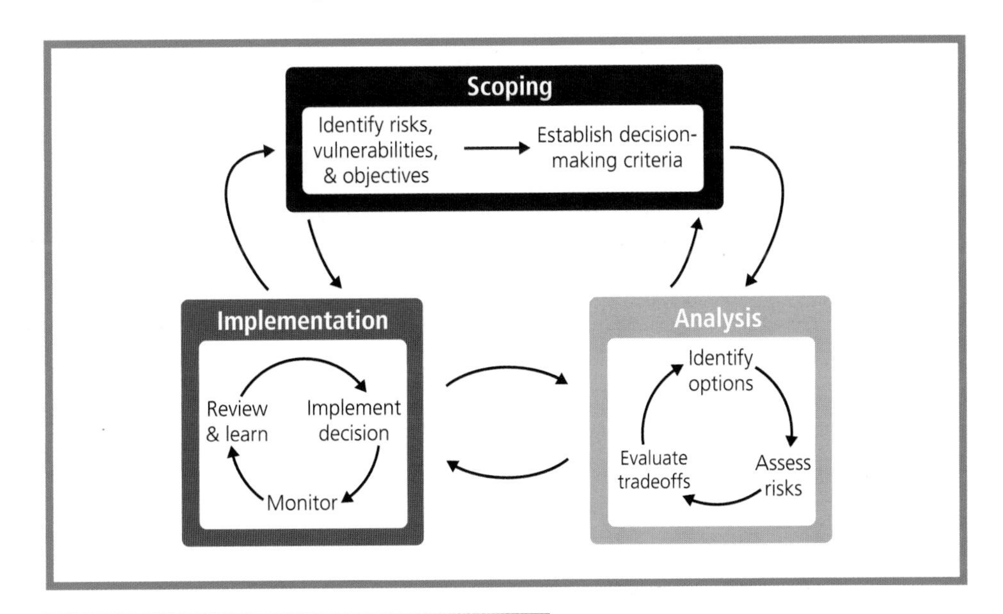

Figure SPM.3 | Climate-change adaptation as an iterative risk management process with multiple feedbacks. People and knowledge shape the process and its outcomes. [Figure 2-1]

[24] 28.2, 28.4
[25] 29.3, 29.6, Table 29-3, Figure 29-1
[26] 30.6
[27] 2.1-4, 3.6, 14.1-3, 15.2-4, 16.2-4, 17.1-3, 17.5, 20.6, 22.4, 25.4, Figure 1-5
[28] WGI AR5 11.3

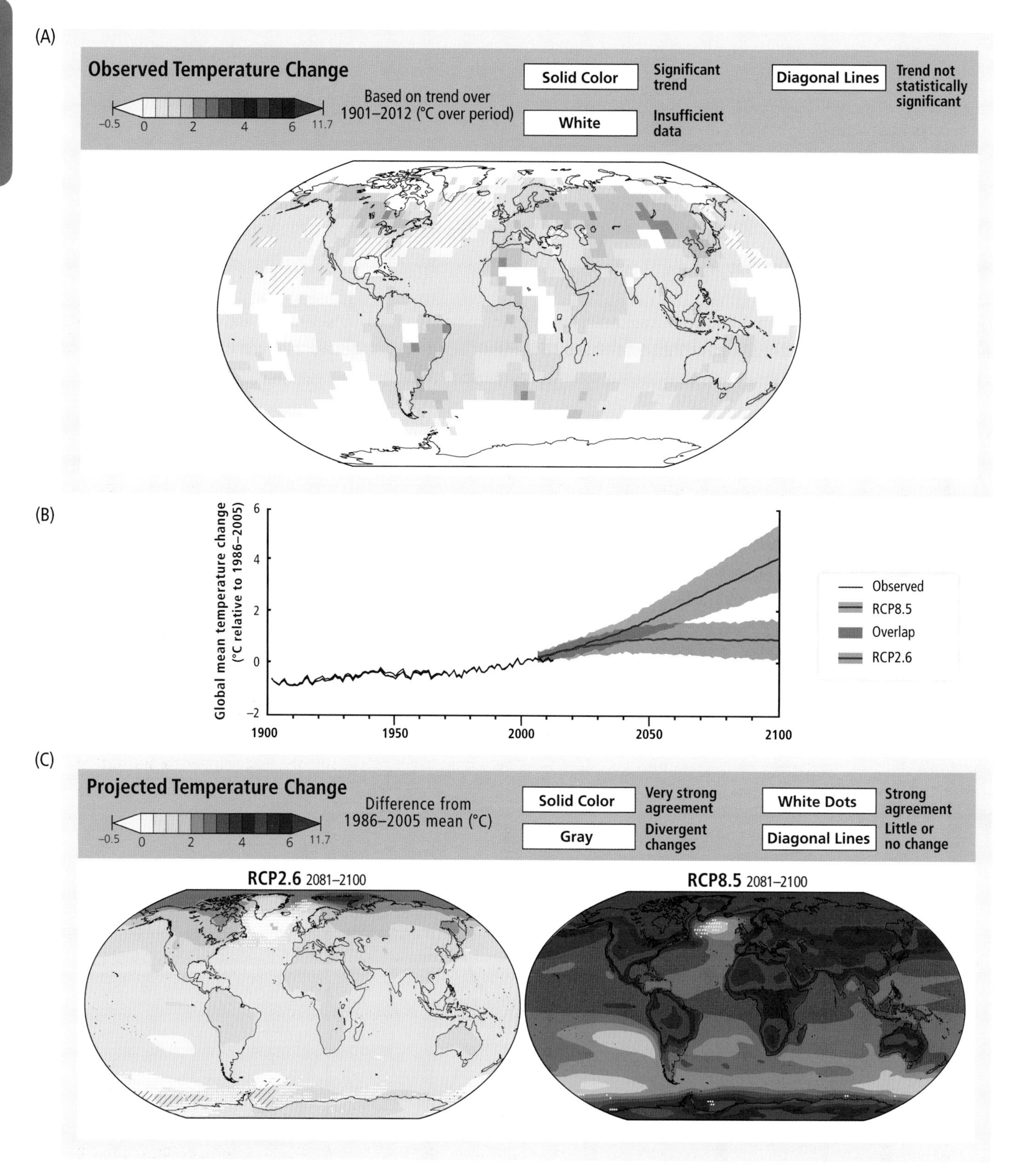

Figure SPM.4 | Observed and projected changes in annual average surface temperature. This figure informs understanding of climate-related risks in the WGII AR5. It illustrates temperature change observed to date and projected warming under continued high emissions and under ambitious mitigation.

Figure SPM.4 Technical Details

(A) Map of observed annual average temperature change from 1901–2012, derived from a linear trend where sufficient data permit a robust estimate; other areas are white. Solid colors indicate areas where trends are significant at the 10% level. Diagonal lines indicate areas where trends are not significant. Observed data (range of grid-point values: –0.53 to 2.50°C over period) are from WGI AR5 Figures SPM.1 and 2.21. (B) Observed and projected future global annual average temperature relative to 1986–2005. Observed warming from 1850–1900 to 1986–2005 is 0.61°C (5–95% confidence interval: 0.55 to 0.67°C). Black lines show temperature estimates from three datasets. Blue and red lines and shading denote the ensemble mean and ±1.64 standard deviation range, based on CMIP5 simulations from 32 models for RCP2.6 and 39 models for RCP8.5. (C) CMIP5 multi-model mean projections of annual average temperature changes for 2081–2100 under RCP2.6 and 8.5, relative to 1986–2005. Solid colors indicate areas with very strong agreement, where the multi-model mean change is greater than twice the baseline variability (natural internal variability in 20-yr means) and ≥90% of models agree on sign of change. Colors with white dots indicate areas with strong agreement, where ≥66% of models show change greater than the baseline variability and ≥66% of models agree on sign of change. Gray indicates areas with divergent changes, where ≥66% of models show change greater than the baseline variability, but <66% agree on sign of change. Colors with diagonal lines indicate areas with little or no change, where <66% of models show change greater than the baseline variability, although there may be significant change at shorter timescales such as seasons, months, or days. Analysis uses model data (range of grid-point values across RCP2.6 and 8.5: 0.06 to 11.71°C) from WGI AR5 Figure SPM.8, with full description of methods in Box CC-RC. See also Annex I of WGI AR5. [Boxes 21-2 and CC-RC; WGI AR5 2.4, Figures SPM.1, SPM.7, and 2.21]

responses, particularly adaptations, will influence near-term outcomes. In the second half of the 21st century and beyond, global temperature increase diverges across emission scenarios (Figure SPM.4B and 4C).[29] For this longer-term period, near-term and longer-term adaptation and mitigation, as well as development pathways, will determine the risks of climate change.[30]

Assessment of risks in the WGII AR5 relies on diverse forms of evidence. Expert judgment is used to integrate evidence into evaluations of risks. Forms of evidence include, for example, empirical observations, experimental results, process-based understanding, statistical approaches, and simulation and descriptive models. Future risks related to climate change vary substantially across plausible alternative development pathways, and the relative importance of development and climate change varies by sector, region, and time period (*high confidence*). Scenarios are useful tools for characterizing possible future socioeconomic pathways, climate change and its risks, and policy implications. Climate-model projections informing evaluations of risks in this report are generally based on the RCPs (Figure SPM.4), as well as the older IPCC *Special Report on Emission Scenarios* (SRES) scenarios.[31]

Uncertainties about future vulnerability, exposure, and responses of interlinked human and natural systems are large (*high confidence*). This motivates exploration of a wide range of socioeconomic futures in assessments of risks. Understanding future vulnerability, exposure, and response capacity of interlinked human and natural systems is challenging due to the number of interacting social, economic, and cultural factors, which have been incompletely considered to date. These factors include wealth and its distribution across society, demographics, migration, access to technology and information, employment patterns, the quality of adaptive responses, societal values, governance structures, and institutions to resolve conflicts. International dimensions such as trade and relations among states are also important for understanding the risks of climate change at regional scales.[32]

B: FUTURE RISKS AND OPPORTUNITIES FOR ADAPTATION

This section presents future risks and more limited potential benefits across sectors and regions, over the next few decades and in the second half of the 21st century and beyond. It examines how they are affected by the magnitude and rate of climate change and by socioeconomic choices. It also assesses opportunities for reducing impacts and managing risks through adaptation and mitigation.

B-1. Key Risks across Sectors and Regions

Key risks are potentially severe impacts relevant to Article 2 of the United Nations Framework Convention on Climate Change, which refers to "dangerous anthropogenic interference with the climate system." Risks are considered key due to high hazard or high vulnerability of societies and systems exposed, or both. Identification of key risks was based on expert judgment using the following specific criteria: large magnitude,

[29] WGI AR5 12.4 and Table SPM.2
[30] 2.5, 21.2-3, 21.5, Box CC-RC
[31] 1.1, 1.3, 2.2-3, 19.6, 20.2, 21.3, 21.5, 26.2, Box CC-RC; WGI AR5 Box SPM.1
[32] 11.3, 12.6, 21.3-5, 25.3-4, 25.11, 26.2

SPM

Assessment Box SPM.1 | Human Interference with the Climate System

Human influence on the climate system is clear.[33] Yet determining whether such influence constitutes "dangerous anthropogenic interference" in the words of Article 2 of the UNFCCC involves both risk assessment and value judgments. This report assesses risks across contexts and through time, providing a basis for judgments about the level of climate change at which risks become dangerous.

Five integrative reasons for concern (RFCs) provide a framework for summarizing key risks across sectors and regions. First identified in the IPCC Third Assessment Report, the RFCs illustrate the implications of warming and of adaptation limits for people, economies, and ecosystems. They provide one starting point for evaluating dangerous anthropogenic interference with the climate system. Risks for each RFC, updated based on assessment of the literature and expert judgments, are presented below and in Assessment Box SPM.1 Figure 1. All temperatures below are given as global average temperature change relative to 1986–2005 ("recent").[34]

1) **Unique and threatened systems**: Some unique and threatened systems, including ecosystems and cultures, are already at risk from climate change (*high confidence*). The number of such systems at risk of severe consequences is higher with additional warming of around 1°C. Many species and systems with limited adaptive capacity are subject to very high risks with additional warming of 2°C, particularly Arctic-sea-ice and coral-reef systems.

2) **Extreme weather events**: Climate-change-related risks from extreme events, such as heat waves, extreme precipitation, and coastal flooding, are already moderate (*high confidence*) and high with 1°C additional warming (*medium confidence*). Risks associated with some types of extreme events (e.g., extreme heat) increase further at higher temperatures (*high confidence*).

3) **Distribution of impacts**: Risks are unevenly distributed and are generally greater for disadvantaged people and communities in countries at all levels of development. Risks are already moderate because of regionally differentiated climate-change impacts on crop production in particular (*medium* to *high confidence*). Based on projected decreases in regional crop yields and water availability, risks of unevenly distributed impacts are high for additional warming above 2°C (*medium confidence*).

4) **Global aggregate impacts**: Risks of global aggregate impacts are moderate for additional warming between 1–2°C, reflecting impacts to both Earth's biodiversity and the overall global economy (*medium confidence*). Extensive biodiversity loss with associated loss of ecosystem goods and services results in high risks around 3°C additional warming (*high confidence*). Aggregate economic damages accelerate with increasing temperature (*limited evidence*, *high agreement*), but few quantitative estimates have been completed for additional warming around 3°C or above.

5) **Large-scale singular events**: With increasing warming, some physical systems or ecosystems may be at risk of abrupt and irreversible changes. Risks associated with such tipping points become moderate between 0–1°C additional warming, due to early warning signs that both warm-water coral reef and Arctic ecosystems are already experiencing irreversible regime shifts (*medium confidence*). Risks increase disproportionately as temperature increases between 1–2°C additional warming and become high above 3°C, due to the potential for a large and irreversible sea level rise from ice sheet loss. For sustained warming greater than some threshold,[35] near-complete loss of the Greenland ice sheet would occur over a millennium or more, contributing up to 7 m of global mean sea level rise.

high probability, or irreversibility of impacts; timing of impacts; persistent vulnerability or exposure contributing to risks; or limited potential to reduce risks through adaptation or mitigation. Key risks are integrated into five complementary and overarching reasons for concern (RFCs) in Assessment Box SPM.1.

The key risks that follow, all of which are identified with *high confidence*, span sectors and regions. Each of these key risks contributes to one or more RFCs.[36]

---

[33] WGI AR5 SPM, 2.2, 6.3, 10.3-6, 10.9

[34] 18.6, 19.6; observed warming from 1850–1900 to 1986–2005 is 0.61°C (5–95% confidence interval: 0.55 to 0.67°C). [WGI AR5 2.4]

[35] Current estimates indicate that this threshold is greater than about 1°C (*low confidence*) but less than about 4°C (*medium confidence*) sustained global mean warming above preindustrial levels. [WGI AR5 SPM, 5.8, 13.4-5]

[36] 19.2-4, 19.6, Table 19-4, Boxes 19-2 and CC-KR

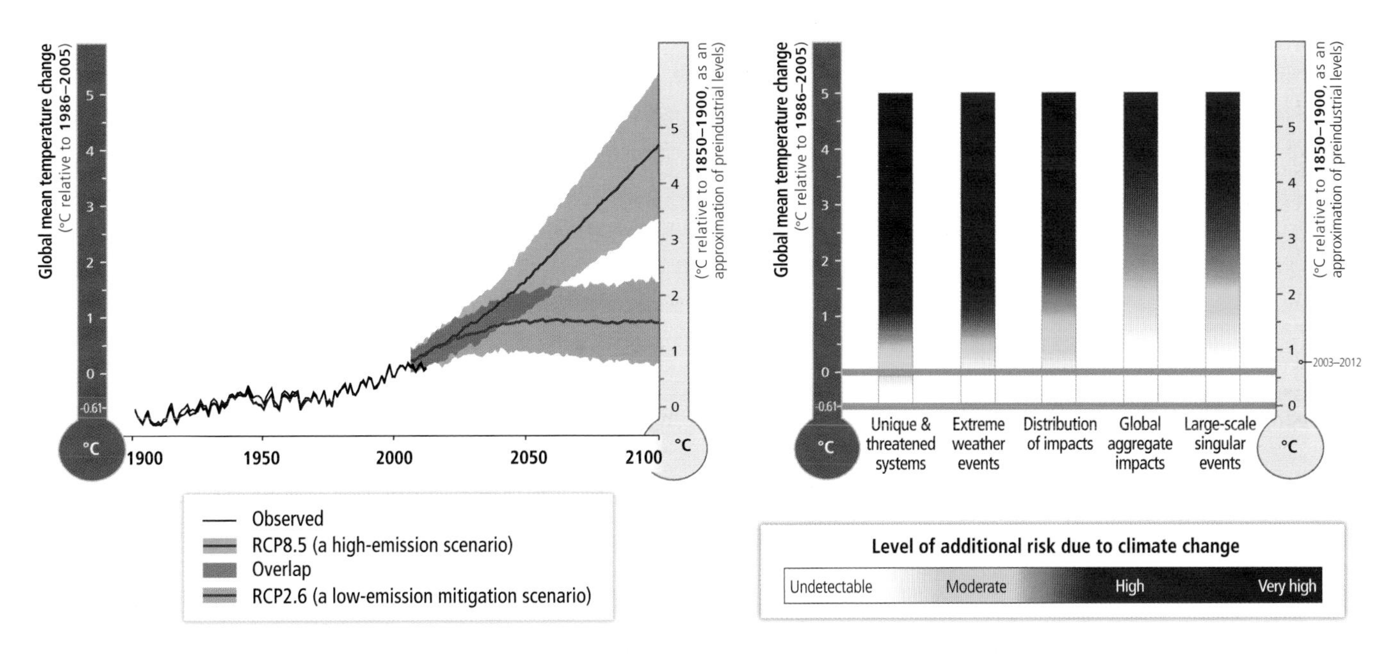

Assessment Box SPM.1 Figure 1 | A global perspective on climate-related risks. Risks associated with reasons for concern are shown at right for increasing levels of climate change. The color shading indicates the additional risk due to climate change when a temperature level is reached and then sustained or exceeded. Undetectable risk (white) indicates no associated impacts are detectable and attributable to climate change. Moderate risk (yellow) indicates that associated impacts are both detectable and attributable to climate change with at least *medium confidence*, also accounting for the other specific criteria for key risks. High risk (red) indicates severe and widespread impacts, also accounting for the other specific criteria for key risks. Purple, introduced in this assessment, shows that very high risk is indicated by all specific criteria for key risks. [Figure 19-4] For reference, past and projected global annual average surface temperature is shown at left, as in Figure SPM.4. [Figure RC-1, Box CC-RC; WGI AR5 Figures SPM.1 and SPM.7] Based on the longest global surface temperature dataset available, the observed change between the average of the period 1850–1900 and of the AR5 reference period (1986–2005) is 0.61°C (5–95% confidence interval: 0.55 to 0.67°C) [WGI AR5 SPM, 2.4], which is used here as an approximation of the change in global mean surface temperature since preindustrial times, referred to as the period before 1750. [WGI and WGII AR5 glossaries]

i) Risk of death, injury, ill-health, or disrupted livelihoods in low-lying coastal zones and small island developing states and other small islands, due to storm surges, coastal flooding, and sea level rise.[37] [RFC 1-5]
ii) Risk of severe ill-health and disrupted livelihoods for large urban populations due to inland flooding in some regions.[38] [RFC 2 and 3]
iii) Systemic risks due to extreme weather events leading to breakdown of infrastructure networks and critical services such as electricity, water supply, and health and emergency services.[39] [RFC 2-4]
iv) Risk of mortality and morbidity during periods of extreme heat, particularly for vulnerable urban populations and those working outdoors in urban or rural areas.[40] [RFC 2 and 3]
v) Risk of food insecurity and the breakdown of food systems linked to warming, drought, flooding, and precipitation variability and extremes, particularly for poorer populations in urban and rural settings.[41] [RFC 2-4]
vi) Risk of loss of rural livelihoods and income due to insufficient access to drinking and irrigation water and reduced agricultural productivity, particularly for farmers and pastoralists with minimal capital in semi-arid regions.[42] [RFC 2 and 3]
vii) Risk of loss of marine and coastal ecosystems, biodiversity, and the ecosystem goods, functions, and services they provide for coastal livelihoods, especially for fishing communities in the tropics and the Arctic.[43] [RFC 1, 2, and 4]
viii) Risk of loss of terrestrial and inland water ecosystems, biodiversity, and the ecosystem goods, functions, and services they provide for livelihoods.[44] [RFC 1, 3, and 4]

Many key risks constitute particular challenges for the least developed countries and vulnerable communities, given their limited ability to cope.

---

[37] 5.4, 8.2, 13.2, 19.2-4, 19.6-7, 24.4-5, 26.7-8, 29.3, 30.3, Tables 19-4 and 26-1, Figure 26-2, Boxes 25-1, 25-7, and CC-KR
[38] 3.4-5, 8.2, 13.2, 19.6, 25.10, 26.3, 26.8, 27.3, Tables 19-4 and 26-1, Boxes 25-8 and CC-KR
[39] 5.4, 8.1-2, 9.3, 10.2-3, 12.6, 19.6, 23.9, 25.10, 26.7-8, 28.3, Table 19-4, Boxes CC-KR and CC-HS
[40] 8.1-2, 11.3-4, 11.6, 13.2, 19.3, 19.6, 23.5, 24.4, 25.8, 26.6, 26.8, Tables 19-4 and 26-1, Boxes CC-KR and CC-HS
[41] 3.5, 7.4-5, 8.2-3, 9.3, 11.3, 11.6, 13.2, 19.3-4, 19.6, 22.3, 24.4, 25.5, 25.7, 26.5, 26.8, 27.3, 28.2, 28.4, Table 19-4, Box CC-KR
[42] 3.4-5, 9.3, 12.2, 13.2, 19.3, 19.6, 24.4, 25.7, 26.8, Table 19-4, Boxes 25-5 and CC-KR
[43] 5.4, 6.3, 7.4, 9.3, 19.5-6, 22.3, 25.6, 27.3, 28.2-3, 29.3, 30.5-7, Table 19-4, Boxes CC-OA, CC-CR, CC-KR, and CC-HS
[44] 4.3, 9.3, 19.3-6, 22.3, 25.6, 27.3, 28.2-3, Table 19-4, Boxes CC-KR and CC-WE

SPM

Increasing magnitudes of warming increase the likelihood of severe, pervasive, and irreversible impacts. Some risks of climate change are considerable at 1 or 2°C above preindustrial levels (as shown in Assessment Box SPM.1). Global climate change risks are high to very high with global mean temperature increase of 4°C or more above preindustrial levels in all reasons for concern (Assessment Box SPM.1), and include severe and widespread impacts on unique and threatened systems, substantial species extinction, large risks to global and regional food security, and the combination of high temperature and humidity compromising normal human activities, including growing food or working outdoors in some areas for parts of the year (*high confidence*). The precise levels of climate change sufficient to trigger tipping points (thresholds for abrupt and irreversible change) remain uncertain, but the risk associated with crossing multiple tipping points in the earth system or in interlinked human and natural systems increases with rising temperature (*medium confidence*).[45]

The overall risks of climate change impacts can be reduced by limiting the rate and magnitude of climate change. Risks are reduced substantially under the assessed scenario with the lowest temperature projections (RCP2.6 – low emissions) compared to the highest temperature projections (RCP8.5 – high emissions), particularly in the second half of the 21st century (*very high confidence*). Reducing climate change can also reduce the scale of adaptation that might be required. Under all assessed scenarios for adaptation and mitigation, some risk from adverse impacts remains (*very high confidence*).[46]

B-2. Sectoral Risks and Potential for Adaptation

Climate change is projected to amplify existing climate-related risks and create new risks for natural and human systems. Some of these risks will be limited to a particular sector or region, and others will have cascading effects. To a lesser extent, climate change is also projected to have some potential benefits.

Freshwater resources

Freshwater-related risks of climate change increase significantly with increasing greenhouse gas concentrations (*robust evidence, high agreement*). The fraction of global population experiencing water scarcity and the fraction affected by major river floods increase with the level of warming in the 21st century.[47]

Climate change over the 21st century is projected to reduce renewable surface water and groundwater resources significantly in most dry subtropical regions (*robust evidence, high agreement*), intensifying competition for water among sectors (*limited evidence, medium agreement*). In presently dry regions, drought frequency will *likely* increase by the end of the 21st century under RCP8.5 (*medium confidence*). In contrast, water resources are projected to increase at high latitudes (*robust evidence, high agreement*). Climate change is projected to reduce raw water quality and pose risks to drinking water quality even with conventional treatment, due to interacting factors: increased temperature; increased sediment, nutrient, and pollutant loadings from heavy rainfall; increased concentration of pollutants during droughts; and disruption of treatment facilities during floods (*medium evidence, high agreement*). Adaptive water management techniques, including scenario planning, learning-based approaches, and flexible and low-regret solutions, can help create resilience to uncertain hydrological changes and impacts due to climate change (*limited evidence, high agreement*).[48]

Terrestrial and freshwater ecosystems

A large fraction of both terrestrial and freshwater species faces increased extinction risk under projected climate change during and beyond the 21st century, especially as climate change interacts with other stressors, such as habitat modification, over-

[45] 4.2-3, 11.8, 19.5, 19.7, 26.5, Box CC-HS
[46] 3.4-5, 16.6, 17.2, 19.7, 20.3, 25.10, Tables 3-2, 8-3, and 8-6, Boxes 16-3 and 25-1
[47] 3.4-5, 26.3, Table 3-2, Box 25-8

exploitation, pollution, and invasive species (*high confidence*). Extinction risk is increased under all RCP scenarios, with risk increasing with both magnitude and rate of climate change. Many species will be unable to track suitable climates under mid- and high-range rates of climate change (i.e., RCP4.5, 6.0, and 8.5) during the 21st century (*medium confidence*). Lower rates of change (i.e., RCP2.6) will pose fewer problems. See Figure SPM.5. Some species will adapt to new climates. Those that cannot adapt sufficiently fast will decrease in abundance or go extinct in part or all of their ranges. Management actions, such as maintenance of genetic diversity, assisted species migration and dispersal, manipulation of disturbance regimes (e.g., fires, floods), and reduction of other stressors, can reduce, but not eliminate, risks of impacts to terrestrial and freshwater ecosystems due to climate change, as well as increase the inherent capacity of ecosystems and their species to adapt to a changing climate (*high confidence*).[49]

Within this century, magnitudes and rates of climate change associated with medium- to high-emission scenarios (RCP4.5, 6.0, and 8.5) pose high risk of abrupt and irreversible regional-scale change in the composition, structure, and function of terrestrial and freshwater ecosystems, including wetlands (*medium confidence*). Examples that could lead to substantial impact on climate are the boreal-tundra Arctic system (*medium confidence*) and the Amazon forest (*low confidence*). Carbon stored in the terrestrial biosphere (e.g., in peatlands, permafrost, and forests) is susceptible to loss to the atmosphere as a result of climate change, deforestation, and ecosystem degradation (*high confidence*). Increased tree mortality and associated forest dieback is projected to occur in many regions over the 21st century, due to increased temperatures and drought (*medium confidence*). Forest dieback poses risks for carbon storage, biodiversity, wood production, water quality, amenity, and economic activity.[50]

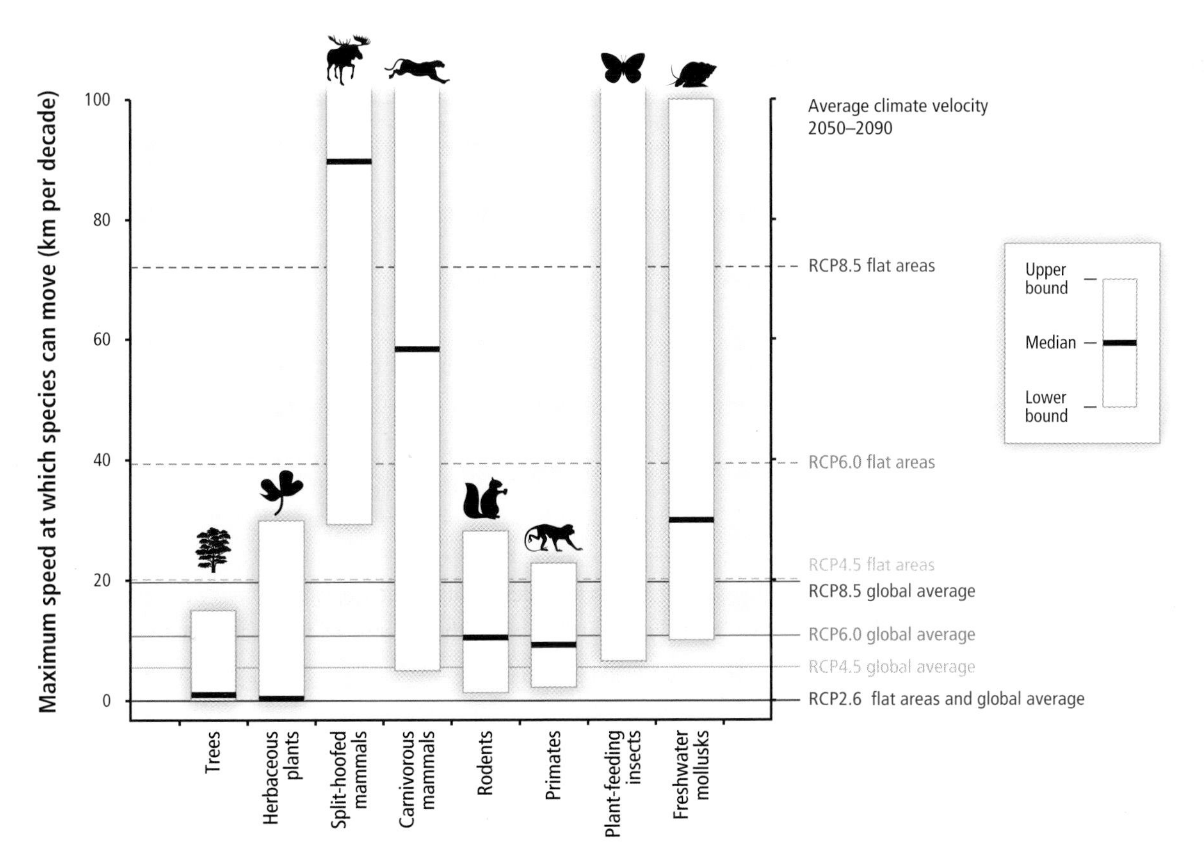

Figure SPM.5 | Maximum speeds at which species can move across landscapes (based on observations and models; vertical axis on left), compared with speeds at which temperatures are projected to move across landscapes (climate velocities for temperature; vertical axis on right). Human interventions, such as transport or habitat fragmentation, can greatly increase or decrease speeds of movement. White boxes with black bars indicate ranges and medians of maximum movement speeds for trees, plants, mammals, plant-feeding insects (median not estimated), and freshwater mollusks. For RCP2.6, 4.5, 6.0, and 8.5 for 2050–2090, horizontal lines show climate velocity for the global-land-area average and for large flat regions. Species with maximum speeds below each line are expected to be unable to track warming in the absence of human intervention. [Figure 4-5]

[48] 3.2, 3.4-6, 22.3, 23.9, 25.5, 26.3, Table 3-2, Table 23-3, Boxes 25-2, CC-RF, and CC-WE; WGI AR5 12.4
[49] 4.3-4, 25.6, 26.4, Box CC-RF
[50] 4.2-3, Figure 4-8, Boxes 4-2, 4-3, and 4-4

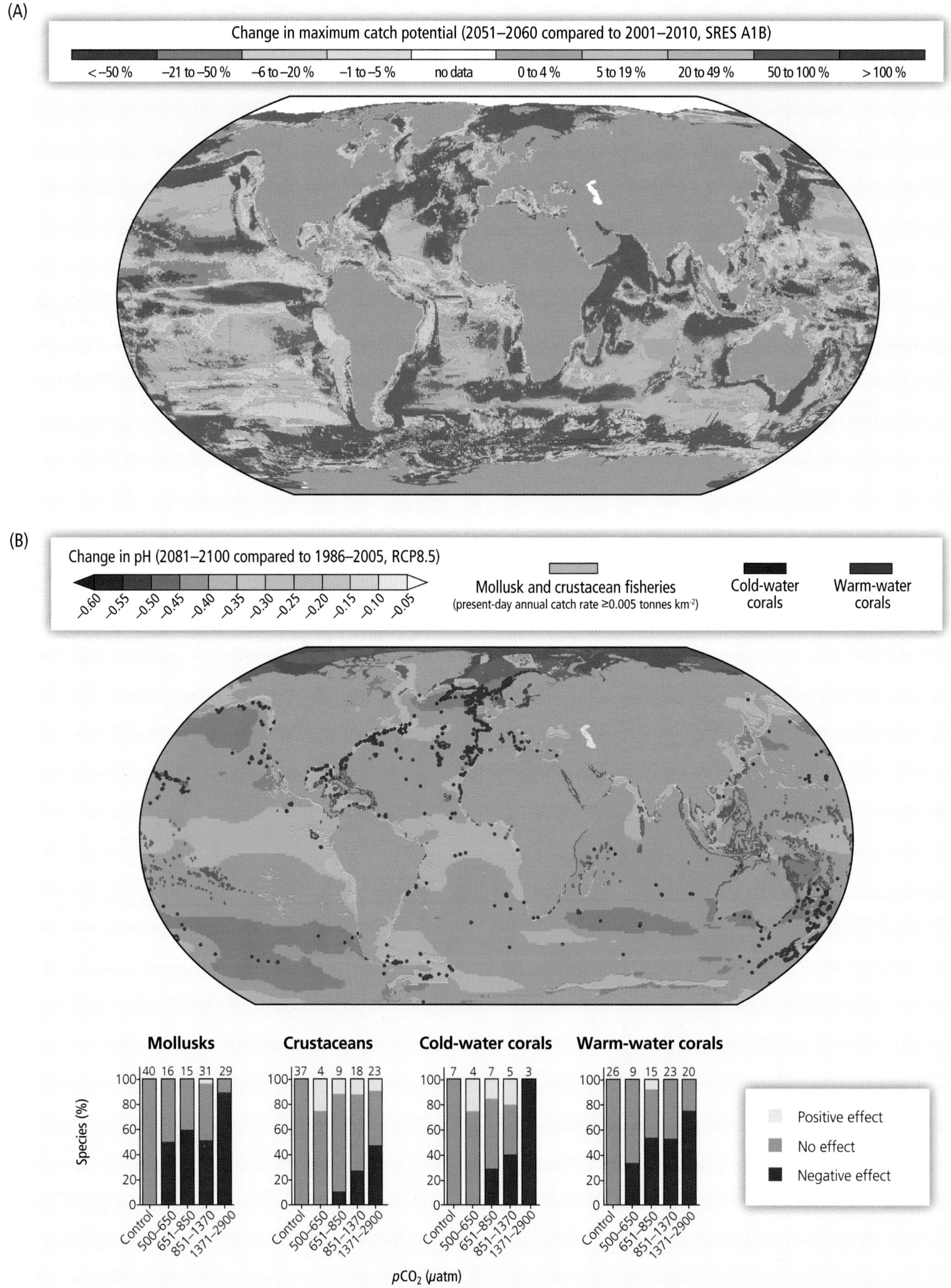
(A)
Change in maximum catch potential (2051–2060 compared to 2001–2010, SRES A1B)
<−50 %
−21 to −50 %
−6 to −20 %
−1 to −5 %
no data
0 to 4 %
5 to 19 %
20 to 49 %
50 to 100 %
> 100 %
(B)
Change in pH (2081–2100 compared to 1986–2005, RCP8.5)
−0.60
−0.55
−0.50
−0.45
−0.40
−0.35
−0.30
−0.25
−0.20
−0.15
−0.10
−0.05
Mollusk and crustacean fisheries
(present-day annual catch rate ≥0.005 tonnes km⁻²)
Cold-water corals
Warm-water corals
Mollusks
Crustaceans
Cold-water corals
Warm-water corals
40 16 15 31 29
37 4 9 18 23
7 4 7 5 3
26 9 15 23 20
Species (%)
100
80
60
40
20
0
Control
500–650
651–850
851–1370
1371–2900
pCO₂ (μatm)
Positive effect
No effect
Negative effect

Figure SPM.6 | Climate change risks for fisheries. (A) Projected global redistribution of maximum catch potential of ~1000 exploited fish and invertebrate species. Projections compare the 10-year averages 2001–2010 and 2051–2060 using SRES A1B, without analysis of potential impacts of overfishing or ocean acidification. (B) Marine mollusk and crustacean fisheries (present-day estimated annual catch rates ≥0.005 tonnes km^{-2}) and known locations of cold- and warm-water corals, depicted on a global map showing the projected distribution of ocean acidification under RCP8.5 (pH change from 1986–2005 to 2081–2100). [WGI AR5 Figure SPM.8] The bottom panel compares sensitivity to ocean acidification across mollusks, crustaceans, and corals, vulnerable animal phyla with socioeconomic relevance (e.g., for coastal protection and fisheries). The number of species analyzed across studies is given for each category of elevated CO_2. For 2100, RCP scenarios falling within each CO_2 partial pressure (pCO_2) category are as follows: RCP4.5 for 500–650 μatm (approximately equivalent to ppm in the atmosphere), RCP6.0 for 651–850 μatm, and RCP8.5 for 851–1370 μatm. By 2150, RCP8.5 falls within the 1371–2900 μatm category. The control category corresponds to 380 μatm. [6.1, 6.3, 30.5, Figures 6-10 and 6-14; WGI AR5 Box SPM.1]

Coastal systems and low-lying areas

Due to sea level rise projected throughout the 21st century and beyond, coastal systems and low-lying areas will increasingly experience adverse impacts such as submergence, coastal flooding, and coastal erosion (*very high confidence*). The population and assets projected to be exposed to coastal risks as well as human pressures on coastal ecosystems will increase significantly in the coming decades due to population growth, economic development, and urbanization (*high confidence*). The relative costs of coastal adaptation vary strongly among and within regions and countries for the 21st century. Some low-lying developing countries and small island states are expected to face very high impacts that, in some cases, could have associated damage and adaptation costs of several percentage points of GDP.[51]

Marine systems

Due to projected climate change by the mid 21st century and beyond, global marine-species redistribution and marine-biodiversity reduction in sensitive regions will challenge the sustained provision of fisheries productivity and other ecosystem services (*high confidence*). Spatial shifts of marine species due to projected warming will cause high-latitude invasions and high local-extinction rates in the tropics and semi-enclosed seas (*medium confidence*). Species richness and fisheries catch potential are projected to increase, on average, at mid and high latitudes (*high confidence*) and decrease at tropical latitudes (*medium confidence*). See Figure SPM.6A. The progressive expansion of oxygen minimum zones and anoxic "dead zones" is projected to further constrain fish habitat. Open-ocean net primary production is projected to redistribute and, by 2100, fall globally under all RCP scenarios. Climate change adds to the threats of over-fishing and other non-climatic stressors, thus complicating marine management regimes (*high confidence*).[52]

For medium- to high-emission scenarios (RCP4.5, 6.0, and 8.5), ocean acidification poses substantial risks to marine ecosystems, especially polar ecosystems and coral reefs, associated with impacts on the physiology, behavior, and population dynamics of individual species from phytoplankton to animals (*medium* to *high confidence*). Highly calcified mollusks, echinoderms, and reef-building corals are more sensitive than crustaceans (*high confidence*) and fishes (*low confidence*), with potentially detrimental consequences for fisheries and livelihoods. See Figure SPM.6B. Ocean acidification acts together with other global changes (e.g., warming, decreasing oxygen levels) and with local changes (e.g., pollution, eutrophication) (*high confidence*). Simultaneous drivers, such as warming and ocean acidification, can lead to interactive, complex, and amplified impacts for species and ecosystems.[53]

Food security and food production systems

For the major crops (wheat, rice, and maize) in tropical and temperate regions, climate change without adaptation is projected to negatively impact production for local temperature increases of 2°C or more above late-20th-century levels, although individual locations may benefit (*medium confidence*). Projected impacts vary across crops and regions and adaptation scenarios, with about 10% of projections for the period 2030–2049 showing yield gains of more than 10%, and about 10% of projections showing yield losses of more than

[51] 5.3-5, 8.2, 22.3, 24.4, 25.6, 26.3, 26.8, Table 26-1, Box 25-1
[52] 6.3-5, 7.4, 25.6, 28.3, 30.6-7, Boxes CC-MB and CC-PP
[53] 5.4, 6.3-5, 22.3, 25.6, 28.3, 30.5, Boxes CC-CR, CC-OA, and TS.7

SPM

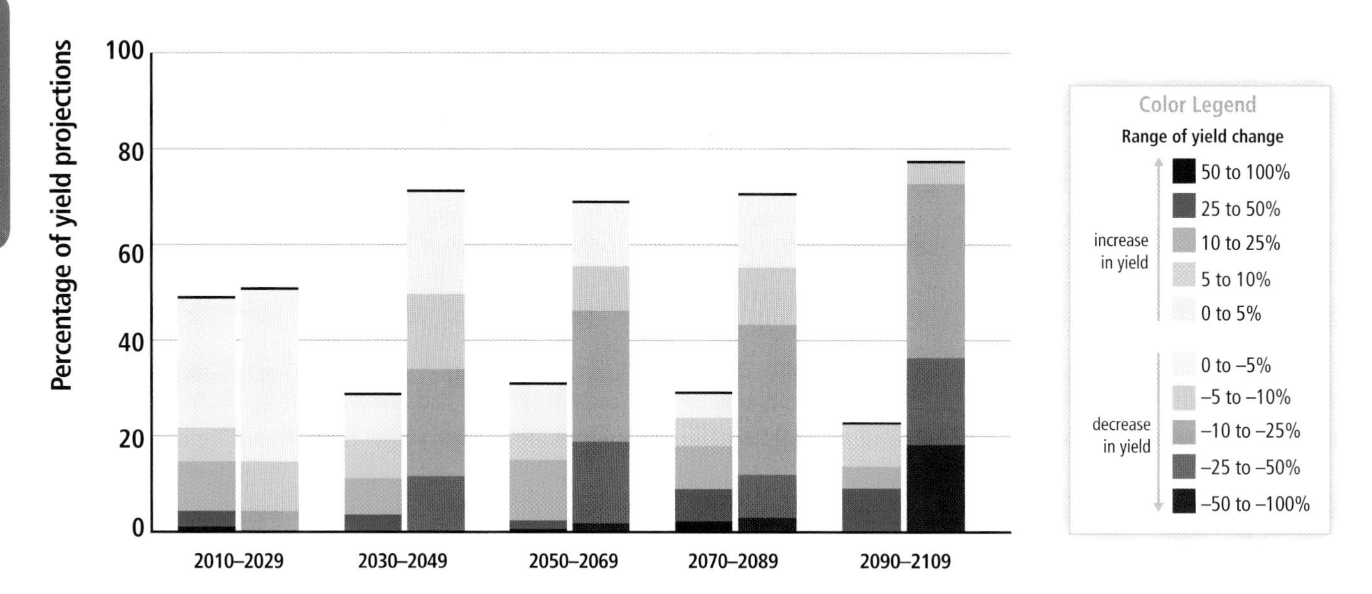

Figure SPM.7 | Summary of projected changes in crop yields, due to climate change over the 21st century. The figure includes projections for different emission scenarios, for tropical and temperate regions, and for adaptation and no-adaptation cases combined. Relatively few studies have considered impacts on cropping systems for scenarios where global mean temperatures increase by 4°C or more. For five timeframes in the near term and long term, data (n=1090) are plotted in the 20-year period on the horizontal axis that includes the midpoint of each future projection period. Changes in crop yields are relative to late-20th-century levels. Data for each timeframe sum to 100%. [Figure 7-5]

25%, compared to the late 20th century. After 2050 the risk of more severe yield impacts increases and depends on the level of warming. See Figure SPM.7. Climate change is projected to progressively increase inter-annual variability of crop yields in many regions. These projected impacts will occur in the context of rapidly rising crop demand.[54]

All aspects of food security are potentially affected by climate change, including food access, utilization, and price stability (*high confidence*). Redistribution of marine fisheries catch potential towards higher latitudes poses risk of reduced supplies, income, and employment in tropical countries, with potential implications for food security (*medium confidence*). Global temperature increases of ~4°C or more above late-20th-century levels, combined with increasing food demand, would pose large risks to food security globally and regionally (*high confidence*). Risks to food security are generally greater in low-latitude areas.[55]

Urban areas

Many global risks of climate change are concentrated in urban areas (*medium confidence*). Steps that build resilience and enable sustainable development can accelerate successful climate-change adaptation globally. Heat stress, extreme precipitation, inland and coastal flooding, landslides, air pollution, drought, and water scarcity pose risks in urban areas for people, assets, economies, and ecosystems (*very high confidence*). Risks are amplified for those lacking essential infrastructure and services or living in poor-quality housing and exposed areas. Reducing basic service deficits, improving housing, and building resilient infrastructure systems could significantly reduce vulnerability and exposure in urban areas. Urban adaptation benefits from effective multi-level urban risk governance, alignment of policies and incentives, strengthened local government and community adaptation capacity, synergies with the private sector, and appropriate financing and institutional development (*medium confidence*). Increased capacity, voice, and influence of low-income groups and vulnerable communities and their partnerships with local governments also benefit adaptation.[56]

[54] 7.4-5, 22.3, 24.4, 25.7, 26.5, Table 7-2, Figures 7-4, 7-5, 7-6, 7-7, and 7-8
[55] 6.3-5, 7.4-5, 9.3, 22.3, 24.4, 25.7, 26.5, Table 7-3, Figures 7-1, 7-4, and 7-7, Box 7-1
[56] 3.5, 8.2-4, 22.3, 24.4-5, 26.8, Table 8-2, Boxes 25-9 and CC-HS

Rural areas

Major future rural impacts are expected in the near term and beyond through impacts on water availability and supply, food security, and agricultural incomes, including shifts in production areas of food and non-food crops across the world (*high confidence*). These impacts are expected to disproportionately affect the welfare of the poor in rural areas, such as female-headed households and those with limited access to land, modern agricultural inputs, infrastructure, and education. Further adaptations for agriculture, water, forestry, and biodiversity can occur through policies taking account of rural decision-making contexts. Trade reform and investment can improve market access for small-scale farms (*medium confidence*).[57]

Key economic sectors and services

For most economic sectors, the impacts of drivers such as changes in population, age structure, income, technology, relative prices, lifestyle, regulation, and governance are projected to be large relative to the impacts of climate change (*medium evidence, high agreement*). Climate change is projected to reduce energy demand for heating and increase energy demand for cooling in the residential and commercial sectors (*robust evidence, high agreement*). Climate change is projected to affect energy sources and technologies differently, depending on resources (e.g., water flow, wind, insolation), technological processes (e.g., cooling), or locations (e.g., coastal regions, floodplains) involved. More severe and/or frequent extreme weather events and/or hazard types are projected to increase losses and loss variability in various regions and challenge insurance systems to offer affordable coverage while raising more risk-based capital, particularly in developing countries. Large-scale public-private risk reduction initiatives and economic diversification are examples of adaptation actions.[58]

Global economic impacts from climate change are difficult to estimate. Economic impact estimates completed over the past 20 years vary in their coverage of subsets of economic sectors and depend on a large number of assumptions, many of which are disputable, and many estimates do not account for catastrophic changes, tipping points, and many other factors.[59] With these recognized limitations, the incomplete estimates of global annual economic losses for additional temperature increases of ~2°C are between 0.2 and 2.0% of income (±1 standard deviation around the mean) (*medium evidence, medium agreement*). Losses are *more likely than not* to be greater, rather than smaller, than this range (*limited evidence, high agreement*). Additionally, there are large differences between and within countries. Losses accelerate with greater warming (*limited evidence, high agreement*), but few quantitative estimates have been completed for additional warming around 3°C or above. Estimates of the incremental economic impact of emitting carbon dioxide lie between a few dollars and several hundreds of dollars per tonne of carbon[60] (*robust evidence, medium agreement*). Estimates vary strongly with the assumed damage function and discount rate.[61]

Human health

Until mid-century, projected climate change will impact human health mainly by exacerbating health problems that already exist (*very high confidence*). Throughout the 21st century, climate change is expected to lead to increases in ill-health in many regions and especially in developing countries with low income, as compared to a baseline without climate change (*high confidence*). Examples include greater likelihood of injury, disease, and death due to more intense heat waves and fires (*very high confidence*); increased likelihood of under-nutrition resulting from diminished food production in poor regions (*high confidence*); risks from lost work capacity and reduced labor productivity in vulnerable populations; and increased risks from food- and water-borne diseases (*very high confidence*) and

---

[57] 9.3, 25.9, 26.8, 28.2, 28.4, Box 25-5

[58] 3.5, 10.2, 10.7, 10.10, 17.4-5, 25.7, 26.7-9, Box 25-7

[59] Disaster loss estimates are lower-bound estimates because many impacts, such as loss of human lives, cultural heritage, and ecosystem services, are difficult to value and monetize, and thus they are poorly reflected in estimates of losses. Impacts on the informal or undocumented economy as well as indirect economic effects can be very important in some areas and sectors, but are generally not counted in reported estimates of losses. [SREX 4.5]

[60] 1 tonne of carbon = 3.667 tonne of CO_2

[61] 10.9

SPM

vector-borne diseases (*medium confidence*). Positive effects are expected to include modest reductions in cold-related mortality and morbidity in some areas due to fewer cold extremes (*low confidence*), geographical shifts in food production (*medium confidence*), and reduced capacity of vectors to transmit some diseases. But globally over the 21st century, the magnitude and severity of negative impacts are projected to increasingly outweigh positive impacts (*high confidence*). The most effective vulnerability reduction measures for health in the near term are programs that implement and improve basic public health measures such as provision of clean water and sanitation, secure essential health care including vaccination and child health services, increase capacity for disaster preparedness and response, and alleviate poverty (*very high confidence*). By 2100 for the high-emission scenario RCP8.5, the combination of high temperature and humidity in some areas for parts of the year is projected to compromise normal human activities, including growing food or working outdoors (*high confidence*).[62]

Human security

Climate change over the 21st century is projected to increase displacement of people (*medium evidence, high agreement*). Displacement risk increases when populations that lack the resources for planned migration experience higher exposure to extreme weather events, in both rural and urban areas, particularly in developing countries with low income. Expanding opportunities for mobility can reduce vulnerability for such populations. Changes in migration patterns can be responses to both extreme weather events and longer-term climate variability and change, and migration can also be an effective adaptation strategy. There is *low confidence* in quantitative projections of changes in mobility, due to its complex, multi-causal nature.[63]

Climate change can indirectly increase risks of violent conflicts in the form of civil war and inter-group violence by amplifying well-documented drivers of these conflicts such as poverty and economic shocks (*medium confidence*). Multiple lines of evidence relate climate variability to these forms of conflict.[64]

The impacts of climate change on the critical infrastructure and territorial integrity of many states are expected to influence national security policies (*medium evidence, medium agreement*). For example, land inundation due to sea level rise poses risks to the territorial integrity of small island states and states with extensive coastlines. Some transboundary impacts of climate change, such as changes in sea ice, shared water resources, and pelagic fish stocks, have the potential to increase rivalry among states, but robust national and intergovernmental institutions can enhance cooperation and manage many of these rivalries.[65]

Livelihoods and poverty

Throughout the 21st century, climate-change impacts are projected to slow down economic growth, make poverty reduction more difficult, further erode food security, and prolong existing and create new poverty traps, the latter particularly in urban areas and emerging hotspots of hunger (*medium confidence*). Climate-change impacts are expected to exacerbate poverty in most developing countries and create new poverty pockets in countries with increasing inequality, in both developed and developing countries. In urban and rural areas, wage-labor-dependent poor households that are net buyers of food are expected to be particularly affected due to food price increases, including in regions with high food insecurity and high inequality (particularly in Africa), although the agricultural self-employed could benefit. Insurance programs, social protection measures, and disaster risk management may enhance long-term livelihood resilience among poor and marginalized people, if policies address poverty and multidimensional inequalities.[66]

B-3. Regional Key Risks and Potential for Adaptation

Risks will vary through time across regions and populations, dependent on myriad factors including the extent of adaptation and mitigation. A selection of key regional risks identified with *medium* to *high confidence* is presented in Assessment Box SPM.2. For extended summary of regional risks and potential benefits, see Technical Summary Section B-3 and WGII AR5 Part B: Regional Aspects.

SPM

Assessment Box SPM.2 | Regional Key Risks

The accompanying Assessment Box SPM.2 Table 1 highlights several representative key risks for each region. Key risks have been identified based on assessment of the relevant scientific, technical, and socioeconomic literature detailed in supporting chapter sections. Identification of key risks was based on expert judgment using the following specific criteria: large magnitude, high probability, or irreversibility of impacts; timing of impacts; persistent vulnerability or exposure contributing to risks; or limited potential to reduce risks through adaptation or mitigation.

For each key risk, risk levels were assessed for three timeframes. For the present, risk levels were estimated for current adaptation and a hypothetical highly adapted state, identifying where current adaptation deficits exist. For two future timeframes, risk levels were estimated for a continuation of current adaptation and for a highly adapted state, representing the potential for and limits to adaptation. The risk levels integrate probability and consequence over the widest possible range of potential outcomes, based on available literature. These potential outcomes result from the interaction of climate-related hazards, vulnerability, and exposure. Each risk level reflects total risk from climatic and non-climatic factors. Key risks and risk levels vary across regions and over time, given differing socioeconomic development pathways, vulnerability and exposure to hazards, adaptive capacity, and risk perceptions. Risk levels are not necessarily comparable, especially across regions, because the assessment considers potential impacts and adaptation in different physical, biological, and human systems across diverse contexts. This assessment of risks acknowledges the importance of differences in values and objectives in interpretation of the assessed risk levels.

Assessment Box SPM.2 Table 1 | Key regional risks from climate change and the potential for reducing risks through adaptation and mitigation. Each key risk is characterized as very low to very high for three timeframes: the present, near term (here, assessed over 2030–2040), and longer term (here, assessed over 2080–2100). In the near term, projected levels of global mean temperature increase do not diverge substantially for different emission scenarios. For the longer term, risk levels are presented for two scenarios of global mean temperature increase (2°C and 4°C above preindustrial levels). These scenarios illustrate the potential for mitigation and adaptation to reduce the risks related to climate change. Climate-related drivers of impacts are indicated by icons.

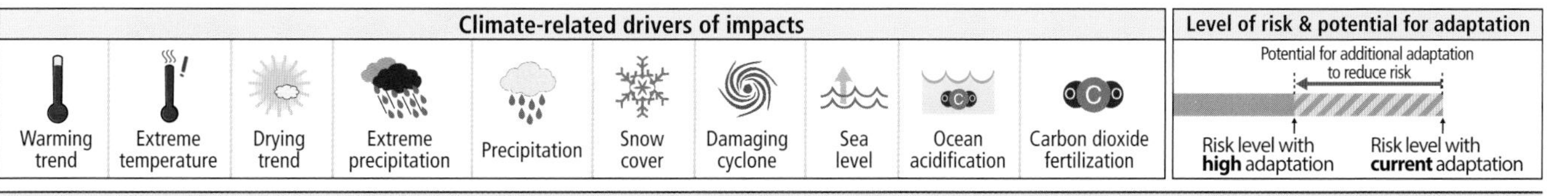

Africa

Key risk	Adaptation issues & prospects	Climatic drivers	Timeframe	Risk & potential for adaptation (Very low – Medium – Very high)
Compounded stress on water resources facing significant strain from overexploitation and degradation at present and increased demand in the future, with drought stress exacerbated in drought-prone regions of Africa (*high confidence*) [22.3-4]	• Reducing non-climate stressors on water resources • Strengthening institutional capacities for demand management, groundwater assessment, integrated water-wastewater planning, and integrated land and water governance • Sustainable urban development	Warming trend, Drying trend, Extreme temperature, Sea level	Present; Near term (2030–2040); Long term (2080–2100) 2°C, 4°C	
Reduced crop productivity associated with heat and drought stress, with strong adverse effects on regional, national, and household livelihood and food security, also given increased pest and disease damage and flood impacts on food system infrastructure (*high confidence*) [22.3-4]	• Technological adaptation responses (e.g., stress-tolerant crop varieties, irrigation, enhanced observation systems) • Enhancing smallholder access to credit and other critical production resources; Diversifying livelihoods • Strengthening institutions at local, national, and regional levels to support agriculture (including early warning systems) and gender-oriented policy • Agronomic adaptation responses (e.g., agroforestry, conservation agriculture)	Warming trend, Drying trend, Extreme temperature, Extreme precipitation	Present; Near term (2030–2040); Long term (2080–2100) 2°C, 4°C	
Changes in the incidence and geographic range of vector- and water-borne diseases due to changes in the mean and variability of temperature and precipitation, particularly along the edges of their distribution (*medium confidence*) [22.3]	• Achieving development goals, particularly improved access to safe water and improved sanitation, and enhancement of public health functions such as surveillance • Vulnerability mapping and early warning systems • Coordination across sectors • Sustainable urban development	Warming trend, Precipitation, Extreme precipitation	Present; Near term (2030–2040); Long term (2080–2100) 2°C, 4°C	

[62] 8.2, 11.3-8, 19.3, 22.3, 25.8, 26.6, Figure 25-5, Box CC-HS

[63] 9.3, 12.4, 19.4, 22.3, 25.9

[64] 12.5, 13.2, 19.4

[65] 12.5-6, 23.9, 25.9

[66] 8.1, 8.3-4, 9.3, 10.9, 13.2-4, 22.3, 26.8

Continued next page →

SPM

Assessment Box SPM.2 Table 1 (continued)

Continued next page →

Europe

Key risk	Adaptation issues & prospects	Climatic drivers	Timeframe	Risk & potential for adaptation
Increased economic losses and people affected by flooding in river basins and coasts, driven by increasing urbanization, increasing sea levels, coastal erosion, and peak river discharges (*high confidence*) [23.2-3, 23.7]	Adaptation can prevent most of the projected damages (*high confidence*). • Significant experience in hard flood-protection technologies and increasing experience with restoring wetlands • High costs for increasing flood protection • Potential barriers to implementation: demand for land in Europe and environmental and landscape concerns		Present Near term (2030–2040) Long term (2080–2100) 2°C / 4°C	Very low — Medium — Very high
Increased water restrictions. Significant reduction in water availability from river abstraction and from groundwater resources, combined with increased water demand (e.g., for irrigation, energy and industry, domestic use) and with reduced water drainage and runoff as a result of increased evaporative demand, particularly in southern Europe (*high confidence*) [23.4, 23.7]	• Proven adaptation potential from adoption of more water-efficient technologies and of water-saving strategies (e.g., for irrigation, crop species, land cover, industries, domestic use) • Implementation of best practices and governance instruments in river basin management plans and integrated water management		Present Near term (2030–2040) Long term (2080–2100) 2°C / 4°C	Very low — Medium — Very high
Increased economic losses and people affected by extreme heat events: impacts on health and well-being, labor productivity, crop production, air quality, and increasing risk of wildfires in southern Europe and in Russian boreal region (*medium confidence*) [23.3-7, Table 23-1]	• Implementation of warning systems • Adaptation of dwellings and workplaces and of transport and energy infrastructure • Reductions in emissions to improve air quality • Improved wildfire management • Development of insurance products against weather-related yield variations		Present Near term (2030–2040) Long term (2080–2100) 2°C / 4°C	Very low — Medium — Very high

Asia

Key risk	Adaptation issues & prospects	Climatic drivers	Timeframe	Risk & potential for adaptation
Increased riverine, coastal, and urban flooding leading to widespread damage to infrastructure, livelihoods, and settlements in Asia (*medium confidence*) [24.4]	• Exposure reduction via structural and non-structural measures, effective land-use planning, and selective relocation • Reduction in the vulnerability of lifeline infrastructure and services (e.g., water, energy, waste management, food, biomass, mobility, local ecosystems, telecommunications) • Construction of monitoring and early warning systems; Measures to identify exposed areas, assist vulnerable areas and households, and diversify livelihoods • Economic diversification		Present Near term (2030–2040) Long term (2080–2100) 2°C / 4°C	Very low — Medium — Very high
Increased risk of heat-related mortality (*high confidence*) [24.4]	• Heat health warning systems • Urban planning to reduce heat islands; Improvement of the built environment; Development of sustainable cities • New work practices to avoid heat stress among outdoor workers		Present Near term (2030–2040) Long term (2080–2100) 2°C / 4°C	Very low — Medium — Very high
Increased risk of drought-related water and food shortage causing malnutrition (*high confidence*) [24.4]	• Disaster preparedness including early-warning systems and local coping strategies • Adaptive/integrated water resource management • Water infrastructure and reservoir development • Diversification of water sources including water re-use • More efficient use of water (e.g., improved agricultural practices, irrigation management, and resilient agriculture)		Present Near term (2030–2040) Long term (2080–2100) 2°C / 4°C	Very low — Medium — Very high

SPM

Assessment Box SPM.2 Table 1 (continued)

Continued next page →

Australasia

Key risk	Adaptation issues & prospects	Climatic drivers	Timeframe	Risk & potential for adaptation
Significant change in community composition and structure of coral reef systems in Australia (*high confidence*) [25.6, 30.5, Boxes CC-CR and CC-OA]	• Ability of corals to adapt naturally appears limited and insufficient to offset the detrimental effects of rising temperatures and acidification. • Other options are mostly limited to reducing other stresses (water quality, tourism, fishing) and early warning systems; direct interventions such as assisted colonization and shading have been proposed but remain untested at scale.		Present Near term (2030–2040) Long term (2080–2100) 2°C / 4°C	Very low – Medium – Very high
Increased frequency and intensity of flood damage to infrastructure and settlements in Australia and New Zealand (*high confidence*) [Table 25-1, Boxes 25-8 and 25-9]	• Significant adaptation deficit in some regions to current flood risk. • Effective adaptation includes land-use controls and relocation as well as protection and accommodation of increased risk to ensure flexibility.		Present Near term (2030–2040) Long term (2080–2100) 2°C / 4°C	Very low – Medium – Very high
Increasing risks to coastal infrastructure and low-lying ecosystems in Australia and New Zealand, with widespread damage towards the upper end of projected sea-level-rise ranges (*high confidence*) [25.6, 25.10, Box 25-1]	• Adaptation deficit in some locations to current coastal erosion and flood risk. Successive building and protection cycles constrain flexible responses. • Effective adaptation includes land-use controls and ultimately relocation as well as protection and accommodation.		Present Near term (2030–2040) Long term (2080–2100) 2°C / 4°C	Very low – Medium – Very high

North America

Key risk	Adaptation issues & prospects	Climatic drivers	Timeframe	Risk & potential for adaptation
Wildfire-induced loss of ecosystem integrity, property loss, human morbidity, and mortality as a result of increased drying trend and temperature trend (*high confidence*) [26.4, 26.8, Box 26-2]	• Some ecosystems are more fire-adapted than others. Forest managers and municipal planners are increasingly incorporating fire protection measures (e.g., prescribed burning, introduction of resilient vegetation). Institutional capacity to support ecosystem adaptation is limited. • Adaptation of human settlements is constrained by rapid private property development in high-risk areas and by limited household-level adaptive capacity. • Agroforestry can be an effective strategy for reduction of slash and burn practices in Mexico.		Present Near term (2030–2040) Long term (2080–2100) 2°C / 4°C	Very low – Medium – Very high
Heat-related human mortality (*high confidence*) [26.6, 26.8]	• Residential air conditioning (A/C) can effectively reduce risk. However, availability and usage of A/C is highly variable and is subject to complete loss during power failures. Vulnerable populations include athletes and outdoor workers for whom A/C is not available. • Community- and household-scale adaptations have the potential to reduce exposure to heat extremes via family support, early heat warning systems, cooling centers, greening, and high-albedo surfaces.		Present Near term (2030–2040) Long term (2080–2100) 2°C / 4°C	Very low – Medium – Very high
Urban floods in riverine and coastal areas, inducing property and infrastructure damage; supply chain, ecosystem, and social system disruption; public health impacts; and water quality impairment, due to sea level rise, extreme precipitation, and cyclones (*high confidence*) [26.2-4, 26.8]	• Implementing management of urban drainage is expensive and disruptive to urban areas. • Low-regret strategies with co-benefits include less impervious surfaces leading to more groundwater recharge, green infrastructure, and rooftop gardens. • Sea level rise increases water elevations in coastal outfalls, which impedes drainage. In many cases, older rainfall design standards are being used that need to be updated to reflect current climate conditions. • Conservation of wetlands, including mangroves, and land-use planning strategies can reduce the intensity of flood events.		Present Near term (2030–2040) Long term (2080–2100) 2°C / 4°C	Very low – Medium – Very high

SPM

Assessment Box SPM.2 Table 1 (continued)

Continued next page →

Central and South America

Key risk	Adaptation issues & prospects	Climatic drivers	Timeframe	Risk & potential for adaptation
Water availability in semi-arid and glacier-melt-dependent regions and Central America; flooding and landslides in urban and rural areas due to extreme precipitation (*high confidence*) [27.3]	• Integrated water resource management • Urban and rural flood management (including infrastructure), early warning systems, better weather and runoff forecasts, and infectious disease control		Present Near term (2030–2040) Long term (2080–2100) 2°C / 4°C	Very low / Medium / Very high
Decreased food production and food quality (*medium confidence*) [27.3]	• Development of new crop varieties more adapted to climate change (temperature and drought) • Offsetting of human and animal health impacts of reduced food quality • Offsetting of economic impacts of land-use change • Strengthening traditional indigenous knowledge systems and practices		Present Near term (2030–2040) Long term (2080–2100) 2°C / 4°C	Very low / Medium / Very high
Spread of vector-borne diseases in altitude and latitude (*high confidence*) [27.3]	• Development of early warning systems for disease control and mitigation based on climatic and other relevant inputs. Many factors augment vulnerability. • Establishing programs to extend basic public health services		Present Near term (2030–2040) Long term (2080–2100) 2°C / 4°C	Very low / Medium / Very high Long term 2°C: not available Long term 4°C: not available

Polar Regions

Key risk	Adaptation issues & prospects	Climatic drivers	Timeframe	Risk & potential for adaptation
Risks for freshwater and terrestrial ecosystems (*high confidence*) and marine ecosystems (*medium confidence*), due to changes in ice, snow cover, permafrost, and freshwater/ocean conditions, affecting species' habitat quality, ranges, phenology, and productivity, as well as dependent economies [28.2-4]	• Improved understanding through scientific and indigenous knowledge, producing more effective solutions and/or technological innovations • Enhanced monitoring, regulation, and warning systems that achieve safe and sustainable use of ecosystem resources • Hunting or fishing for different species, if possible, and diversifying income sources		Present Near term (2030–2040) Long term (2080–2100) 2°C / 4°C	Very low / Medium / Very high
Risks for the health and well-being of Arctic residents, resulting from injuries and illness from the changing physical environment, food insecurity, lack of reliable and safe drinking water, and damage to infrastructure, including infrastructure in permafrost regions (*high confidence*) [28.2-4]	• Co-production of more robust solutions that combine science and technology with indigenous knowledge • Enhanced observation, monitoring, and warning systems • Improved communications, education, and training • Shifting resource bases, land use, and/or settlement areas		Present Near term (2030–2040) Long term (2080–2100) 2°C / 4°C	Very low / Medium / Very high
Unprecedented challenges for northern communities due to complex inter-linkages between climate-related hazards and societal factors, particularly if rate of change is faster than social systems can adapt (*high confidence*) [28.2-4]	• Co-production of more robust solutions that combine science and technology with indigenous knowledge • Enhanced observation, monitoring, and warning systems • Improved communications, education, and training • Adaptive co-management responses developed through the settlement of land claims		Present Near term (2030–2040) Long term (2080–2100) 2°C / 4°C	Very low / Medium / Very high

Small Islands

Key risk	Adaptation issues & prospects	Climatic drivers	Timeframe	Risk & potential for adaptation
Loss of livelihoods, coastal settlements, infrastructure, ecosystem services, and economic stability (*high confidence*) [29.6, 29.8, Figure 29-4]	• Significant potential exists for adaptation in islands, but additional external resources and technologies will enhance response. • Maintenance and enhancement of ecosystem functions and services and of water and food security • Efficacy of traditional community coping strategies is expected to be substantially reduced in the future.		Present Near term (2030–2040) Long term (2080–2100) 2°C / 4°C	Very low / Medium / Very high
The interaction of rising global mean sea level in the 21st century with high-water-level events will threaten low-lying coastal areas (*high confidence*) [29.4, Table 29-1; WGI AR5 13.5, Table 13.5]	• High ratio of coastal area to land mass will make adaptation a significant financial and resource challenge for islands. • Adaptation options include maintenance and restoration of coastal landforms and ecosystems, improved management of soils and freshwater resources, and appropriate building codes and settlement patterns.		Present Near term (2030–2040) Long term (2080–2100) 2°C / 4°C	Very low / Medium / Very high

SPM

Assessment Box SPM.2 Table 1 (continued)

The Ocean				
Key risk	**Adaptation issues & prospects**	**Climatic drivers**	**Timeframe**	**Risk & potential for adaptation**
Distributional shift in fish and invertebrate species, and decrease in fisheries catch potential at low latitudes, e.g., in equatorial upwelling and coastal boundary systems and sub-tropical gyres (*high confidence*) [6.3, 30.5-6, Tables 6-6 and 30-3, Box CC-MB]	• Evolutionary adaptation potential of fish and invertebrate species to warming is limited as indicated by their changes in distribution to maintain temperatures. • Human adaptation options: Large-scale translocation of industrial fishing activities following the regional decreases (low latitude) vs. possibly transient increases (high latitude) in catch potential; Flexible management that can react to variability and change; Improvement of fish resilience to thermal stress by reducing other stressors such as pollution and eutrophication; Expansion of sustainable aquaculture and the development of alternative livelihoods in some regions.		Present; Near term (2030–2040); Long term (2080–2100) 2°C, 4°C	Very low — Medium — Very high
Reduced biodiversity, fisheries abundance, and coastal protection by coral reefs due to heat-induced mass coral bleaching and mortality increases, exacerbated by ocean acidification, e.g., in coastal boundary systems and sub-tropical gyres (*high confidence*) [5.4, 6.4, 30.3, 30.5-6, Tables 6-6 and 30-3, Box CC-CR]	• Evidence of rapid evolution by corals is very limited. Some corals may migrate to higher latitudes, but entire reef systems are not expected to be able to track the high rates of temperature shifts. • Human adaptation options are limited to reducing other stresses, mainly by enhancing water quality, and limiting pressures from tourism and fishing. These options will delay human impacts of climate change by a few decades, but their efficacy will be severely reduced as thermal stress increases.		Present; Near term (2030–2040); Long term (2080–2100) 2°C, 4°C	Very low — Medium — Very high
Coastal inundation and habitat loss due to sea level rise, extreme events, changes in precipitation, and reduced ecological resilience, e.g., in coastal boundary systems and sub-tropical gyres (*medium* to *high confidence*) [5.5, 30.5-6, Tables 6-6 and 30-3, Box CC-CR]	• Human adaptation options are limited to reducing other stresses, mainly by reducing pollution and limiting pressures from tourism, fishing, physical destruction, and unsustainable aquaculture. • Reducing deforestation and increasing reforestation of river catchments and coastal areas to retain sediments and nutrients • Increased mangrove, coral reef, and seagrass protection, and restoration to protect numerous ecosystem goods and services such as coastal protection, tourist value, and fish habitat		Present; Near term (2030–2040); Long term (2080–2100) 2°C, 4°C	Very low — Medium — Very high

C: MANAGING FUTURE RISKS AND BUILDING RESILIENCE

Managing the risks of climate change involves adaptation and mitigation decisions with implications for future generations, economies, and environments. This section evaluates adaptation as a means to build resilience and to adjust to climate-change impacts. It also considers limits to adaptation, climate-resilient pathways, and the role of transformation. See Figure SPM.8 for an overview of responses for addressing risk related to climate change.

C-1. Principles for Effective Adaptation

Adaptation is place- and context-specific, with no single approach for reducing risks appropriate across all settings (*high confidence*). Effective risk reduction and adaptation strategies consider the dynamics of vulnerability and exposure and their linkages with socioeconomic processes, sustainable development, and climate change. Specific examples of responses to climate change are presented in Table SPM.1.[67]

Adaptation planning and implementation can be enhanced through complementary actions across levels, from individuals to governments (*high confidence*). National governments can coordinate adaptation efforts of local and subnational governments, for example by protecting vulnerable groups, by supporting economic diversification, and by providing information, policy and legal frameworks, and financial support (*robust evidence*, *high agreement*). Local government and the private sector are increasingly recognized as critical to progress in adaptation, given their roles in scaling up adaptation of communities, households, and civil society and in managing risk information and financing (*medium evidence*, *high agreement*).[68]

A first step towards adaptation to future climate change is reducing vulnerability and exposure to present climate variability (*high confidence*). Strategies include actions with co-benefits for other objectives. Available strategies and actions can increase resilience across a range of possible future climates while helping to improve human health, livelihoods, social and economic well-being, and

[67] 2.1, 8.3-4, 13.1, 13.3-4, 15.2-3, 15.5, 16.2-3, 16.5, 17.2, 17.4, 19.6, 21.3, 22.4, 26.8-9, 29.6, 29.8

[68] 2.1-4, 3.6, 5.5, 8.3-4, 9.3-4, 14.2, 15.2-3, 15.5, 16.2-5, 17.2-3, 22.4, 24.4, 25.4, 26.8-9, 30.7, Tables 21-1, 21-5, & 21-6, Box 16-2

SPM

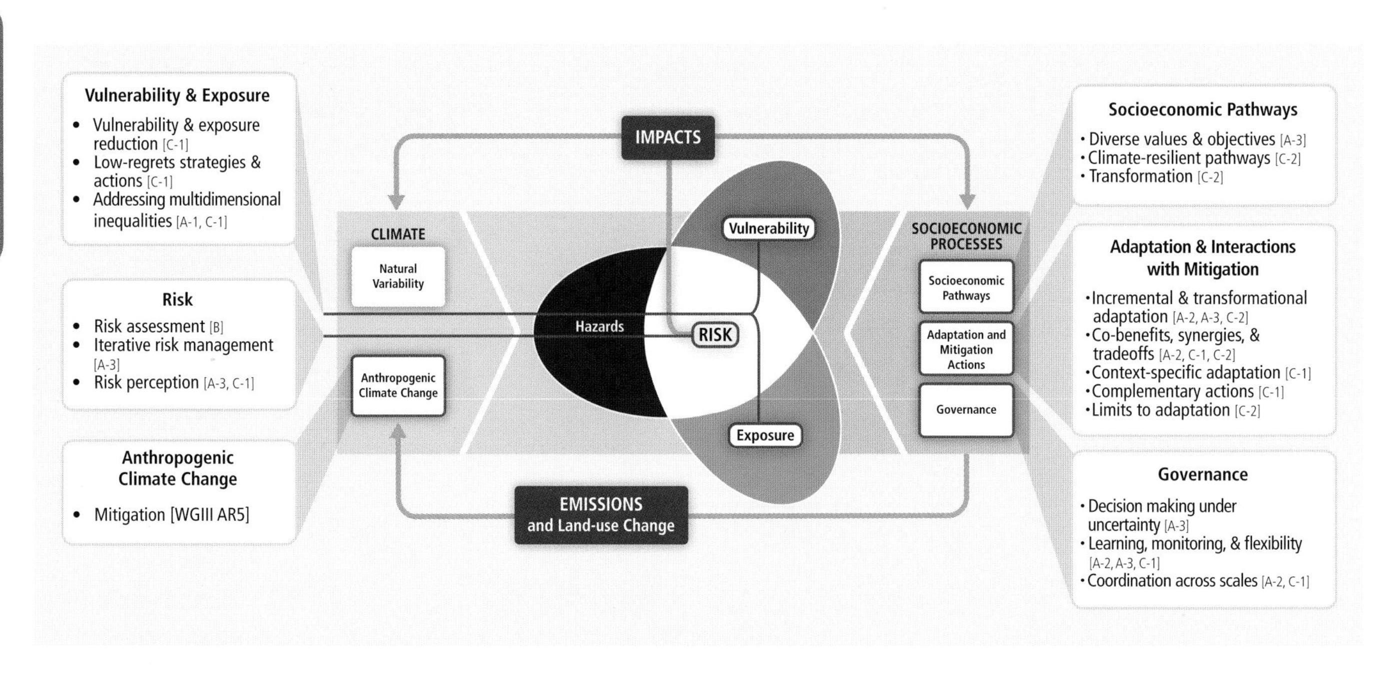

Figure SPM.8 | The solution space. Core concepts of the WGII AR5, illustrating overlapping entry points and approaches, as well as key considerations, in managing risks related to climate change, as assessed in this report and presented throughout this SPM. Bracketed references indicate sections of this summary with corresponding assessment findings.

environmental quality. See Table SPM.1. Integration of adaptation into planning and decision making can promote synergies with development and disaster risk reduction.[69]

Adaptation planning and implementation at all levels of governance are contingent on societal values, objectives, and risk perceptions (*high confidence*). Recognition of diverse interests, circumstances, social-cultural contexts, and expectations can benefit decision-making processes. Indigenous, local, and traditional knowledge systems and practices, including indigenous peoples' holistic view of community and environment, are a major resource for adapting to climate change, but these have not been used consistently in existing adaptation efforts. Integrating such forms of knowledge with existing practices increases the effectiveness of adaptation.[70]

Decision support is most effective when it is sensitive to context and the diversity of decision types, decision processes, and constituencies (*robust evidence, high agreement*). Organizations bridging science and decision making, including climate services, play an important role in the communication, transfer, and development of climate-related knowledge, including translation, engagement, and knowledge exchange (*medium evidence, high agreement*).[71]

Existing and emerging economic instruments can foster adaptation by providing incentives for anticipating and reducing impacts (*medium confidence*). Instruments include public-private finance partnerships, loans, payments for environmental services, improved resource pricing, charges and subsidies, norms and regulations, and risk sharing and transfer mechanisms. Risk financing mechanisms in the public and private sector, such as insurance and risk pools, can contribute to increasing resilience, but without attention to major design challenges, they can also provide disincentives, cause market failure, and decrease equity. Governments often play key roles as regulators, providers, or insurers of last resort.[72]

Constraints can interact to impede adaptation planning and implementation (*high confidence*). Common constraints on implementation arise from the following: limited financial and human resources; limited integration or coordination of governance; uncertainties

---

[69] 3.6, 8.3, 9.4, 14.3, 15.2-3, 17.2, 20.4, 20.6, 22.4, 24.4-5, 25.4, 25.10, 27.3-5, 29.6, Boxes 25-2 and 25-6
[70] 2.2-4, 9.4, 12.3, 13.2, 15.2, 16.2-4, 16.7, 17.2-3, 21.3, 22.4, 24.4, 24.6, 25.4, 25.8, 26.9, 28.2, 28.4, Table 15-1, Box 25-7
[71] 2.1-4, 8.4, 14.4, 16.2-3, 16.5, 21.2-3, 21.5, 22.4, Box 9-4
[72] 10.7, 10.9, 13.3, 17.4-5, Box 25-7

Table SPM.1 | Approaches for managing the risks of climate change. These approaches should be considered overlapping rather than discrete, and they are often pursued simultaneously. Mitigation is considered essential for managing the risks of climate change. It is not addressed in this table as mitigation is the focus of WGIII AR5. Examples are presented in no specific order and can be relevant to more than one category. [14.2-3, Table 14-1]

Overlapping Approaches	Category	Examples	Chapter Reference(s)
Vulnerability & Exposure Reduction through development, planning, & practices including many low-regrets measures	Human development	Improved access to education, nutrition, health facilities, energy, safe housing & settlement structures, & social support structures; Reduced gender inequality & marginalization in other forms.	8.3, 9.3, 13.1-3, 14.2-3, 22.4
	Poverty alleviation	Improved access to & control of local resources; Land tenure; Disaster risk reduction; Social safety nets & social protection; Insurance schemes.	8.3-4, 9.3, 13.1-3
	Livelihood security	Income, asset, & livelihood diversification; Improved infrastructure; Access to technology & decision-making fora; Increased decision-making power; Changed cropping, livestock, & aquaculture practices; Reliance on social networks.	7.5, 9.4, 13.1-3, 22.3-4, 23.4, 26.5, 27.3, 29.6, Table SM24-7
	Disaster risk management	Early warning systems; Hazard & vulnerability mapping; Diversifying water resources; Improved drainage; Flood & cyclone shelters; Building codes & practices; Storm & wastewater management; Transport & road infrastructure improvements.	8.2-4, 11.7, 14.3, 15.4, 22.4, 24.4, 26.6, 28.4, Box 25-1, Table 3-3
	Ecosystem management	Maintaining wetlands & urban green spaces; Coastal afforestation; Watershed & reservoir management; Reduction of other stressors on ecosystems & of habitat fragmentation; Maintenance of genetic diversity; Manipulation of disturbance regimes; Community-based natural resource management.	4.3-4, 8.3, 22.4, Table 3-3, Boxes 4-3, 8-2, 15-1, 25-8, 25-9, & CC-EA
	Spatial or land-use planning	Provisioning of adequate housing, infrastructure, & services; Managing development in flood prone & other high risk areas; Urban planning & upgrading programs; Land zoning laws; Easements; Protected areas.	4.4, 8.1-4, 22.4, 23.7-8, 27.3, Box 25-8
Adaptation including incremental & transformational adjustments	Structural/physical	***Engineered & built-environment options***: Sea walls & coastal protection structures; Flood levees; Water storage; Improved drainage; Flood & cyclone shelters; Building codes & practices; Storm & wastewater management; Transport & road infrastructure improvements; Floating houses; Power plant & electricity grid adjustments.	3.5-6, 5.5, 8.2-3, 10.2, 11.7, 23.3, 24.4, 25.7, 26.3, 26.8, Boxes 15-1, 25-1, 25-2, & 25-8
		Technological options: New crop & animal varieties; Indigenous, traditional, & local knowledge, technologies, & methods; Efficient irrigation; Water-saving technologies; Desalinization; Conservation agriculture; Food storage & preservation facilities; Hazard & vulnerability mapping & monitoring; Early warning systems; Building insulation; Mechanical & passive cooling; Technology development, transfer, & diffusion.	7.5, 8.3, 9.4, 10.3, 15.4, 22.4, 24.4, 26.3, 26.5, 27.3, 28.2, 28.4, 29.6-7, Boxes 20-5 & 25-2, Tables 3-3 & 15-1
		Ecosystem-based options: Ecological restoration; Soil conservation; Afforestation & reforestation; Mangrove conservation & replanting; Green infrastructure (e.g., shade trees, green roofs); Controlling overfishing; Fisheries co-management; Assisted species migration & dispersal; Ecological corridors; Seed banks, gene banks, & other *ex situ* conservation; Community-based natural resource management.	4.4, 5.5, 6.4, 8.3, 9.4, 11.7, 15.4, 22.4, 23.6-7, 24.4, 25.6, 27.3, 28.2, 29.7, 30.6, Boxes 15-1, 22-2, 25-9, 26-2, & CC-EA
		Services: Social safety nets & social protection; Food banks & distribution of food surplus; Municipal services including water & sanitation; Vaccination programs; Essential public health services; Enhanced emergency medical services.	3.5-6, 8.3, 9.3, 11.7, 11.9, 22.4, 29.6, Box 13-2
	Institutional	***Economic options***: Financial incentives; Insurance; Catastrophe bonds; Payments for ecosystem services; Pricing water to encourage universal provision and careful use; Microfinance; Disaster contingency funds; Cash transfers; Public-private partnerships.	8.3-4, 9.4, 10.7, 11.7, 13.3, 15.4, 17.5, 22.4, 26.7, 27.6, 29.6, Box 25-7
		Laws & regulations: Land zoning laws; Building standards & practices; Easements; Water regulations & agreements; Laws to support disaster risk reduction; Laws to encourage insurance purchasing; Defined property rights & land tenure security; Protected areas; Fishing quotas; Patent pools & technology transfer.	4.4, 8.3, 9.3, 10.5, 10.7, 15.2, 15.4, 17.5, 22.4, 23.4, 23.7, 24.4, 25.4, 26.3, 27.3, 30.6, Table 25-2, Box CC-CR
		National & government policies & programs: National & regional adaptation plans including mainstreaming; Sub-national & local adaptation plans; Economic diversification; Urban upgrading programs; Municipal water management programs; Disaster planning & preparedness; Integrated water resource management; Integrated coastal zone management; Ecosystem-based management; Community-based adaptation.	2.4, 3.6, 4.4, 5.5, 6.4, 7.5, 8.3, 11.7, 15.2-5, 22.4, 23.7, 25.4, 25.8, 26.8-9, 27.3-4, 29.6, Boxes 25-1, 25-2, & 25-9, Tables 9-2 & 17-1
	Social	***Educational options***: Awareness raising & integrating into education; Gender equity in education; Extension services; Sharing indigenous, traditional, & local knowledge; Participatory action research & social learning; Knowledge-sharing & learning platforms.	8.3-4, 9.4, 11.7, 12.3, 15.2-4, 22.4, 25.4, 28.4, 29.6, Tables 15-1 & 25-2
		Informational options: Hazard & vulnerability mapping; Early warning & response systems; Systematic monitoring & remote sensing; Climate services; Use of indigenous climate observations; Participatory scenario development; Integrated assessments.	2.4, 5.5, 8.3-4, 9.4, 11.7, 15.2-4, 22.4, 23.5, 24.4, 25.8, 26.6, 26.8, 27.3, 28.2, 28.5, 30.6, Table 25-2, Box 26-3
		Behavioral options: Household preparation & evacuation planning; Migration; Soil & water conservation; Storm drain clearance; Livelihood diversification; Changed cropping, livestock, & aquaculture practices; Reliance on social networks.	5.5, 7.5, 9.4, 12.4, 22.3-4, 23.4, 23.7, 25.7, 26.5, 27.3, 29.6, Table SM24-7, Box 25-5
Transformation	Spheres of change	***Practical***: Social & technical innovations, behavioral shifts, or institutional & managerial changes that produce substantial shifts in outcomes.	8.3, 17.3, 20.5, Box 25-5
		Political: Political, social, cultural, & ecological decisions & actions consistent with reducing vulnerability & risk & supporting adaptation, mitigation, & sustainable development.	14.2-3, 20.5, 25.4, 30.7, Table 14-1
		Personal: Individual & collective assumptions, beliefs, values, & worldviews influencing climate-change responses.	14.2-3, 20.5, 25.4, Table 14-1

SPM

about projected impacts; different perceptions of risks; competing values; absence of key adaptation leaders and advocates; and limited tools to monitor adaptation effectiveness. Another constraint includes insufficient research, monitoring, and observation and the finance to maintain them. Underestimating the complexity of adaptation as a social process can create unrealistic expectations about achieving intended adaptation outcomes.[73]

Poor planning, overemphasizing short-term outcomes, or failing to sufficiently anticipate consequences can result in maladaptation (*medium evidence, high agreement*). Maladaptation can increase the vulnerability or exposure of the target group in the future, or the vulnerability of other people, places, or sectors. Some near-term responses to increasing risks related to climate change may also limit future choices. For example, enhanced protection of exposed assets can lock in dependence on further protection measures.[74]

Limited evidence indicates a gap between global adaptation needs and the funds available for adaptation (*medium confidence*). There is a need for a better assessment of global adaptation costs, funding, and investment. Studies estimating the global cost of adaptation are characterized by shortcomings in data, methods, and coverage (*high confidence*).[75]

Significant co-benefits, synergies, and trade-offs exist between mitigation and adaptation and among different adaptation responses; interactions occur both within and across regions (*very high confidence*). Increasing efforts to mitigate and adapt to climate change imply an increasing complexity of interactions, particularly at the intersections among water, energy, land use, and biodiversity, but tools to understand and manage these interactions remain limited. Examples of actions with co-benefits include (i) improved energy efficiency and cleaner energy sources, leading to reduced emissions of health-damaging climate-altering air pollutants; (ii) reduced energy and water consumption in urban areas through greening cities and recycling water; (iii) sustainable agriculture and forestry; and (iv) protection of ecosystems for carbon storage and other ecosystem services.[76]

C-2. Climate-resilient Pathways and Transformation

Climate-resilient pathways are sustainable-development trajectories that combine adaptation and mitigation to reduce climate change and its impacts. They include iterative processes to ensure that effective risk management can be implemented and sustained. See Figure SPM.9.[77]

Prospects for climate-resilient pathways for sustainable development are related fundamentally to what the world accomplishes with climate-change mitigation (*high confidence*). Since mitigation reduces the rate as well as the magnitude of warming, it also increases the time available for adaptation to a particular level of climate change, potentially by several decades. Delaying mitigation actions may reduce options for climate-resilient pathways in the future.[78]

Greater rates and magnitude of climate change increase the likelihood of exceeding adaptation limits (*high confidence*). Limits to adaptation occur when adaptive actions to avoid intolerable risks for an actor's objectives or for the needs of a system are not possible or are not currently available. Value-based judgments of what constitutes an intolerable risk may differ. Limits to adaptation emerge from the interaction among climate change and biophysical and/or socioeconomic constraints. Opportunities to take advantage of positive synergies between adaptation and mitigation may decrease with time, particularly if limits to adaptation are exceeded. In some parts of the world, insufficient responses to emerging impacts are already eroding the basis for sustainable development.[79]

---

[73] 3.6, 4.4, 5.5, 8.4, 9.4, 13.2-3, 14.2, 14.5, 15.2-3, 15.5, 16.2-3, 16.5, 17.2-3, 22.4, 23.7, 24.5, 25.4, 25.10, 26.8-9, 30.6, Table 16-3, Boxes 16-1 and 16-3

[74] 5.5, 8.4, 14.6, 15.5, 16.3, 17.2-3, 20.2, 22.4, 24.4, 25.10, 26.8, Table 14-4, Box 25-1

[75] 14.2, 17.4, Tables 17-2 and 17-3

[76] 2.4-5, 3.7, 4.2, 4.4, 5.4-5, 8.4, 9.3, 11.9, 13.3, 17.2, 19.3-4, 20.2-5, 21.4, 22.6, 23.8, 24.6, 25.6-7, 25.9, 26.8-9, 27.3, 29.6-8, Boxes 25-2, 25-9, 25-10, 30.6-7, CC-WE, and CC-RF

[77] 2.5, 20.3-4

[78] 1.1, 19.7, 20.2-3, 20.6, Figure 1-5

[79] 1.1, 11.8, 13.4, 16.2-7, 17.2, 20.2-3, 20.5-6, 25.10, 26.5, Boxes 16-1, 16-3, and 16-4

Transformations in economic, social, technological, and political decisions and actions can enable climate-resilient pathways (*high confidence*). Specific examples are presented in Table SPM.1. Strategies and actions can be pursued now that will move towards climate-resilient pathways for sustainable development, while at the same time helping to improve livelihoods, social and economic well-being, and responsible environmental management. At the national level, transformation is considered most effective when it reflects a country's own visions and approaches to achieving sustainable development in accordance with its national circumstances and priorities. Transformations to sustainability are considered to benefit from iterative learning, deliberative processes, and innovation.[80]

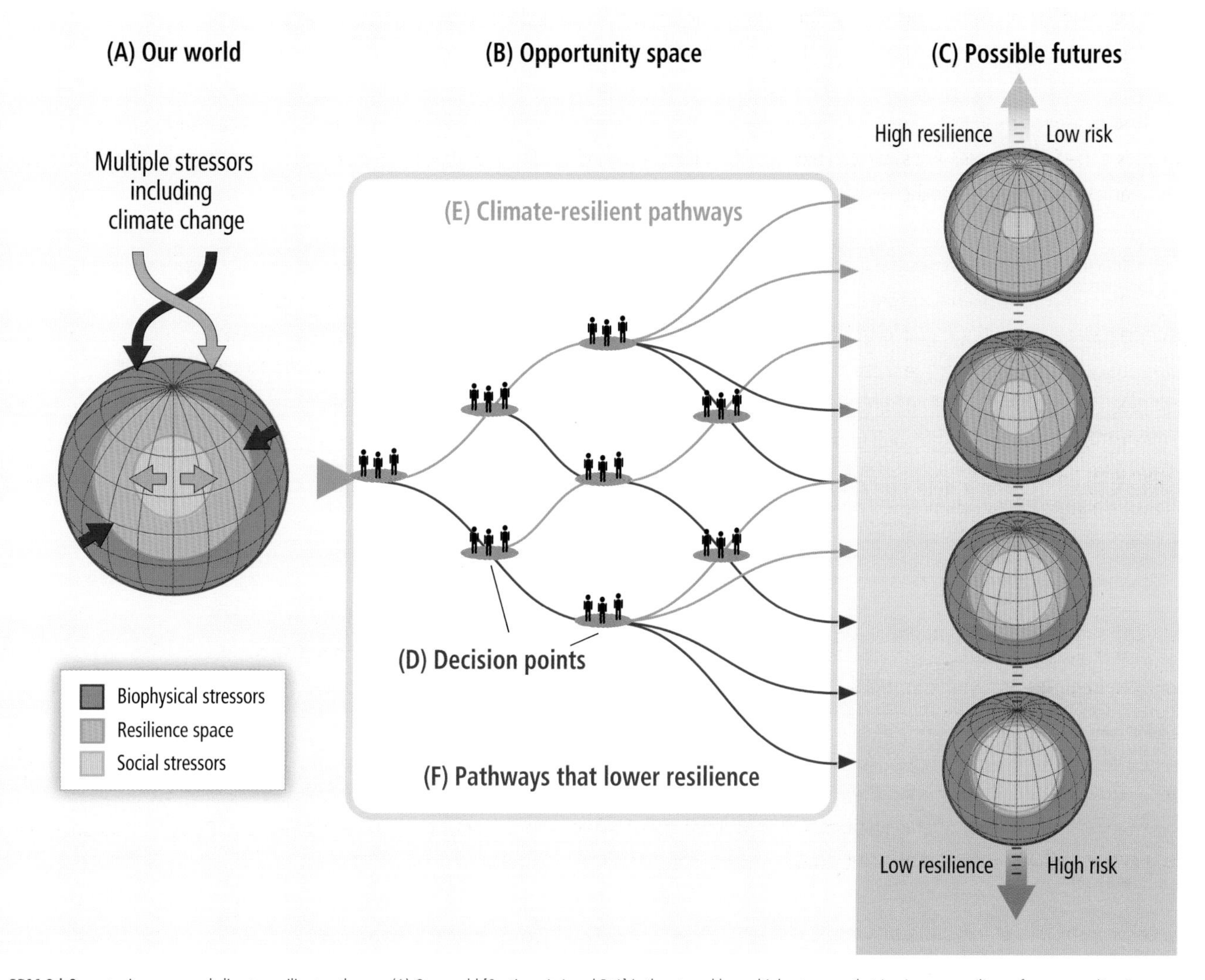

Figure SPM.9 | Opportunity space and climate-resilient pathways. (A) Our world [Sections A-1 and B-1] is threatened by multiple stressors that impinge on resilience from many directions, represented here simply as biophysical and social stressors. Stressors include climate change, climate variability, land-use change, degradation of ecosystems, poverty and inequality, and cultural factors. (B) Opportunity space [Sections A-2, A-3, B-2, C-1, and C-2] refers to decision points and pathways that lead to a range of (C) possible futures [Sections C and B-3] with differing levels of resilience and risk. (D) Decision points result in actions or failures-to-act throughout the opportunity space, and together they constitute the process of managing or failing to manage risks related to climate change. (E) Climate-resilient pathways (in green) within the opportunity space lead to a more resilient world through adaptive learning, increasing scientific knowledge, effective adaptation and mitigation measures, and other choices that reduce risks. (F) Pathways that lower resilience (in red) can involve insufficient mitigation, maladaptation, failure to learn and use knowledge, and other actions that lower resilience; and they can be irreversible in terms of possible futures.

[80] 1.1, 2.1, 2.5, 8.4, 14.1, 14.3, 16.2-7, 20.5, 22.4, 25.4, 25.10, Figure 1-5, Boxes 16-1, 16-4, and TS.8

SUPPLEMENTARY MATERIAL

Table SPM.A1 | Observed impacts attributed to climate change reported in the scientific literature since the AR4. These impacts have been attributed to climate change with *very low*, *low*, *medium*, or *high confidence*, with the relative contribution of climate change to the observed change indicated (major or minor), for natural and human systems across eight major world regions over the past several decades. [Tables 18-5, 18-6, 18-7, 18-8, and 18-9] Absence from the table of additional impacts attributed to climate change does not imply that such impacts have not occurred.

	Africa
Snow & Ice, Rivers & Lakes, Floods & Drought	• Retreat of tropical highland glaciers in East Africa (*high confidence*, major contribution from climate change) • Reduced discharge in West African rivers (*low confidence*, major contribution from climate change) • Lake surface warming and water column stratification increases in the Great Lakes and Lake Kariba (*high confidence*, major contribution from climate change) • Increased soil moisture drought in the Sahel since 1970, partially wetter conditions since 1990 (*medium confidence*, major contribution from climate change) [22.2-3, Tables 18-5, 18-6, and 22-3]
Terrestrial Ecosystems	• Tree density decreases in western Sahel and semi-arid Morocco, beyond changes due to land use (*medium confidence*, major contribution from climate change) • Range shifts of several southern plants and animals, beyond changes due to land use (*medium confidence*, major contribution from climate change) • Increases in wildfires on Mt. Kilimanjaro (*low confidence*, major contribution from climate change) [22.3, Tables 18-7 and 22-3]
Coastal Erosion & Marine Ecosystems	• Decline in coral reefs in tropical African waters, beyond decline due to human impacts (*high confidence*, major contribution from climate change) [Table 18-8]
Food Production & Livelihoods	• Adaptive responses to changing rainfall by South African farmers, beyond changes due to economic conditions (*very low confidence*, major contribution from climate change) • Decline in fruit-bearing trees in Sahel (*low confidence*, major contribution from climate change) • Malaria increases in Kenyan highlands, beyond changes due to vaccination, drug resistance, demography, and livelihoods (*low confidence*, minor contribution from climate change) • Reduced fisheries productivity of Great Lakes and Lake Kariba, beyond changes due to fisheries management and land use (*low confidence*, minor contribution from climate change) [7.2, 11.5, 13.2, 22.3, Table 18-9]
	Europe
Snow & Ice, Rivers & Lakes, Floods & Drought	• Retreat of Alpine, Scandinavian, and Icelandic glaciers (*high confidence*, major contribution from climate change) • Increase in rock slope failures in western Alps (*medium confidence*, major contribution from climate change) • Changed occurrence of extreme river discharges and floods (*very low confidence*, minor contribution from climate change) [18.3, 23.2-3, Tables 18-5 and 18-6; WGI AR5 4.3]
Terrestrial Ecosystems	• Earlier greening, leaf emergence, and fruiting in temperate and boreal trees (*high confidence*, major contribution from climate change) • Increased colonization of alien plant species in Europe, beyond a baseline of some invasion (*medium confidence*, major contribution from climate change) • Earlier arrival of migratory birds in Europe since 1970 (*medium confidence*, major contribution from climate change) • Upward shift in tree-line in Europe, beyond changes due to land use (*low confidence*, major contribution from climate change) • Increasing burnt forest areas during recent decades in Portugal and Greece, beyond some increase due to land use (*high confidence*, major contribution from climate change) [4.3, 18.3, Tables 18-7 and 23-6]
Coastal Erosion & Marine Ecosystems	• Northward distributional shifts of zooplankton, fishes, seabirds, and benthic invertebrates in northeast Atlantic (*high confidence*, major contribution from climate change) • Northward and depth shift in distribution of many fish species across European seas (*medium confidence*, major contribution from climate change) • Plankton phenology changes in northeast Atlantic (*medium confidence*, major contribution from climate change) • Spread of warm water species into the Mediterranean, beyond changes due to invasive species and human impacts (*medium confidence*, major contribution from climate change) [6.3, 23.6, 30.5, Tables 6-2 and 18-8, Boxes 6-1 and CC-MB]
Food Production & Livelihoods	• Shift from cold-related mortality to heat-related mortality in England and Wales, beyond changes due to exposure and health care (*low confidence*, major contribution from climate change) • Impacts on livelihoods of Sámi people in northern Europe, beyond effects of economic and sociopolitical changes (*medium confidence*, major contribution from climate change) • Stagnation of wheat yields in some countries in recent decades, despite improved technology (*medium confidence*, minor contribution from climate change) • Positive yield impacts for some crops mainly in northern Europe, beyond increase due to improved technology (*medium confidence*, minor contribution from climate change) • Spread of bluetongue virus in sheep and of ticks across parts of Europe (*medium confidence*, minor contribution from climate change) [18.4, 23.4-5, Table 18-9, Figure 7-2]

Continued next page →

Table SPM.A1 (continued)

	Asia
Snow & Ice, Rivers & Lakes, Floods & Drought	• Permafrost degradation in Siberia, Central Asia, and Tibetan Plateau (*high confidence*, major contribution from climate change) • Shrinking mountain glaciers across most of Asia (*medium confidence*, major contribution from climate change) • Changed water availability in many Chinese rivers, beyond changes due to land use (*low confidence*, minor contribution from climate change) • Increased flow in several rivers due to shrinking glaciers (*high confidence*, major contribution from climate change) • Earlier timing of maximum spring flood in Russian rivers (*medium confidence*, major contribution from climate change) • Reduced soil moisture in north-central and northeast China (1950–2006) (*medium confidence*, major contribution from climate change) • Surface water degradation in parts of Asia, beyond changes due to land use (*medium confidence*, minor contribution from climate change) [24.3-4, 28.2, Tables 18-5, 18-6, and SM24-4, Box 3-1; WGI AR5 4.3, 10.5]
Terrestrial Ecosystems	• Changes in plant phenology and growth in many parts of Asia (earlier greening), particularly in the north and east (*medium confidence*, major contribution from climate change) • Distribution shifts of many plant and animal species upwards in elevation or polewards, particularly in the north of Asia (*medium confidence*, major contribution from climate change) • Invasion of Siberian larch forests by pine and spruce during recent decades (*low confidence*, major contribution from climate change) • Advance of shrubs into the Siberian tundra (*high confidence*, major contribution from climate change) [4.3, 24.4, 28.2, Table 18-7, Figure 4-4]
Coastal Erosion & Marine Ecosystems	• Decline in coral reefs in tropical Asian waters, beyond decline due to human impacts (*high confidence*, major contribution from climate change) • Northward range extension of corals in the East China Sea and western Pacific, and of a predatory fish in the Sea of Japan (*medium confidence*, major contribution from climate change) • Shift from sardines to anchovies in the western North Pacific, beyond fluctuations due to fisheries (*low confidence*, major contribution from climate change) • Increased coastal erosion in Arctic Asia (*low confidence*, major contribution from climate change) [6.3, 24.4, 30.5, Tables 6-2 and 18-8]
Food Production & Livelihoods	• Impacts on livelihoods of indigenous groups in Arctic Russia, beyond economic and sociopolitical changes (*low confidence*, major contribution from climate change) • Negative impacts on aggregate wheat yields in South Asia, beyond increase due to improved technology (*medium confidence*, minor contribution from climate change) • Negative impacts on aggregate wheat and maize yields in China, beyond increase due to improved technology (*low confidence*, minor contribution from climate change) • Increases in a water-borne disease in Israel (*low confidence*, minor contribution from climate change) [7.2, 13.2, 18.4, 28.2, Tables 18-4 and 18-9, Figure 7-2]
	Australasia
Snow & Ice, Rivers & Lakes, Floods & Drought	• Significant decline in late-season snow depth at 3 of 4 alpine sites in Australia (1957–2002) (*medium confidence*, major contribution from climate change) • Substantial reduction in ice and glacier ice volume in New Zealand (*medium confidence*, major contribution from climate change) • Intensification of hydrological drought due to regional warming in southeast Australia (*low confidence*, minor contribution from climate change) • Reduced inflow in river systems in southwestern Australia (since the mid-1970s) (*high confidence*, major contribution from climate change) [25.5, Tables 18-5, 18-6, and 25-1; WGI AR5 4.3]
Terrestrial Ecosystems	• Changes in genetics, growth, distribution, and phenology of many species, in particular birds, butterflies, and plants in Australia, beyond fluctuations due to variable local climates, land use, pollution, and invasive species (*high confidence*, major contribution from climate change) • Expansion of some wetlands and contraction of adjacent woodlands in southeast Australia (*low confidence*, major contribution from climate change) • Expansion of monsoon rainforest at expense of savannah and grasslands in northern Australia (*medium confidence*, major contribution from climate change) • Migration of glass eels advanced by several weeks in Waikato River, New Zealand (*low confidence*, major contribution from climate change) [Tables 18-7 and 25-3]
Coastal Erosion & Marine Ecosystems	• Southward shifts in the distribution of marine species near Australia, beyond changes due to short-term environmental fluctuations, fishing, and pollution (*medium confidence*, major contribution from climate change) • Change in timing of migration of seabirds in Australia (*low confidence*, major contribution from climate change) • Increased coral bleaching in Great Barrier Reef and western Australian reefs, beyond effects from pollution and physical disturbance (*high confidence*, major contribution from climate change) • Changed coral disease patterns at Great Barrier Reef, beyond effects from pollution (*medium confidence*, major contribution from climate change) [6.3, 25.6, Tables 18-8 and 25-3]
Food Production & Livelihoods	• Advanced timing of wine-grape maturation in recent decades, beyond advance due to improved management (*medium confidence*, major contribution from climate change) • Shift in winter vs. summer human mortality in Australia, beyond changes due to exposure and health care (*low confidence*, major contribution from climate change) • Relocation or diversification of agricultural activities in Australia, beyond changes due to policy, markets, and short-term climate variability (*low confidence*, minor contribution from climate change) [11.4, 18.4, 25.7-8, Tables 18-9 and 25-3, Box 25-5]
	North America
Snow & Ice, Rivers & Lakes, Floods & Drought	• Shrinkage of glaciers across western and northern North America (*high confidence*, major contribution from climate change) • Decreasing amount of water in spring snowpack in western North America (1960–2002) (*high confidence*, major contribution from climate change) • Shift to earlier peak flow in snow dominated rivers in western North America (*high confidence*, major contribution from climate change) • Increased runoff in the midwestern and northeastern US (*medium confidence*, minor contribution from climate change) [Tables 18-5 and 18-6; WGI AR5 2.6, 4.3]
Terrestrial Ecosystems	• Phenology changes and species distribution shifts upward in elevation and northward across multiple taxa (*medium confidence*, major contribution from climate change) • Increased wildfire frequency in subarctic conifer forests and tundra (*medium confidence*, major contribution from climate change) • Regional increases in tree mortality and insect infestations in forests (*low confidence*, minor contribution from climate change) • Increase in wildfire activity, fire frequency and duration, and burnt area in forests of the western US and boreal forests in Canada, beyond changes due to land use and fire management (*medium confidence*, minor contribution from climate change) [26.4, 28.2, Table 18-7, Box 26-2]
Coastal Erosion & Marine Ecosystems	• Northward distributional shifts of northwest Atlantic fish species (*high confidence*, major contribution from climate change) • Changes in musselbeds along the west coast of US (*high confidence*, major contribution from climate change) • Changed migration and survival of salmon in northeast Pacific (*high confidence*, major contribution from climate change) • Increased coastal erosion in Alaska and Canada (*medium confidence*, major contribution from climate change) [18.3, 30.5, Tables 6-2 and 18-8]
Food Production & Livelihoods	• Impacts on livelihoods of indigenous groups in the Canadian Arctic, beyond effects of economic and sociopolitical changes (*medium confidence*, major contribution from climate change) [18.4, 28.2, Tables 18-4 and 18-9]

Continued next page →

SPM

Table SPM.A1 (continued)

	Central and South America
Snow & Ice, Rivers & Lakes, Floods & Drought	• Shrinkage of Andean glaciers (*high confidence*, major contribution from climate change) • Changes in extreme flows in Amazon River (*medium confidence*, major contribution from climate change) • Changing discharge patterns in rivers in the western Andes (*medium confidence*, major contribution from climate change) • Increased streamflow in sub-basins of the La Plata River, beyond increase due to land-use change (*high confidence*, major contribution from climate change) [27.3, Tables 18-5, 18-6, and 27-3; WGI AR5 4.3]
Terrestrial Ecosystems	• Increased tree mortality and forest fire in the Amazon (*low confidence*, minor contribution from climate change) • Rainforest degradation and recession in the Amazon, beyond reference trends in deforestation and land degradation (*low confidence*, minor contribution from climate change) [4.3, 18.3, 27.2-3, Table 18-7]
Coastal Erosion & Marine Ecosystems	• Increased coral bleaching in western Caribbean, beyond effects from pollution and physical disturbance (*high confidence*, major contribution from climate change) • Mangrove degradation on north coast of South America, beyond degradation due to pollution and land use (*low confidence*, minor contribution from climate change) [27.3, Table 18-8]
Food Production & Livelihoods	• More vulnerable livelihood trajectories for indigenous Aymara farmers in Bolivia due to water shortage, beyond effects of increasing social and economic stress (*medium confidence*, major contribution from climate change) • Increase in agricultural yields and expansion of agricultural areas in southeastern South America, beyond increase due to improved technology (*medium confidence*, major contribution from climate change) [13.1, 27.3, Table 18-9]
	Polar Regions
Snow & Ice, Rivers & Lakes, Floods & Drought	• Decreasing Arctic sea ice cover in summer (*high confidence*, major contribution from climate change) • Reduction in ice volume in Arctic glaciers (*high confidence*, major contribution from climate change) • Decreasing snow cover extent across the Arctic (*medium confidence*, major contribution from climate change) • Widespread permafrost degradation, especially in the southern Arctic (*high confidence*, major contribution from climate change) • Ice mass loss along coastal Antarctica (*medium confidence*, major contribution from climate change) • Increased river discharge for large circumpolar rivers (1997–2007) (*low confidence*, major contribution from climate change) • Increased winter minimum river flow in most of the Arctic (*medium confidence*, major contribution from climate change) • Increased lake water temperatures 1985–2009 and prolonged ice-free seasons (*medium confidence*, major contribution from climate change) • Disappearance of thermokarst lakes due to permafrost degradation in the low Arctic. New lakes created in areas of formerly frozen peat (*high confidence*, major contribution from climate change) [28.2, Tables 18-5 and 18-6; WGI AR5 4.2-4, 4.6, 10.5]
Terrestrial Ecosystems	• Increased shrub cover in tundra in North America and Eurasia (*high confidence*, major contribution from climate change) • Advance of Arctic tree-line in latitude and altitude (*medium confidence*, major contribution from climate change) • Changed breeding area and population size of subarctic birds, due to snowbed reduction and/or tundra shrub encroachment (*medium confidence*, major contribution from climate change) • Loss of snow-bed ecosystems and tussock tundra (*high confidence*, major contribution from climate change) • Impacts on tundra animals from increased ice layers in snow pack, following rain-on-snow events (*medium confidence*, major contribution from climate change) • Increased plant species ranges in the West Antarctic Peninsula and nearby islands over the past 50 years (*high confidence*, major contribution from climate change) • Increased phytoplankton productivity in Signy Island lake waters (*high confidence*, major contribution from climate change) [28.2, Table 18-7]
Coastal Erosion & Marine Ecosystems	• Increased coastal erosion across Arctic (*medium confidence*, major contribution from climate change) • Negative effects on non-migratory Arctic species (*high confidence*, major contribution from climate change) • Decreased reproductive success in Arctic seabirds (*medium confidence*, major contribution from climate change) • Decline in Southern Ocean seals and seabirds (*medium confidence*, major contribution from climate change) • Reduced thickness of foraminiferal shells in southern oceans, due to ocean acidification (*medium confidence*, major contribution from climate change) • Reduced krill density in Scotia Sea (*medium confidence*, major contribution from climate change) [6.3, 18.3, 28.2-3, Table 18-8]
Food Production & Livelihoods	• Impact on livelihoods of Arctic indigenous peoples, beyond effects of economic and sociopolitical changes (*medium confidence*, major contribution from climate change) • Increased shipping traffic across the Bering Strait (*medium confidence*, major contribution from climate change) [18.4, 28.2, Tables 18-4 and 18-9, Figure 28-4]
	Small Islands
Snow & Ice, Rivers & Lakes, Floods & Drought	• Increased water scarcity in Jamaica, beyond increase due to water use (*very low confidence*, minor contribution from climate change) [Table 18-6]
Terrestrial Ecosystems	• Tropical bird population changes in Mauritius (*medium confidence*, major contribution from climate change) • Decline of an endemic plant in Hawai'i (*medium confidence*, major contribution from climate change) • Upward trend in tree-lines and associated fauna on high-elevation islands (*low confidence*, minor contribution from climate change) [29.3, Table 18-7]
Coastal Erosion & Marine Ecosystems	• Increased coral bleaching near many tropical small islands, beyond effects of degradation due to fishing and pollution (*high confidence*, major contribution from climate change) • Degradation of mangroves, wetlands, and seagrass around small islands, beyond degradation due to other disturbances (*very low confidence*, minor contribution from climate change) • Increased flooding and erosion, beyond erosion due to human activities, natural erosion, and accretion (*low confidence*, minor contribution from climate change) • Degradation of groundwater and freshwater ecosystems due to saline intrusion, beyond degradation due to pollution and groundwater pumping (*low confidence*, minor contribution from climate change) [29.3, Table 18-8]
Food Production & Livelihoods	• Increased degradation of coastal fisheries due to direct effects and effects of increased coral reef bleaching, beyond degradation due to overfishing and pollution (*low confidence*, minor contribution from climate change) [18.3-4, 29.3, 30.6, Table 18-9, Box CC-CR]

Technical Summary

Technical Summary

Prepared under the leadership of the Working Group II Bureau:

Amjad Abdulla (Maldives), Vicente R. Barros (Argentina), Eduardo Calvo (Peru), Christopher B. Field (USA), José M. Moreno (Spain), Nirivololona Raholijao (Madagascar), Sergey Semenov (Russian Federation), Neville Smith (Australia)

Coordinating Lead Authors:

Christopher B. Field (USA), Vicente R. Barros (Argentina), Katharine J. Mach (USA), Michael D. Mastrandrea (USA)

Lead Authors:

Maarten K. van Aalst (Netherlands), W. Neil Adger (UK), Douglas J. Arent (USA), Jonathon Barnett (Australia), Richard A. Betts (UK), T. Eren Bilir (USA), Joern Birkmann (Germany), JoAnn Carmin (USA), Dave D. Chadee (Trinidad and Tobago), Andrew J. Challinor (UK), Monalisa Chatterjee (USA/India), Wolfgang Cramer (Germany/France), Debra J. Davidson (Canada), Yuka Otsuki Estrada (USA/Japan), Jean-Pierre Gattuso (France), Yasuaki Hijioka (Japan), Ove Hoegh-Guldberg (Australia), He-Qing Huang (China), Gregory E. Insarov (Russian Federation), Roger N. Jones (Australia), R. Sari Kovats (UK), Joan Nymand Larsen (Iceland), Iñigo J. Losada (Spain), José A. Marengo (Brazil), Roger F. McLean (Australia), Linda O. Mearns (USA), Reinhard Mechler (Germany/Austria), John F. Morton (UK), Isabelle Niang (Senegal), Taikan Oki (Japan), Jane Mukarugwiza Olwoch (South Africa), Maggie Opondo (Kenya), Elvira S. Poloczanska (Australia), Hans-O. Pörtner (Germany), Margaret Hiza Redsteer (USA), Andy Reisinger (New Zealand), Aromar Revi (India), Patricia Romero-Lankao (Mexico), Daniela N. Schmidt (UK), M. Rebecca Shaw (USA), William Solecki (USA), Dáithí A. Stone (Canada/South Africa/USA), John M.R. Stone (Canada), Kenneth M. Strzepek (UNU/USA), Avelino G. Suarez (Cuba), Petra Tschakert (USA), Riccardo Valentini (Italy), Sebastián Vicuña (Chile), Alicia Villamizar (Venezuela), Katharine E. Vincent (South Africa), Rachel Warren (UK), Leslie L. White (USA), Thomas J. Wilbanks (USA), Poh Poh Wong (Singapore), Gary W. Yohe (USA)

Review Editors:

Paulina Aldunce (Chile), Jean Pierre Ometto (Brazil), Nirivololona Raholijao (Madagascar), Kazuya Yasuhara (Japan)

This Technical Summary should be cited as:

Field, C.B., V.R. Barros, K.J. Mach, M.D. Mastrandrea, M. van Aalst, W.N. Adger, D.J. Arent, J. Barnett, R. Betts, T.E. Bilir, J. Birkmann, J. Carmin, D.D. Chadee, A.J. Challinor, M. Chatterjee, W. Cramer, D.J. Davidson, Y.O. Estrada, J.-P. Gattuso, Y. Hijioka, O. Hoegh-Guldberg, H.Q. Huang, G.E. Insarov, R.N. Jones, R.S. Kovats, P. Romero-Lankao, J.N. Larsen, I.J. Losada, J.A. Marengo, R.F. McLean, L.O. Mearns, R. Mechler, J.F. Morton, I. Niang, T. Oki, J.M. Olwoch, M. Opondo, E.S. Poloczanska, H.-O. Pörtner, M.H. Redsteer, A. Reisinger, A. Revi, D.N. Schmidt, M.R. Shaw, W. Solecki, D.A. Stone, J.M.R. Stone, K.M. Strzepek, A.G. Suarez, P. Tschakert, R. Valentini, S. Vicuña, A. Villamizar, K.E. Vincent, R. Warren, L.L. White, T.J. Wilbanks, P.P. Wong, and G.W. Yohe, 2014: Technical summary. In: *Climate Change 2014: Impacts, Adaptation, and Vulnerability. Summaries, Frequently Asked Questions, and Cross-Chapter Boxes. A Contribution of Working Group II to the Fifth Assessment Report of the Intergovernmental Panel on Climate Change* [Field, C.B., V.R. Barros, D.J. Dokken, K.J. Mach, M.D. Mastrandrea, T.E. Bilir, M. Chatterjee, K.L. Ebi, Y.O. Estrada, R.C. Genova, B. Girma, E.S. Kissel, A.N. Levy, S. MacCracken, P.R. Mastrandrea, and L.L. White (eds.)]. World Meteorological Organization, Geneva, Switzerland, pp. 35-92. (in Arabic, Chinese, English, French, Russian, and Spanish)

Contents

ASSESSING AND MANAGING THE RISKS OF CLIMATE CHANGE

Human interference with the climate system is occurring (WGI AR5 SPM Section D.3; WGI AR5 Sections 2.2, 6.3, 10.3 to 10.6, 10.9). Climate change poses risks for human and natural systems (Figure TS.1). The assessment of impacts, adaptation, and vulnerability in the Working Group II contribution to the IPCC's Fifth Assessment Report (WGII AR5) evaluates how patterns of risks and potential benefits are shifting due to climate change. It considers how impacts and risks related to climate change can be reduced and managed through adaptation and mitigation. The report assesses needs, options, opportunities, constraints, resilience, limits, and other aspects associated with adaptation. It recognizes that risks of climate change will vary across regions and populations, through space and time, dependent on myriad factors including the extent of adaptation and mitigation.

Climate change involves complex interactions and changing likelihoods of diverse impacts. A focus on risk, which is new in this report, supports decision making in the context of climate change and complements other elements of the report. People and societies may perceive or rank risks and potential benefits differently, given diverse values and goals.

Compared to past WGII reports, the WGII AR5 assesses a substantially larger knowledge base of relevant scientific, technical, and socioeconomic literature. Increased literature has facilitated comprehensive assessment across a broader set of topics and sectors, with expanded coverage of human systems, adaptation, and the ocean. See Box TS.1.

Section A of this summary characterizes observed impacts, vulnerability and exposure, and adaptive responses to date. Section B examines future risks and potential benefits across sectors and regions, highlighting where choices matter for reducing risks through mitigation and adaptation. Section C considers principles for effective adaptation and the broader interactions among adaptation, mitigation, and sustainable development.

Box TS.2 defines central concepts. To convey the degree of certainty in key findings, the report relies on the consistent use of calibrated uncertainty language, introduced in Box TS.3. Chapter references in brackets indicate support for findings, figures, and tables in this summary.

A: OBSERVED IMPACTS, VULNERABILITY, AND ADAPTATION IN A COMPLEX AND CHANGING WORLD

This section presents observed effects of climate change, building from understanding of vulnerability, exposure, and climate-related hazards as determinants of impacts. The section considers the factors, including development and non-climatic stressors, that influence vulnerability and

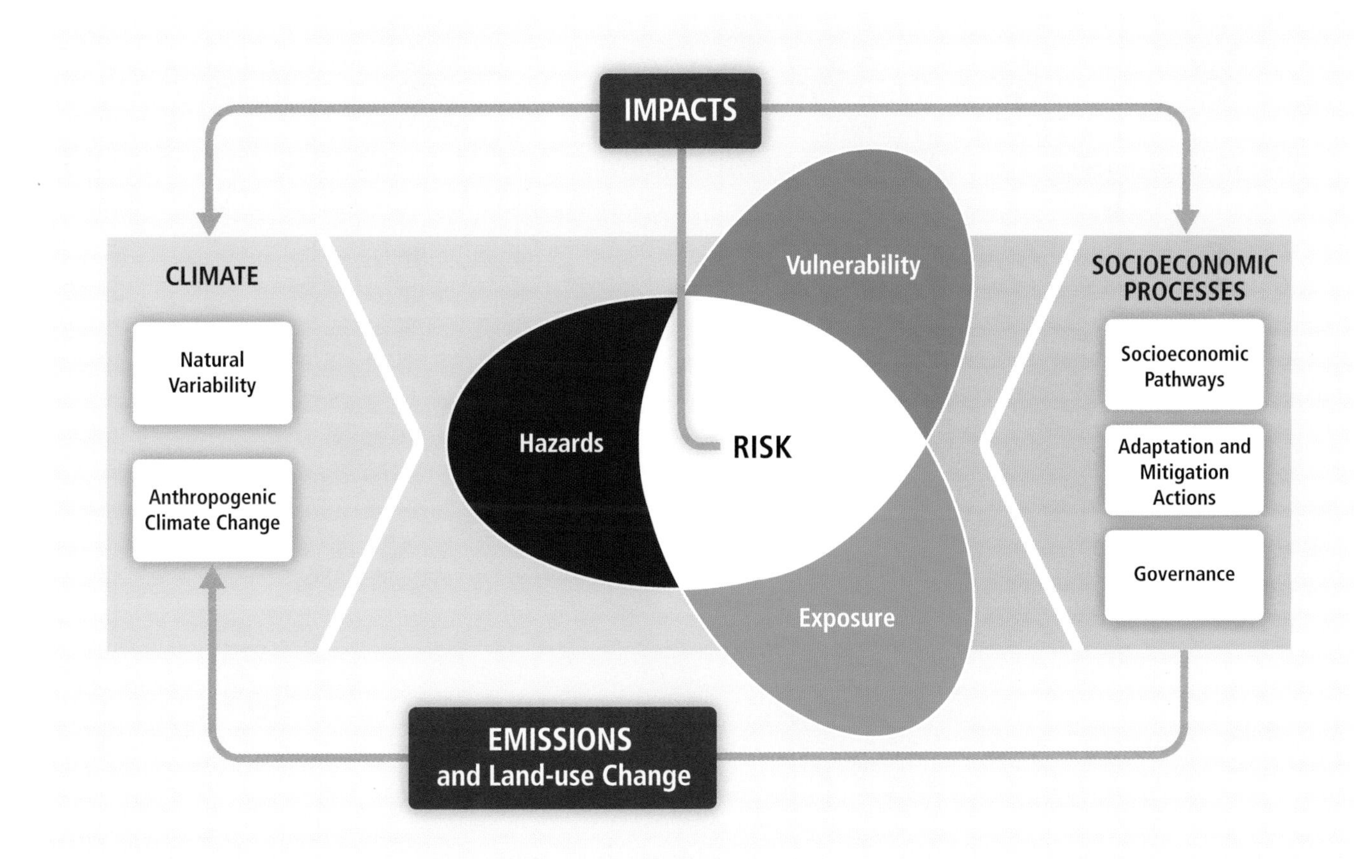

Figure TS.1 | Illustration of the core concepts of the WGII AR5. Risk of climate-related impacts results from the interaction of climate-related hazards (including hazardous events and trends) with the vulnerability and exposure of human and natural systems. Changes in both the climate system (left) and socioeconomic processes including adaptation and mitigation (right) are drivers of hazards, exposure, and vulnerability. [19.2, Figure 19-1]

Box TS.1 | Context for the Assessment

For the past 2 decades, IPCC's Working Group II has developed assessments of climate change impacts, adaptation, and vulnerability. The WGII AR5 builds from the WGII contribution to the IPCC's Fourth Assessment Report (WGII AR4), published in 2007, and the *Special Report on Managing the Risks of Extreme Events and Disasters to Advance Climate Change Adaptation* (SREX), published in 2012. It follows the Working Group I contribution to the AR5 (WGI AR5). The WGII AR5 is presented in two parts (Part A: Global and Sectoral Aspects, and Part B: Regional Aspects), reflecting the expanded literature basis and multidisciplinary approach, increased focus on societal impacts and responses, and continued regionally comprehensive coverage. [1.1 to 1.3]

The number of scientific publications available for assessing climate change impacts, adaptation, and vulnerability more than doubled between 2005 and 2010, with especially rapid increases in publications related to adaptation, allowing for a more robust assessment that supports policymaking (*high confidence*). The diversity of the topics and regions covered has similarly expanded, as has the geographic distribution of authors contributing to the knowledge base for climate change assessments (Box TS.1 Figure 1). Authorship of climate change publications from developing countries has increased, although it still represents a small fraction of the total. The unequal distribution of publications presents a challenge to the production of a comprehensive and balanced global assessment. [1.1, Figure 1-1]

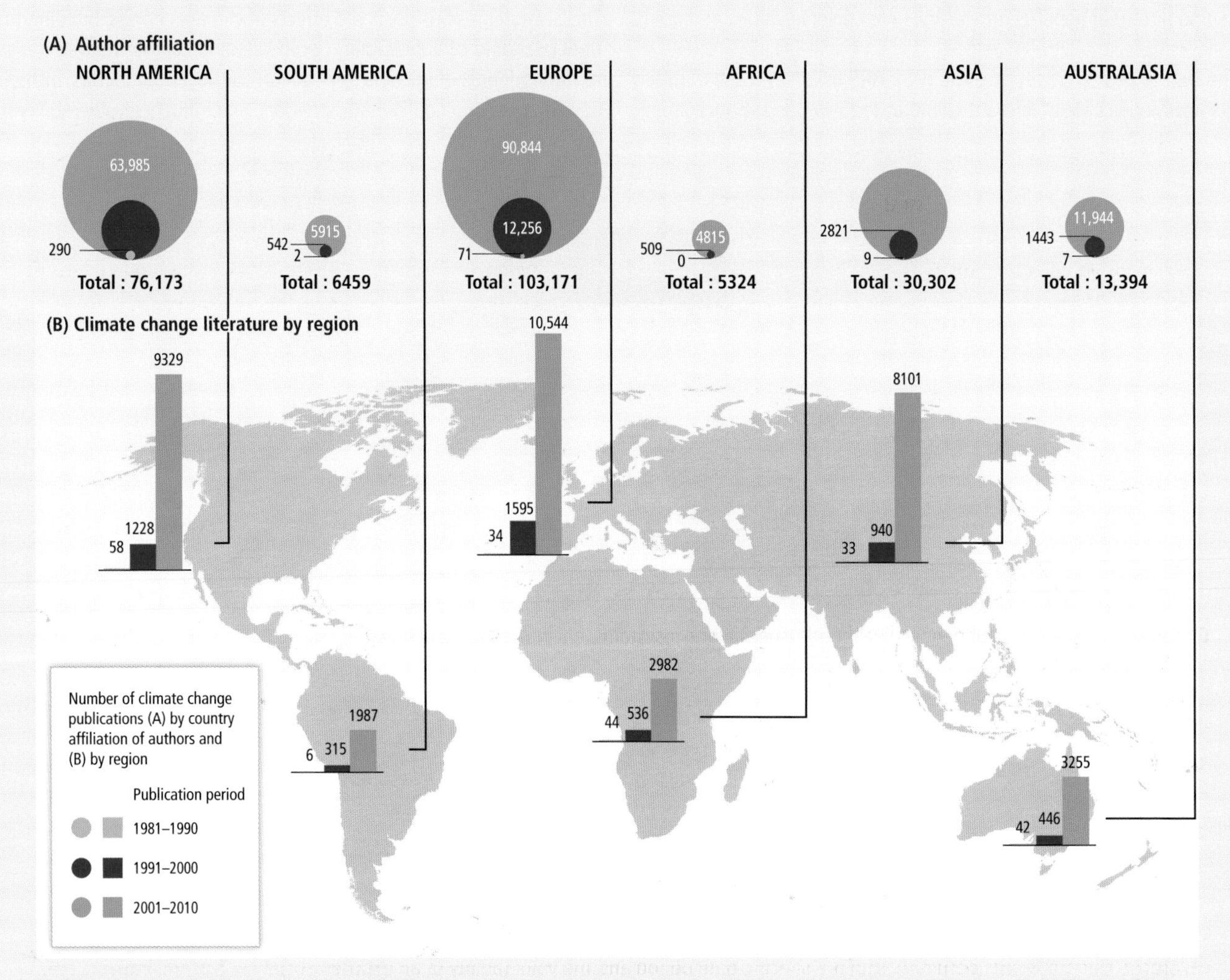

Box TS.1 Figure 1 | Number of climate change publications listed in the Scopus bibliographic database. (A) Number of climate change publications in English (as of July 2011) summed by country affiliation of all authors of the publications and sorted by region. Each publication can be counted multiple times (i.e., the number of different countries in the author affiliation list). (B) Number of climate change publications in English with individual countries mentioned in title, abstract, or key words (as of July 2011) sorted by region for the decades 1981–1990, 1991–2000, and 2001–2010. Each publication can be counted multiple times if more than one country is listed. [Figure 1-1]

Continued next page →

Box TS.1 (continued)

Adaptation has emerged as a central area in climate change research, in country-level planning, and in implementation of climate change strategies (*high confidence*). The body of literature, including government and private sector reports, shows an increased focus on adaptation opportunities and the interrelations between adaptation, mitigation, and alternative sustainable pathways. The literature shows an emergence of studies on transformative processes that take advantage of synergies between adaptation planning, development strategies, social protection, and disaster risk reduction and management. [1.1]

As a core feature and innovation of IPCC assessment, major findings are presented with defined, calibrated language that communicates the strength of scientific understanding, including uncertainties and areas of disagreement (Box TS.3). Each finding is supported by a traceable account of the evaluation of evidence and agreement. [1.1, Box 1-1]

Box TS.2 | Terms Central for Understanding the Summary

Central concepts defined in the WGII AR5 glossary and used throughout the report include the following terms. Reflecting progress in science, some definitions differ in breadth and focus from the definitions used in the AR4 and other IPCC reports.

Climate change: Climate change refers to a change in the state of the climate that can be identified (e.g., by using statistical tests) by changes in the mean and/or the variability of its properties, and that persists for an extended period, typically decades or longer. Climate change may be due to natural internal processes or external forcings such as modulations of the solar cycles, volcanic eruptions, and persistent anthropogenic changes in the composition of the atmosphere or in land use. Note that the Framework Convention on Climate Change (UNFCCC), in its Article 1, defines climate change as: "a change of climate which is attributed directly or indirectly to human activity that alters the composition of the global atmosphere and which is in addition to natural climate variability observed over comparable time periods." The UNFCCC thus makes a distinction between climate change attributable to human activities altering the atmospheric composition, and climate variability attributable to natural causes.

Hazard: The potential occurrence of a natural or human-induced physical event or trend or physical impact that may cause loss of life, injury, or other health impacts, as well as damage and loss to property, infrastructure, livelihoods, service provision, ecosystems, and environmental resources. In this report, the term *hazard* usually refers to climate-related physical events or trends or their physical impacts.

Exposure: The presence of people, livelihoods, species or ecosystems, environmental functions, services, and resources, infrastructure, or economic, social, or cultural assets in places and settings that could be adversely affected.

Vulnerability: The propensity or predisposition to be adversely affected. Vulnerability encompasses a variety of concepts including sensitivity or susceptibility to harm and lack of capacity to cope and adapt.

Impacts: Effects on natural and human systems. In this report, the term *impacts* is used primarily to refer to the effects on natural and human systems of extreme weather and climate events and of climate change. Impacts generally refer to effects on lives, livelihoods, health, ecosystems, economies, societies, cultures, services, and infrastructure due to the interaction of climate changes or hazardous climate events occurring within a specific time period and the vulnerability of an exposed society or system. Impacts are also referred to as *consequences* and *outcomes*. The impacts of climate change on geophysical systems, including floods, droughts, and sea level rise, are a subset of impacts called physical impacts.

Continued next page →

Box TS.2 (continued)

Risk: The potential for consequences where something of value is at stake and where the outcome is uncertain, recognizing the diversity of values. Risk is often represented as probability of occurrence of hazardous events or trends multiplied by the impacts if these events or trends occur. Risk results from the interaction of vulnerability, exposure, and hazard (see Figure TS.1). In this report, the term *risk* is used primarily to refer to the risks of climate-change impacts.

Adaptation: The process of adjustment to actual or expected climate and its effects. In human systems, adaptation seeks to moderate or avoid harm or exploit beneficial opportunities. In some natural systems, human intervention may facilitate adjustment to expected climate and its effects.

Incremental adaptation: Adaptation actions where the central aim is to maintain the essence and integrity of a system or process at a given scale.

Transformational adaptation: Adaptation that changes the fundamental attributes of a system in response to climate and its effects.

Transformation: A change in the fundamental attributes of natural and human systems.

Resilience: The capacity of social, economic, and environmental systems to cope with a hazardous event or trend or disturbance, responding or reorganizing in ways that maintain their essential function, identity, and structure, while also maintaining the capacity for adaptation, learning, and transformation.

exposure, evaluating the sensitivity of systems to climate change. The section also identifies challenges and options based on adaptation experience, looking at what has motivated previous adaptation actions in the context of climate change and broader objectives. It examines current understanding of decision making as relevant to climate change.

A-1. Observed Impacts, Vulnerability, and Exposure

In recent decades, changes in climate have caused impacts on natural and human systems on all continents and across the oceans. This conclusion is strengthened by more numerous and improved observations and analyses since the AR4. Evidence of climate-change impacts is strongest and most comprehensive for natural systems. Some impacts on human systems have also been attributed to climate change, with a major or minor contribution of climate change distinguishable from other influences such as changing social and economic factors. In many regions, impacts on natural and human systems are now detected even in the presence of strong confounding factors such as pollution or land use change. See Figure TS.2 and Table TS.1 for a summary of observed impacts, illustrating broader trends presented in this section. Attribution of observed impacts in the WGII AR5 generally links responses of natural and human systems to observed climate change, regardless of its cause. Most reported impacts of climate change are attributed to warming and/or to shifts in precipitation patterns. There is also emerging evidence of impacts of ocean acidification. Relatively few robust attribution studies and meta-analyses have linked impacts in physical and biological systems to anthropogenic climate change. [18.1, 18.3 to 18.6]

Differences in vulnerability and exposure arise from non-climatic factors and from multidimensional inequalities often produced by uneven development processes (*very high confidence*). These differences shape differential risks from climate change. See Figure TS.1 and Box TS.4. Vulnerability and exposure vary over time and across geographic contexts. Changes in poverty or socioeconomic status, ethnic composition, age structure, and governance have had a significant influence on the outcome of past crises associated with climate-related hazards. [8.2, 9.3, 12.2, 13.1, 13.2, 14.1 to 14.3, 19.2, 19.6, 26.8, Box CC-GC]

Impacts from recent climate-related extremes, such as heat waves, droughts, floods, cyclones, and wildfires, reveal significant vulnerability and exposure of some ecosystems and many human systems to current climate variability (*very high confidence*). Impacts of such climate-related extremes include alteration of ecosystems, disruption of food production and water supply, damage to infrastructure and settlements, morbidity and mortality, and consequences for mental health and human well-being. For countries at all levels of development, these impacts are consistent with a significant lack of preparedness for current climate variability in some sectors. The following examples

Box TS.3 | Communication of the Degree of Certainty in Assessment Findings

Based on the Guidance Note for Lead Authors of the IPCC Fifth Assessment Report on Consistent Treatment of Uncertainties, the WGII AR5 relies on two metrics for communicating the degree of certainty in key findings:

- Confidence in the validity of a finding, based on the type, amount, quality, and consistency of evidence (e.g., data, mechanistic understanding, theory, models, expert judgment) and the degree of agreement. Confidence is expressed qualitatively.
- Quantified measures of uncertainty in a finding expressed probabilistically (based on statistical analysis of observations or model results, or both, and expert judgment).

Each finding has its foundation in evaluation of associated evidence and agreement. The summary terms to describe evidence are: *limited*, *medium*, or *robust*; and agreement: *low*, *medium*, or *high*. These terms are presented with some key findings. In many cases, assessment authors in addition evaluate their confidence about the validity of a finding, providing a synthesis of the evaluation of evidence and agreement. Levels of confidence include five qualifiers: *very low*, *low*, *medium*, *high*, and *very high*. Box TS.3 Figure 1 illustrates the flexible relationship between the summary terms for evidence and agreement and the confidence metric. For a given evidence and agreement statement, different confidence levels could be assigned, but increasing levels of evidence and degrees of agreement are correlated with increasing confidence.

Agreement ↑			
	High agreement Limited evidence	*High agreement Medium evidence*	*High agreement Robust evidence*
	Medium agreement Limited evidence	*Medium agreement Medium evidence*	*Medium agreement Robust evidence*
	Low agreement Limited evidence	*Low agreement Medium evidence*	*Low agreement Robust evidence*
	Evidence (type, amount, quality, consistency) →		

Confidence Scale

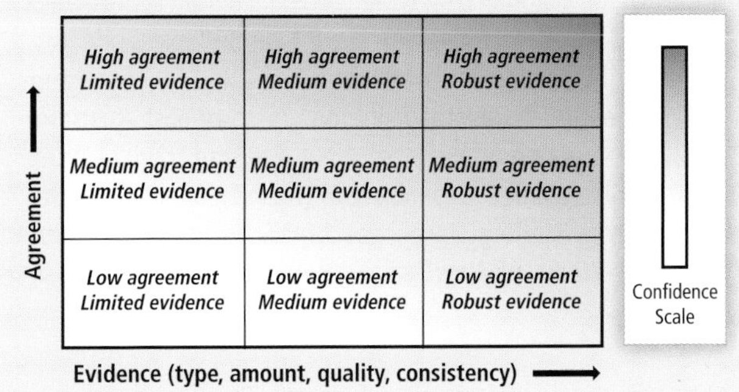

Box TS.3 Figure 1 | Evidence and agreement statements and their relationship to confidence. The shading increasing toward the top right corner indicates increasing confidence. Generally, evidence is most robust when there are multiple, consistent independent lines of high-quality evidence. [Figure 1-3]

When assessment authors evaluate the likelihood, or probability, of some well-defined outcome having occurred or occurring in the future, a finding can include likelihood terms (see below) or a more precise presentation of probability. Use of likelihood is not an alternative to use of confidence. Unless otherwise indicated, findings assigned a likelihood term are associated with *high* or *very high* confidence.

Term	Likelihood of the outcome
Virtually certain	99–100% probability
Extremely likely	95–100% probability
Very likely	90–100% probability
Likely	66–100% probability
More likely than not	>50–100% probability
About as likely as not	33–66% probability
Unlikely	0–33% probability
Very unlikely	0–10% probability
Extremely unlikely	0–5% probability
Exceptionally unlikely	0–1% probability

Where appropriate, findings are also formulated as statements of fact without using uncertainty qualifiers.

Within paragraphs of this summary, the confidence, evidence, and agreement terms given for a key finding apply to subsequent statements in the paragraph, unless additional terms are provided.

[1.1, Box 1-1]

TS

illustrate impacts of extreme weather and climate events experienced across regional contexts:

- In Africa, extreme weather and climate events including droughts and floods have significant impacts on economic sectors, natural resources, ecosystems, livelihoods, and human health. The floods of the Zambezi River in Mozambique in 2008, for example, displaced 90,000 people, and along the Zambezi River Valley, with approximately 1 million people living in the flood-affected areas, temporary displacement is taking on permanent characteristics. [22.3, 22.4, 22.6]
- Recent floods in Australia and New Zealand caused severe damage to infrastructure and settlements and 35 deaths in Queensland alone (2011). The Victorian heat wave (2009) increased heat-related morbidity and was associated with more than 300 excess deaths, while intense bushfires destroyed more than 2000 buildings and led to 173 deaths. Widespread drought in southeast Australia (1997–2009) and many parts of New Zealand (2007–2009; 2012–2013) resulted in economic losses (e.g., regional GDP in the southern Murray-Darling Basin was below forecast by about 5.7% in 2007–2008, and New Zealand lost about NZ$3.6 billion in direct and off-farm output in 2007–2009). [13.2, 25.6, 25.8, Table 25-1, Boxes 25-5, 25-6, and 25-8]
- In Europe, extreme weather events currently have significant impacts in multiple economic sectors as well as adverse social and health effects (*high confidence*). [Table 23-1]
- In North America, most economic sectors and human systems have been affected by and have responded to extreme weather, including hurricanes, flooding, and intense rainfall (*high confidence*). Extreme heat events currently result in increases in mortality and morbidity (*very high confidence*), with impacts that vary by age, location, and socioeconomic factors (*high confidence*). Extreme coastal storm events have caused excess mortality and morbidity, particularly along the east coast of the United States, and the gulf coast of both Mexico and the United States. Much North American infrastructure is currently vulnerable to extreme weather events (*medium confidence*), with deteriorating water-resource and transportation infrastructure particularly vulnerable (*high confidence*). [26.6, 26.7, Figure 26-2]
- In the Arctic, extreme weather events have had direct and indirect adverse health effects for residents (*high confidence*). [28.2]

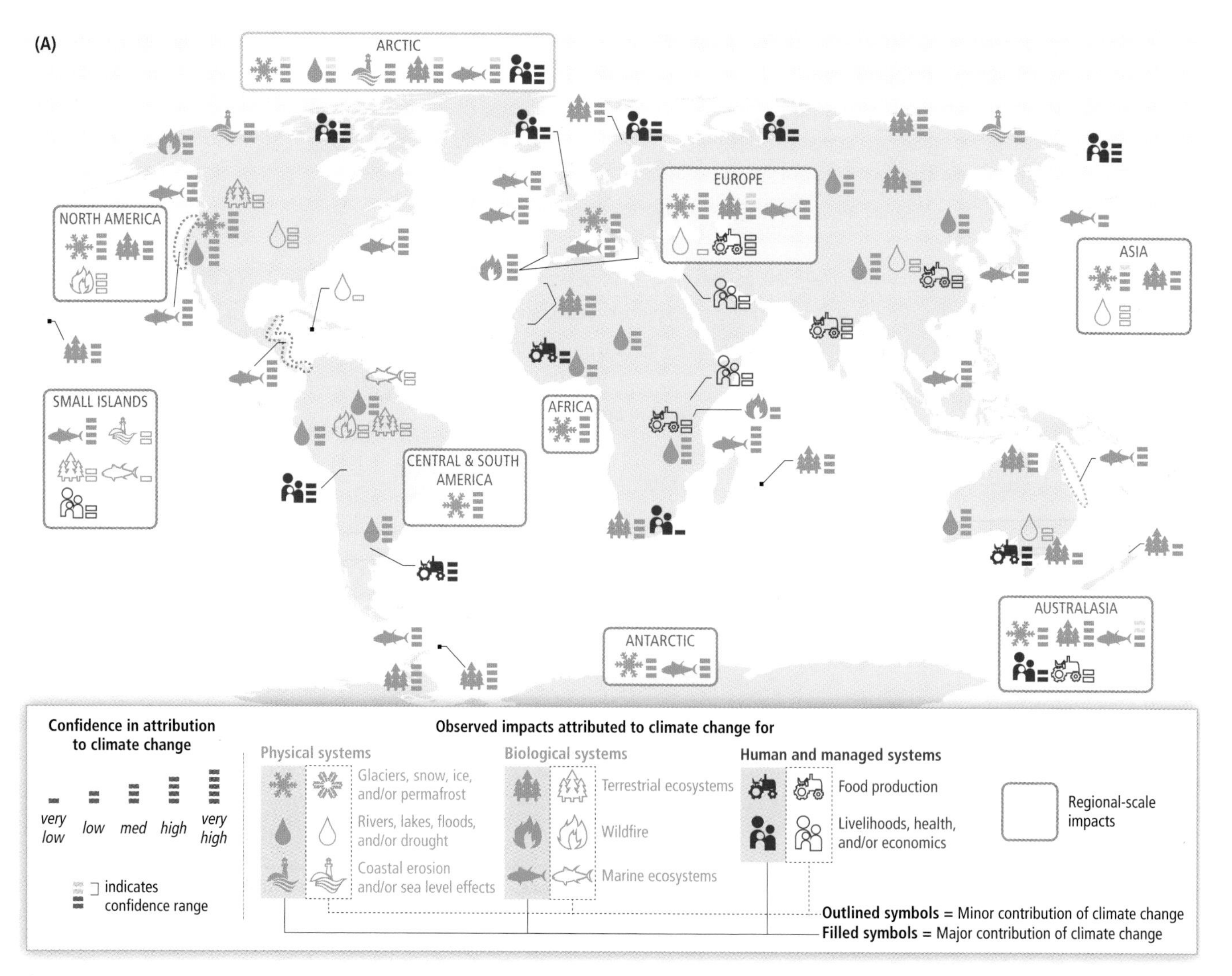

Figure TS.2

Continued next page →

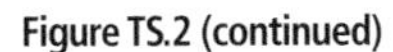

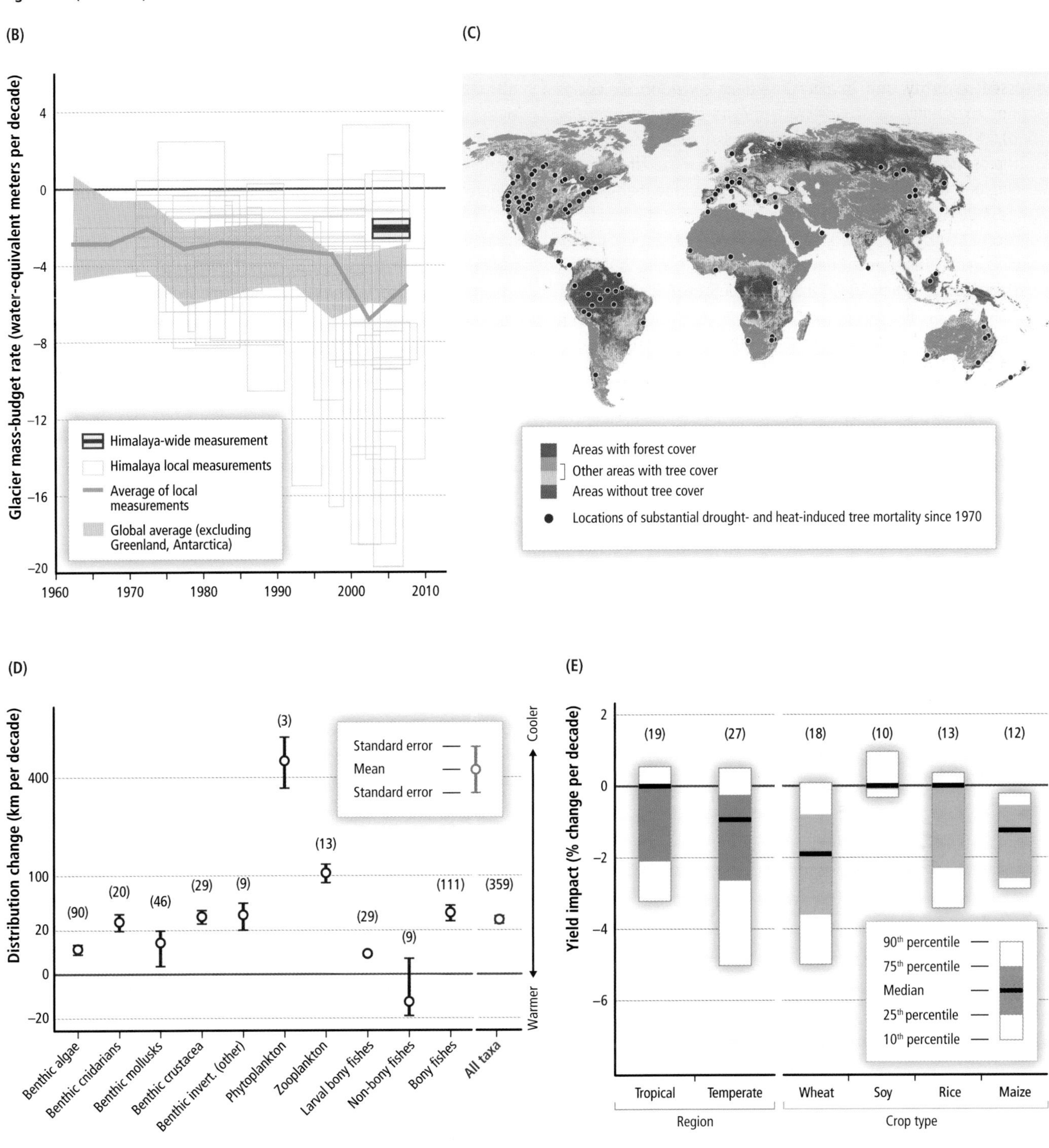

Figure TS.2 | Widespread impacts in a changing world. (A) Global patterns of impacts in recent decades attributed to climate change, based on studies since the AR4. Impacts are shown at a range of geographic scales. Symbols indicate categories of attributed impacts, the relative contribution of climate change (major or minor) to the observed impact, and confidence in attribution. See Table TS.1 for descriptions of the impacts. (B) Changes in glacier mass from all published measurements for Himalayan glaciers. Negative values indicate loss of glacier mass. Local measurements are mostly for small, accessible Himalayan glaciers. The blue box for each local Himalaya measurement is centered vertically on its average, and has a height of ±1 standard deviation for annual measurements and a height of ±1 standard error for multiannual measurements. Himalaya-wide measurement (red) was made by satellite laser altimetry. For reference, global average glacier mass change estimates from WGI AR5 4.3 are also shown, with shading indicating ±1 standard deviation. (C) Locations of substantial drought- and heat-induced tree mortality around the globe over 1970–2011. (D) Average rates of change in distribution (km per decade) for marine taxonomic groups based on observations over 1900–2010. Positive distribution changes are consistent with warming (moving into previously cooler waters, generally poleward). The number of responses analyzed is given within parentheses for each category. (E) Summary of estimated impacts of observed climate changes on yields over 1960–2013 for four major crops in temperate and tropical regions, with the number of data points analyzed given within parentheses for each category. [Figures 3-3, 4-7, 7-2, 18-3, and MB-2]

Freshwater Resources

In many regions, changing precipitation or melting snow and ice are altering hydrological systems, affecting water resources in terms of quantity and quality (*medium confidence*). Glaciers continue to shrink almost worldwide due to climate change (*high confidence*) (e.g., Figure TS.2B), affecting runoff and water resources downstream (*medium confidence*). Climate change is causing permafrost warming and thawing in high-latitude regions and in high-elevation regions (*high confidence*). There is no evidence that surface water and groundwater drought frequency has changed over the last few decades, although impacts of drought have increased mostly due to increased water demand. [3.2, 4.3, 18.3, 18.5, 24.4, 25.5, 26.2, 28.2, Tables 3-1 and 25-1, Figures 18-2 and 26-1]

Terrestrial and Freshwater Ecosystems

Many terrestrial and freshwater plant and animal species have shifted their geographic ranges and seasonal activities and altered their abundance in response to observed climate change over recent decades, and they are doing so now in many regions (*high confidence*). Increased tree mortality, observed in many places worldwide, has been attributed to climate change in some regions (Figure TS.2C). Increases in the frequency or intensity of ecosystem disturbances such as droughts, wind storms, fires, and pest outbreaks have been detected in many parts of the world and in some cases are attributed to climate change (*medium confidence*). While recent climate change contributed to the extinction of some species of Central American amphibians (*medium confidence*), most recent observed terrestrial

Table TS.1 | Observed impacts attributed to climate change reported in the scientific literature since the AR4. These impacts have been attributed to climate change with *very low*, *low*, *medium*, or *high confidence*, with the relative contribution of climate change to the observed change indicated (major or minor), for natural and human systems across eight major world regions over the past several decades. [Tables 18-5 to 18-9] Absence from the table of additional impacts attributed to climate change does not imply that such impacts have not occurred.

	Africa
Snow & Ice, Rivers & Lakes, Floods & Drought	• Retreat of tropical highland glaciers in East Africa (*high confidence*, major contribution from climate change) • Reduced discharge in West African rivers (*low confidence*, major contribution from climate change) • Lake surface warming and water column stratification increases in the Great Lakes and Lake Kariba (*high confidence*, major contribution from climate change) • Increased soil moisture drought in the Sahel since 1970, partially wetter conditions since 1990 (*medium confidence*, major contribution from climate change) [22.2, 22.3, Tables 18-5, 18-6, and 22-3]
Terrestrial Ecosystems	• Tree density decreases in western Sahel and semi-arid Morocco, beyond changes due to land use (*medium confidence*, major contribution from climate change) • Range shifts of several southern plants and animals, beyond changes due to land use (*medium confidence*, major contribution from climate change) • Increases in wildfires on Mt. Kilimanjaro (*low confidence*, major contribution from climate change) [22.3, Tables 18-7 and 22-3]
Coastal Erosion & Marine Ecosystems	• Decline in coral reefs in tropical African waters, beyond decline due to human impacts (*high confidence*, major contribution from climate change) [Table 18-8]
Food Production & Livelihoods	• Adaptive responses to changing rainfall by South African farmers, beyond changes due to economic conditions (*very low confidence*, major contribution from climate change) • Decline in fruit-bearing trees in Sahel (*low confidence*, major contribution from climate change) • Malaria increases in Kenyan highlands, beyond changes due to vaccination, drug resistance, demography, and livelihoods (*low confidence*, minor contribution from climate change) • Reduced fisheries productivity of Great Lakes and Lake Kariba, beyond changes due to fisheries management and land use (*low confidence*, minor contribution from climate change) [7.2, 11.5, 13.2, 22.3, Table 18-9]
	Europe
Snow & Ice, Rivers & Lakes, Floods & Drought	• Retreat of Alpine, Scandinavian, and Icelandic glaciers (*high confidence*, major contribution from climate change) • Increase in rock slope failures in western Alps (*medium confidence*, major contribution from climate change) • Changed occurrence of extreme river discharges and floods (*very low confidence*, minor contribution from climate change) [18.3, 23.2, 23.3, Tables 18-5 and 18-6; WGI AR5 4.3]
Terrestrial Ecosystems	• Earlier greening, leaf emergence, and fruiting in temperate and boreal trees (*high confidence*, major contribution from climate change) • Increased colonization of alien plant species in Europe, beyond a baseline of some invasion (*medium confidence*, major contribution from climate change) • Earlier arrival of migratory birds in Europe since 1970 (*medium confidence*, major contribution from climate change) • Upward shift in tree-line in Europe, beyond changes due to land use (*low confidence*, major contribution from climate change) • Increasing burnt forest areas during recent decades in Portugal and Greece, beyond some increase due to land use (*high confidence*, major contribution from climate change) [4.3, 18.3, Tables 18-7 and 23-6]
Coastal Erosion & Marine Ecosystems	• Northward distributional shifts of zooplankton, fishes, seabirds, and benthic invertebrates in northeast Atlantic (*high confidence*, major contribution from climate change) • Northward and depth shift in distribution of many fish species across European seas (*medium confidence*, major contribution from climate change) • Plankton phenology changes in northeast Atlantic (*medium confidence*, major contribution from climate change) • Spread of warm water species into the Mediterranean, beyond changes due to invasive species and human impacts (*medium confidence*, major contribution from climate change) [6.3, 23.6, 30.5, Tables 6-2 and 18-8, Boxes 6-1 and CC-MB]
Food Production & Livelihoods	• Shift from cold-related mortality to heat-related mortality in England and Wales, beyond changes due to exposure and health care (*low confidence*, major contribution from climate change) • Impacts on livelihoods of Sámi people in northern Europe, beyond effects of economic and sociopolitical changes (*medium confidence*, major contribution from climate change) • Stagnation of wheat yields in some countries in recent decades, despite improved technology (*medium confidence*, minor contribution from climate change) • Positive yield impacts for some crops mainly in northern Europe, beyond increase due to improved technology (*medium confidence*, minor contribution from climate change) • Spread of bluetongue virus in sheep and of ticks across parts of Europe (*medium confidence*, minor contribution from climate change) [18.4, 23.4, 23.5, Table 18-9, Figure 7-2]

Continued next page →

Table TS.1 (continued)

	Asia
Snow & Ice, Rivers & Lakes, Floods & Drought	• Permafrost degradation in Siberia, Central Asia, and Tibetan Plateau (*high confidence*, major contribution from climate change) • Shrinking mountain glaciers across most of Asia (*medium confidence*, major contribution from climate change) • Changed water availability in many Chinese rivers, beyond changes due to land use (*low confidence*, minor contribution from climate change) • Increased flow in several rivers due to shrinking glaciers (*high confidence*, major contribution from climate change) • Earlier timing of maximum spring flood in Russian rivers (*medium confidence*, major contribution from climate change) • Reduced soil moisture in north-central and northeast China (1950–2006) (*medium confidence*, major contribution from climate change) • Surface water degradation in parts of Asia, beyond changes due to land use (*medium confidence*, minor contribution from climate change) [24.3, 24.4, 28.2, Tables 18-5, 18-6, and SM24-4, Box 3-1; WGI AR5 4.3, 10.5]
Terrestrial Ecosystems	• Changes in plant phenology and growth in many parts of Asia (earlier greening), particularly in the north and east (*medium confidence*, major contribution from climate change) • Distribution shifts of many plant and animal species upwards in elevation or polewards, particularly in the north of Asia (*medium confidence*, major contribution from climate change) • Invasion of Siberian larch forests by pine and spruce during recent decades (*low confidence*, major contribution from climate change) • Advance of shrubs into the Siberian tundra (*high confidence*, major contribution from climate change) [4.3, 24.4, 28.2, Table 18-7, Figure 4-4]
Coastal Erosion & Marine Ecosystems	• Decline in coral reefs in tropical Asian waters, beyond decline due to human impacts (*high confidence*, major contribution from climate change) • Northward range extension of corals in the East China Sea and western Pacific, and of a predatory fish in the Sea of Japan (*medium confidence*, major contribution from climate change) • Shift from sardines to anchovies in the western North Pacific, beyond fluctuations due to fisheries (*low confidence*, major contribution from climate change) • Increased coastal erosion in Arctic Asia (*low confidence*, major contribution from climate change) [6.3, 24.4, 30.5, Tables 6-2 and 18-8]
Food Production & Livelihoods	• Impacts on livelihoods of indigenous groups in Arctic Russia, beyond economic and sociopolitical changes (*low confidence*, major contribution from climate change) • Negative impacts on aggregate wheat yields in South Asia, beyond increase due to improved technology (*medium confidence*, minor contribution from climate change) • Negative impacts on aggregate wheat and maize yields in China, beyond increase due to improved technology (*low confidence*, minor contribution from climate change) • Increases in a water-borne disease in Israel (*low confidence*, minor contribution from climate change) [7.2, 13.2, 18.4, 28.2, Tables 18-4 and 18-9, Figure 7-2]
	Australasia
Snow & Ice, Rivers & Lakes, Floods & Drought	• Significant decline in late-season snow depth at 3 of 4 alpine sites in Australia (1957–2002) (*medium confidence*, major contribution from climate change) • Substantial reduction in ice and glacier ice volume in New Zealand (*medium confidence*, major contribution from climate change) • Intensification of hydrological drought due to regional warming in southeast Australia (*low confidence*, minor contribution from climate change) • Reduced inflow in river systems in southwestern Australia (since the mid-1970s) (*high confidence*, major contribution from climate change) [25.5, Tables 18-5, 18-6, and 25-1; WGI AR5 4.3]
Terrestrial Ecosystems	• Changes in genetics, growth, distribution, and phenology of many species, in particular birds, butterflies, and plants in Australia, beyond fluctuations due to variable local climates, land use, pollution, and invasive species (*high confidence*, major contribution from climate change) • Expansion of some wetlands and contraction of adjacent woodlands in southeast Australia (*low confidence*, major contribution from climate change) • Expansion of monsoon rainforest at expense of savannah and grasslands in northern Australia (*medium confidence*, major contribution from climate change) • Migration of glass eels advanced by several weeks in Waikato River, New Zealand (*low confidence*, major contribution from climate change) [Tables 18-7 and 25-3]
Coastal Erosion & Marine Ecosystems	• Southward shifts in the distribution of marine species near Australia, beyond changes due to short-term environmental fluctuations, fishing, and pollution (*medium confidence*, major contribution from climate change) • Change in timing of migration of seabirds in Australia (*low confidence*, major contribution from climate change) • Increased coral bleaching in Great Barrier Reef and western Australian reefs, beyond effects from pollution and physical disturbance (*high confidence*, major contribution from climate change) • Changed coral disease patterns at Great Barrier Reef, beyond effects from pollution (*medium confidence*, major contribution from climate change) [6.3, 25.6, Tables 18-8 and 25-3]
Food Production & Livelihoods	• Advanced timing of wine-grape maturation in recent decades, beyond advance due to improved management (*medium confidence*, major contribution from climate change) • Shift in winter vs. summer human mortality in Australia, beyond changes due to exposure and health care (*low confidence*, major contribution from climate change) • Relocation or diversification of agricultural activities in Australia, beyond changes due to policy, markets, and short-term climate variability (*low confidence*, minor contribution from climate change) [11.4, 18.4, 25.7, 25.8, Tables 18-9 and 25-3, Box 25-5]
	North America
Snow & Ice, Rivers & Lakes, Floods & Drought	• Shrinkage of glaciers across western and northern North America (*high confidence*, major contribution from climate change) • Decreasing amount of water in spring snowpack in western North America (1960–2002) (*high confidence*, major contribution from climate change) • Shift to earlier peak flow in snow dominated rivers in western North America (*high confidence*, major contribution from climate change) • Increased runoff in the midwestern and northeastern US (*medium confidence*, minor contribution from climate change) [Tables 18-5 and 18-6; WGI AR5 2.6, 4.3]
Terrestrial Ecosystems	• Phenology changes and species distribution shifts upward in elevation and northward across multiple taxa (*medium confidence*, major contribution from climate change) • Increased wildfire frequency in subarctic conifer forests and tundra (*medium confidence*, major contribution from climate change) • Regional increases in tree mortality and insect infestations in forests (*low confidence*, minor contribution from climate change) • Increase in wildfire activity, fire frequency and duration, and burnt area in forests of the western US and boreal forests in Canada, beyond changes due to land use and fire management (*medium confidence*, minor contribution from climate change) [26.4, 28.2, Table 18-7, Box 26-2]
Coastal Erosion & Marine Ecosystems	• Northward distributional shifts of northwest Atlantic fish species (*high confidence*, major contribution from climate change) • Changes in musselbeds along the west coast of US (*high confidence*, major contribution from climate change) • Changed migration and survival of salmon in northeast Pacific (*high confidence*, major contribution from climate change) • Increased coastal erosion in Alaska and Canada (*medium confidence*, major contribution from climate change) [18.3, 30.5, Tables 6-2 and 18-8]
Food Production & Livelihoods	• Impacts on livelihoods of indigenous groups in the Canadian Arctic, beyond effects of economic and sociopolitical changes (*medium confidence*, major contribution from climate change) [18.4, 28.2, Tables 18-4 and 18-9]

Continued next page →

Table TS.1 (continued)

	Central and South America
Snow & Ice, Rivers & Lakes, Floods & Drought	• Shrinkage of Andean glaciers (*high confidence*, major contribution from climate change) • Changes in extreme flows in Amazon River (*medium confidence*, major contribution from climate change) • Changing discharge patterns in rivers in the western Andes (*medium confidence*, major contribution from climate change) • Increased streamflow in sub-basins of the La Plata River, beyond increase due to land-use change (*high confidence*, major contribution from climate change) [27.3, Tables 18-5, 18-6, and 27-3; WGI AR5 4.3]
Terrestrial Ecosystems	• Increased tree mortality and forest fire in the Amazon (*low confidence*, minor contribution from climate change) • Rainforest degradation and recession in the Amazon, beyond reference trends in deforestation and land degradation (*low confidence*, minor contribution from climate change) [4.3, 18.3, 27.2, 27.3, Table 18-7]
Coastal Erosion & Marine Ecosystems	• Increased coral bleaching in western Caribbean, beyond effects from pollution and physical disturbance (*high confidence*, major contribution from climate change) • Mangrove degradation on north coast of South America, beyond degradation due to pollution and land use (*low confidence*, minor contribution from climate change) [27.3, Table 18-8]
Food Production & Livelihoods	• More vulnerable livelihood trajectories for indigenous Aymara farmers in Bolivia due to water shortage, beyond effects of increasing social and economic stress (*medium confidence*, major contribution from climate change) • Increase in agricultural yields and expansion of agricultural areas in southeastern South America, beyond increase due to improved technology (*medium confidence*, major contribution from climate change) [13.1, 27.3, Table 18-9]
	Polar Regions
Snow & Ice, Rivers & Lakes, Floods & Drought	• Decreasing Arctic sea ice cover in summer (*high confidence*, major contribution from climate change) • Reduction in ice volume in Arctic glaciers (*high confidence*, major contribution from climate change) • Decreasing snow cover extent across the Arctic (*medium confidence*, major contribution from climate change) • Widespread permafrost degradation, especially in the southern Arctic (*high confidence*, major contribution from climate change) • Ice mass loss along coastal Antarctica (*medium confidence*, major contribution from climate change) • Increased river discharge for large circumpolar rivers (1997–2007) (*low confidence*, major contribution from climate change) • Increased winter minimum river flow in most of the Arctic (*medium confidence*, major contribution from climate change) • Increased lake water temperatures 1985–2009 and prolonged ice-free seasons (*medium confidence*, major contribution from climate change) • Disappearance of thermokarst lakes due to permafrost degradation in the low Arctic. New lakes created in areas of formerly frozen peat (*high confidence*, major contribution from climate change) [28.2, Tables 18-5 and 18-6; WGI AR5 4.2 to 4.4, 4.6, 10.5]
Terrestrial Ecosystems	• Increased shrub cover in tundra in North America and Eurasia (*high confidence*, major contribution from climate change) • Advance of Arctic tree-line in latitude and altitude (*medium confidence*, major contribution from climate change) • Changed breeding area and population size of subarctic birds, due to snowbed reduction and/or tundra shrub encroachment (*medium confidence*, major contribution from climate change) • Loss of snow-bed ecosystems and tussock tundra (*high confidence*, major contribution from climate change) • Impacts on tundra animals from increased ice layers in snow pack, following rain-on-snow events (*medium confidence*, major contribution from climate change) • Increased plant species ranges in the West Antarctic Peninsula and nearby islands over the past 50 years (*high confidence*, major contribution from climate change) • Increased phytoplankton productivity in Signy Island lake waters (*high confidence*, major contribution from climate change) [28.2, Table 18-7]
Coastal Erosion & Marine Ecosystems	• Increased coastal erosion across Arctic (*medium confidence*, major contribution from climate change) • Negative effects on non-migratory Arctic species (*high confidence*, major contribution from climate change) • Decreased reproductive success in Arctic seabirds (*medium confidence*, major contribution from climate change) • Decline in Southern Ocean seals and seabirds (*medium confidence*, major contribution from climate change) • Reduced thickness of foraminiferal shells in southern oceans, due to ocean acidification (*medium confidence*, major contribution from climate change) • Reduced krill density in Scotia Sea (*medium confidence*, major contribution from climate change) [6.3, 18.3, 28.2, 28.3, Table 18-8]
Food Production & Livelihoods	• Impact on livelihoods of Arctic indigenous peoples, beyond effects of economic and sociopolitical changes (*medium confidence*, major contribution from climate change) • Increased shipping traffic across the Bering Strait (*medium confidence*, major contribution from climate change) [18.4, 28.2, Tables 18-4 and 18-9, Figure 28-4]
	Small Islands
Snow & Ice, Rivers & Lakes, Floods & Drought	• Increased water scarcity in Jamaica, beyond increase due to water use (*very low confidence*, minor contribution from climate change) [Table 18-6]
Terrestrial Ecosystems	• Tropical bird population changes in Mauritius (*medium confidence*, major contribution from climate change) • Decline of an endemic plant in Hawai'i (*medium confidence*, major contribution from climate change) • Upward trend in tree-lines and associated fauna on high-elevation islands (*low confidence*, minor contribution from climate change) [29.3, Table 18-7]
Coastal Erosion & Marine Ecosystems	• Increased coral bleaching near many tropical small islands, beyond effects of degradation due to fishing and pollution (*high confidence*, major contribution from climate change) • Degradation of mangroves, wetlands, and seagrass around small islands, beyond degradation due to other disturbances (*very low confidence*, minor contribution from climate change) • Increased flooding and erosion, beyond erosion due to human activities, natural erosion, and accretion (*low confidence*, minor contribution from climate change) • Degradation of groundwater and freshwater ecosystems due to saline intrusion, beyond degradation due to pollution and groundwater pumping (*low confidence*, minor contribution from climate change) [29.3, Table 18-8]
Food Production & Livelihoods	• Increased degradation of coastal fisheries due to direct effects and effects of increased coral reef bleaching, beyond degradation due to overfishing and pollution (*low confidence*, minor contribution from climate change) [18.3, 18.4, 29.3, 30.6, Table 18-9, Box CC-CR]

species extinctions have not been attributed to climate change (*high confidence*). [4.2, 4.4, 18.3, 18.5, 22.3, 25.6, 26.4, 28.2, Figure 4-10, Boxes 4-2, 4-3, 4-4, and 25-3]

Coastal Systems and Low-lying Areas

Coastal systems are particularly sensitive to changes in sea level and ocean temperature and to ocean acidification (*very high confidence*). Coral bleaching and species range shifts have been attributed to changes in ocean temperature. For many other coastal changes, the impacts of climate change are difficult to identify given other human-related drivers (e.g. land use change, coastal development, pollution) (*robust evidence, high agreement*). [5.3 to 5.5, 18.3, 25.6, 26.4, Box 25-3]

Marine Systems

Warming has caused and will continue to cause shifts in the abundance, geographic distribution, migration patterns, and timing of seasonal activities of marine species (*very high confidence*), paralleled by reduction in maximum body sizes (*medium confidence*). This has resulted and will further result in changing interactions between species, including competition and predator-prey dynamics (*high confidence*). Numerous observations over the last decades in all ocean basins show global-scale changes including large-scale distribution shifts of species (*very high confidence*) and altered ecosystem composition (*high confidence*) on multi-decadal time scales, tracking climate trends. Many fishes, invertebrates, and phytoplankton have shifted their distribution and/or abundance poleward and/or to deeper, cooler waters (Figure TS.2D). Some warm-water corals and their reefs have responded to warming with species replacement, bleaching, and decreased coral cover causing habitat loss. Few field observations to date demonstrate biological responses attributable to anthropogenic ocean acidification, as in many places these responses are not yet outside their natural variability and may be influenced by confounding local or regional factors. See also Box TS.7. Natural global climate change at rates slower than current anthropogenic climate change caused significant ecosystem shifts, including species emergences and extinctions, during the past millions of years. [5.4, 6.1, 6.3 to 6.5, 18.3, 18.5, 22.3, 25.6, 26.4, 30.4, 30.5, Boxes 25-3, CC-OA, CC-CR, and CC-MB]

Vulnerability of most marine organisms to warming is set by their physiology, which defines their limited temperature ranges and hence their thermal sensitivity (*high confidence*). See Figure TS.3. Temperature defines the geographic distribution of many species and their responses to climate change. Shifting temperature means and extremes alter habitat (e.g., sea ice and coastal habitat), and cause changes in species abundances through local extinctions and latitudinal distribution expansions or shifts of up to hundreds of kilometers per decade (*very high confidence*). Although genetic adaptation occurs (*medium confidence*), the capacity of fauna and flora to compensate for or keep up with the rate of ongoing thermal change is limited (*low confidence*). [6.3, 6.5, 30.5]

Oxygen minimum zones are progressively expanding in the tropical Pacific, Atlantic, and Indian Oceans, due to reduced ventilation and O_2 solubilities in more stratified oceans at higher temperatures (*high confidence*). In combination with human activities that increase the productivity of coastal systems, hypoxic areas ("dead zones") are increasing in number and size. Regional exacerbation of hypoxia causes shifts to hypoxia-tolerant biota and reduces habitat for commercially relevant species, with implications for fisheries. [6.1, 6.3, 30.3, 30.5, 30.6; WGI AR5 3.8]

Food Security and Food Production Systems

Based on many studies covering a wide range of regions and crops, negative impacts of climate change on crop yields have been more common than positive impacts (*high confidence*). The smaller number of studies showing positive impacts relate mainly to high-latitude regions, though it is not yet clear whether the balance of impacts has been negative or positive in these regions. Climate change has negatively affected wheat and maize yields for many regions and in the global aggregate (*medium confidence*). Effects on rice and soybean yield have been smaller in major production regions and globally, with a median change of zero across all available data, which are fewer for soy compared to the other crops. Observed impacts relate mainly to production aspects of food security rather than access or other components of food security. See Figure TS.2E. Since AR4, several periods of rapid food and cereal price increases following climate extremes in key producing regions indicate a sensitivity of current markets to climate extremes among other factors (*medium confidence*). Crop yields have a large negative sensitivity to extreme daytime temperatures around 30°C, throughout the growing season (*high confidence*). CO_2 has stimulatory effects on crop yields in most cases, and elevated tropospheric ozone has damaging effects. Interactions among CO_2 and ozone, mean temperature, extremes, water, and nitrogen are non-linear and difficult to predict (*medium confidence*). [7.2, 7.3, 18.4, 22.3, 26.5, Figures 7-2, 7-3, and 7-7, Box 25-3]

Urban Areas

Urban areas hold more than half the world's population and most of its built assets and economic activities. A high proportion of the population and economic activities at risk from climate change are in urban areas, and a high proportion of global greenhouse gas emissions are generated by urban-based activities and residents. Cities are composed of complex inter-dependent systems that can be leveraged to support climate change adaptation via effective city governments supported by cooperative multilevel governance (*medium confidence*). This can enable synergies with infrastructure investment and maintenance, land use management, livelihood creation, and ecosystem services protection. [8.1, 8.3, 8.4]

Rapid urbanization and growth of large cities in developing countries have been accompanied by expansion of highly vulnerable urban communities living in informal settlements, many of which are on land exposed to extreme weather (*medium confidence*). [8.2, 8.3]

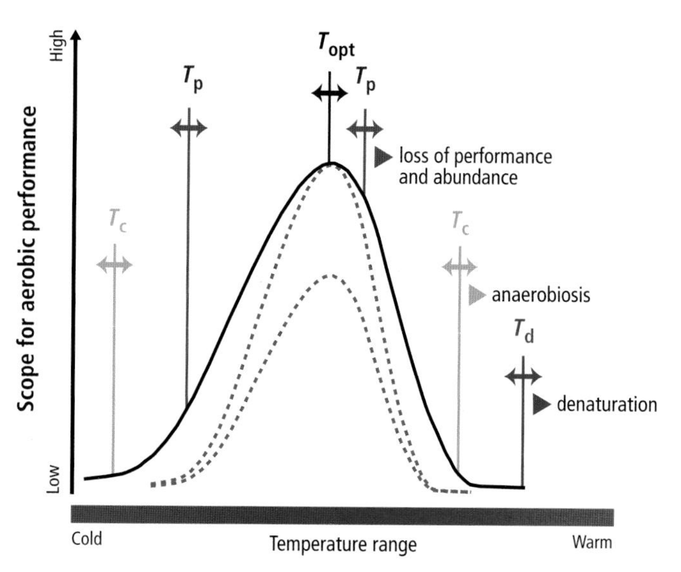

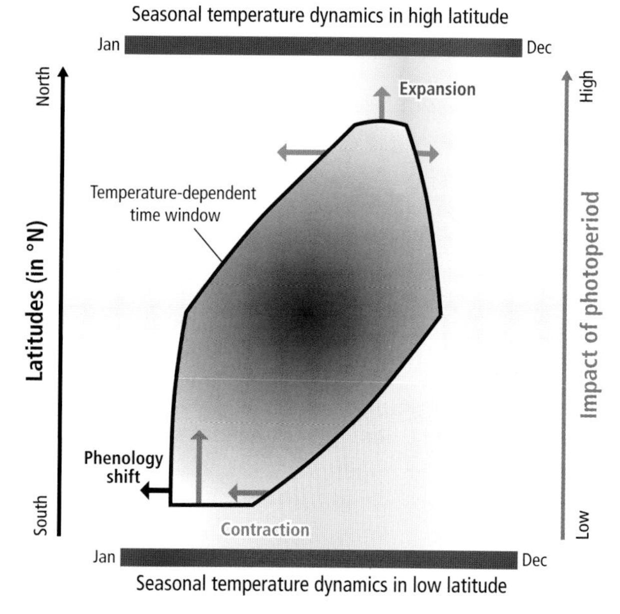

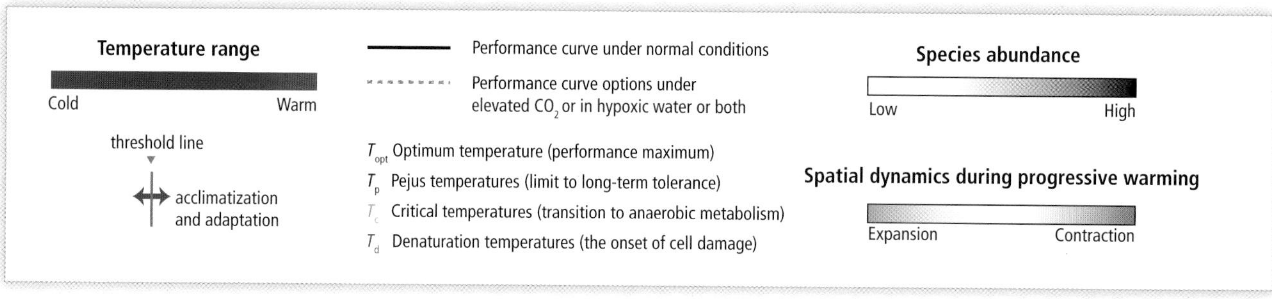

(C)

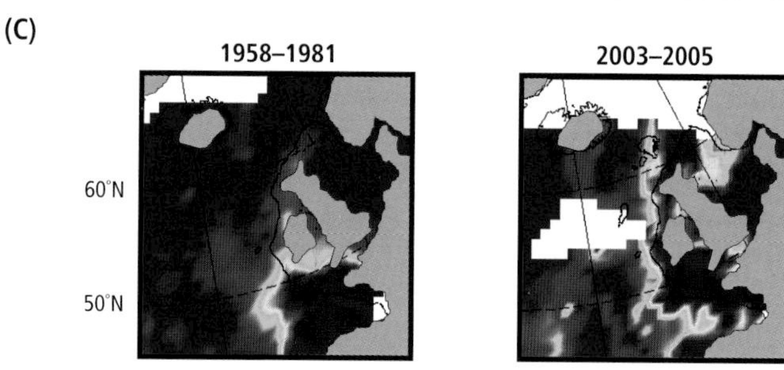

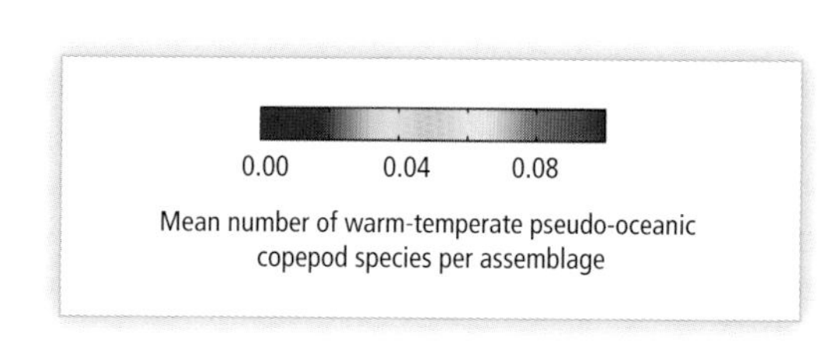

Figure TS.3 | Temperature specialization of species (A), which is influenced by other factors such as oxygen, causes warming-induced distribution shifts (B), for example, the northward expansion of warm-temperate species in the northeast Atlantic (C). These distribution changes depend on species-specific physiology and ecology. Detailed introduction of each panel follows: (A) The temperature tolerance range and performance levels of an organism are described by its performance curve. Each performance (e.g., exercise, growth, reproduction) is highest at optimum temperature (T_{opt}) and lower at cooler or warmer temperatures. Surpassing temperature thresholds (T_p) means going into time-limited tolerance, and more extreme temperature changes lead to exceedance of thresholds that cause metabolic disturbances (T_c) and ultimately onset of cell damage (T_d). These thresholds for an individual can shift (horizontal arrows), within limits, between summer and winter (seasonal acclimatization) or when the species adapts to a cooler or warmer climate over generations (evolutionary adaptation). Under elevated CO_2 levels (ocean acidification) or low oxygen, thermal windows narrow (dashed gray curves). (B) During climate warming, a species follows its normal temperatures as it moves or is displaced, typically resulting in a poleward shift of the biogeographic range (exemplified for the Northern Hemisphere). The polygon delineates the distribution range in space and seasonal time; the level of gray denotes abundance. (C) Long-term changes in the mean number of warm-temperate pseudo-oceanic copepod species in the northeast Atlantic from 1958 to 2005. [Figures 6-5, 6-7, and 6-8]

Rural Areas

Climate change in rural areas will take place in the context of many important economic, social, and land use trends (*very high confidence*). In different regions, absolute rural populations have peaked or will peak in the next few decades. The proportion of the rural population depending on agriculture is varied across regions, but declining everywhere. Poverty rates in rural areas are higher than overall poverty rates, but also falling more sharply, and the proportions of population in extreme poverty accounted for by rural people are also falling: in both cases with the exception of sub-Saharan Africa, where these rates are rising. Accelerating globalization, through migration, labor linkages, regional and international trade, and new information and communication technologies, is bringing about economic transformation in rural areas of developing and developed countries. [9.3, Figure 9-2]

For rural households and communities, access to land and natural resources, flexible local institutions, knowledge and information, and livelihood strategies can contribute to resilience to climate change (*high confidence*). Especially in developing countries, rural people are subject to multiple non-climatic stressors, including underinvestment in agriculture, problems with land and natural resource policy, and processes of environmental degradation (*very high confidence*). In developed countries, there are important shifts toward multiple uses of rural areas, especially leisure uses, and new rural policies based on the collaboration of multiple stakeholders, the targeting of multiple sectors, and a change from subsidy-based to investment-based policy. [9.3, 22.4, Table 9-3]

Key Economic Sectors and Services

Economic losses due to extreme weather events have increased globally, mostly due to increase in wealth and exposure, with a possible influence of climate change (*low confidence* in attribution to climate change). Flooding can have major economic costs, both in term of impacts (e.g., capital destruction, disruption) and adaptation (e.g., construction, defensive investment) (*robust evidence, high agreement*). Since the mid-20th century, socioeconomic losses from flooding have increased mainly due to greater exposure and vulnerability (*high confidence*). [3.2, 3.4, 10.3, 18.4, 23.2, 23.3, 26.7, Figure 26-2, Box 25-7]

Human Health

At present the worldwide burden of human ill-health from climate change is relatively small compared with effects of other stressors and is not well quantified. However, there has been increased heat-related mortality and decreased cold-related mortality in some regions as a result of warming (*medium confidence*). Local changes in temperature and rainfall have altered the distribution of some waterborne illnesses and disease vectors (*medium confidence*). [11.4 to 11.6, 18.4, 25.8]

The health of human populations is sensitive to shifts in weather patterns and other aspects of climate change (*very high confidence*). These effects occur directly, due to changes in temperature and precipitation and in the occurrence of heat waves, floods, droughts, and fires. Health may be damaged indirectly by climate change-related ecological disruptions, such as crop failures or shifting patterns of disease vectors, or by social responses to climate change, such as displacement of populations following prolonged drought. Variability in temperatures is a risk factor in its own right, over and above the influence of average temperatures on heat-related deaths. [11.4, 28.2]

Human Security

Challenges for vulnerability reduction and adaptation actions are particularly high in regions that have shown severe difficulties in governance (*high confidence*). Violent conflict increases vulnerability to climate change (*medium evidence, high agreement*). Large-scale violent conflict harms assets that facilitate adaptation, including infrastructure, institutions, natural resources, social capital, and livelihood opportunities. [12.5, 19.2, 19.6]

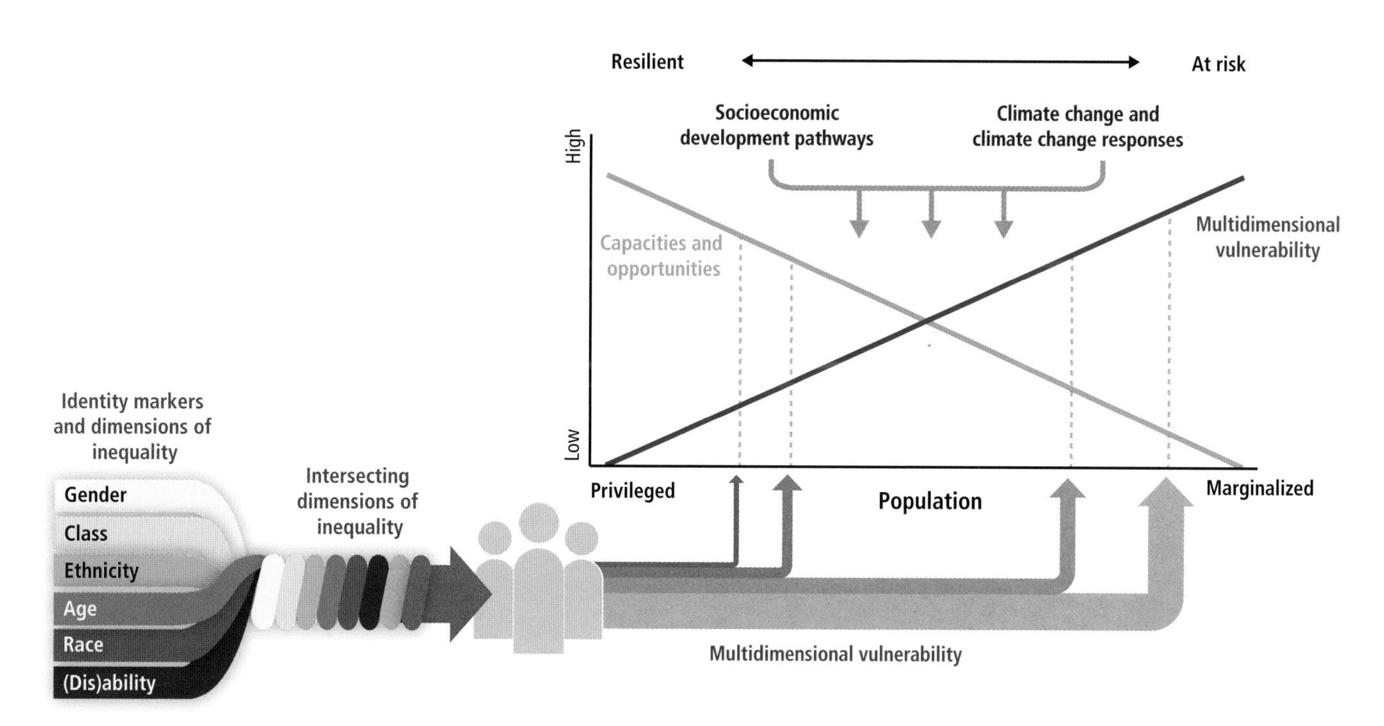

Box TS.4 Figure 1 | Multidimensional vulnerability driven by intersecting dimensions of inequality. Vulnerability increases when people's capacities and opportunities to adapt to climate change and adjust to climate change responses are diminished. [Figure 13-5]

Box TS.4 | Multidimensional Inequality and Vulnerability to Climate Change

People who are socially, economically, culturally, politically, institutionally, or otherwise marginalized in society are especially vulnerable to climate change and also to some adaptation and mitigation responses (*medium evidence*, *high agreement*). This heightened vulnerability is rarely due to a single cause. Rather, it is the product of intersecting social processes that result in inequalities in socioeconomic status and income, as well as in exposure. Such social processes include, for example, discrimination on the basis of gender, class, race/ethnicity, age, and (dis)ability. See Box TS.4 Figure 1 on previous page. Understanding differential capacities and opportunities of individuals, households, and communities requires knowledge of these intersecting social drivers, which may be context-specific and clustered in diverse ways (e.g., class and ethnicity in one case, gender and age in another). Few studies depict the full spectrum of these intersecting social processes and the ways in which they shape multidimensional vulnerability to climate change.

Examples of inequality-driven impacts and risks of climate change and climate change responses (*medium evidence*, *high agreement*):

- Privileged members of society can benefit from climate change impacts and response strategies, given their flexibility in mobilizing and accessing resources and positions of power, often to the detriment of others. [13.2, 13.3, 22.4, 26.8]
- Differential impacts on men and women arise from distinct roles in society, the way these roles are enhanced or constrained by other dimensions of inequality, risk perceptions, and the nature of response to hazards. [8.2, 9.3, 11.3, 12.2, 13.2, 18.4, 19.6, 22.4, Box CC-GC]
- Both male and female deaths are recorded after flooding, affected by socioeconomic disadvantage, occupation, and culturally imposed expectations to save lives. Although women are generally more sensitive to heat stress, more male workers are reported to have died largely as a result of responsibilities related to outdoor and indoor work. [11.3, 13.2, Box CC-GC]
- Women often experience additional duties as laborers and caregivers as a result of extreme weather events and climate change, as well as responses (e.g., male outmigration), while facing more psychological and emotional distress, reduced food intake, adverse mental health outcomes due to displacement, and in some cases increasing incidences of domestic violence. [9.3, 9.4, 12.4, 13.2, Box CC-GC]
- Children and the elderly are often at higher risk due to narrow mobility, susceptibility to infectious diseases, reduced caloric intake, and social isolation. While adults and older children are more severely affected by some climate-sensitive vector-borne diseases such as dengue, young children are more likely to die from or be severely compromised by diarrheal diseases and floods. The elderly face disproportional physical harm and death from heat stress, droughts, and wildfires. [8.2, 10.9, 11.1, 11.4, 11.5, 13.2, 22.4, 23.5, 26.6]
- In most urban areas, low-income groups, including migrants, face large climate change risks because of poor-quality, insecure, and clustered housing, inadequate infrastructure, and lack of provision for health care, emergency services, flood exposure, and measures for disaster risk reduction. [8.1, 8.2, 8.4, 8.5, 12.4, 22.3, 26.8]
- People disadvantaged by race or ethnicity, especially in developed countries, experience more harm from heat stress, often due to low economic status and poor health conditions, and displacement after extreme events. [11.3, 12.4, 13.2]
- Livelihoods and lifestyles of indigenous peoples, pastoralists, and fisherfolk, often dependent on natural resources, are highly sensitive to climate change and climate change policies, especially those that marginalize their knowledge, values, and activities. [9.3, 11.3, 12.3, 14.2, 22.4, 25.8, 26.8, 28.2]
- Disadvantaged groups without access to land and labor, including female-headed households, tend to benefit less from climate change response mechanisms (e.g., Clean Development Mechanism (CDM), Reduction of Emissions from Deforestation and Forest Degradation (REDD+), large-scale land acquisition for biofuels, and planned agricultural adaptation projects). [9.3, 12.2, 12.5, 13.3, 22.4, 22.6]

Livelihoods and Poverty

Climate-related hazards exacerbate other stressors, often with negative outcomes for livelihoods, especially for people living in poverty (*high confidence*). Climate-related hazards affect poor people's lives directly through impacts on livelihoods, reductions in crop yields, or destruction of homes and indirectly through, for example, increased food prices and food insecurity. Urban and rural transient poor who face multiple deprivations can slide into chronic poverty as a result of extreme events, or a series of events, when unable to rebuild their eroded assets (*limited evidence, high agreement*). Observed positive effects for poor and marginalized people, which are limited and often indirect, include examples such as diversification of social networks and of agricultural practices. [8.2, 8.3, 9.3, 11.3, 13.1 to 13.3, 22.3, 24.4, 26.8]

Livelihoods of indigenous peoples in the Arctic have been altered by climate change, through impacts on food security and traditional and cultural values (*medium confidence*). There is emerging evidence of climate change impacts on livelihoods of indigenous people in other regions. [18.4, Table 18-9, Box 18-5]

A-2. Adaptation Experience

Throughout history, people and societies have adjusted to and coped with climate, climate variability, and extremes, with varying degrees of success. This section focuses on adaptive human responses to observed and projected climate-change impacts, which can also address broader risk-reduction and development objectives.

Adaptation is becoming embedded in some planning processes, with more limited implementation of responses (*high confidence*). Engineered and technological options are commonly implemented adaptive responses, often integrated within existing programs such as disaster risk management and water management. There is increasing recognition of the value of social, institutional, and ecosystem-based measures and of the extent of constraints to adaptation. Adaptation options adopted to date continue to emphasize incremental adjustments and co-benefits and are starting to emphasize flexibility and learning (*medium evidence, medium agreement*). [4.4, 5.5, 6.4, 8.3, 9.4, 11.7, 14.1, 14.3, 15.2 to 15.5, 17.2, 17.3, 22.4, 23.7, 25.4, 25.10, 26.8, 26.9, 27.3, 30.6, Boxes 25-1, 25-2, 25-9, and CC-EA]

Most assessments of adaptation have been restricted to impacts, vulnerability, and adaptation planning, with very few assessing the processes of implementation or the effects of adaptation actions (*medium evidence, high agreement*). Vulnerability indicators define, quantify, and weight aspects of vulnerability across regional units, but methods of constructing indices are subjective, often lack transparency, and can be difficult to interpret. There are conflicting views on the choice of adaptation metrics, given differing values placed on needs and outcomes, many of which cannot be captured in a comparable way by metrics. Indicators proving most useful for policy learning are those that track not just process and implementation, but also the extent to which targeted outcomes are occurring. Multi-metric evaluations including risk and uncertainty are increasingly used, an evolution from a previous focus on cost-benefit analysis and identification of "best economic adaptations" (*high confidence*). Adaptation assessments best suited to delivering effective adaptation measures often include both top-down assessments of biophysical climate changes and bottom-up assessments of vulnerability targeted toward local solutions to globally derived risks and toward particular decisions. [4.4, 14.4, 14.5, 15.2, 15.3, 17.2, 17.3, 21.3, 21.5, 22.4, 25.4, 25.10, 26.8, 26.9, Box CC-EA]

Adaptation experience is accumulating across regions in the public and private sector and within communities (*high confidence*). Governments at various levels are starting to develop adaptation plans and policies and to integrate climate-change considerations into broader development plans. Examples of adaptation across regions and contexts include the following:

- Urban adaptation has emphasized city-based disaster risk management such as early warning systems and infrastructure investments; ecosystem-based adaptation and green roofs; enhanced storm and wastewater management; urban and peri-urban agriculture improving food security; enhanced social protection; and good-quality, affordable, and well-located housing (*high confidence*). [8.3, 8.4, 15.4, 26.8, Boxes 25-9, CC-UR, and CC-EA]
- There is a growing body of literature on adaptation practices in both developed and developing country rural areas, including documentation of practical experience in agriculture, water, forestry, and biodiversity and, to a lesser extent, fisheries (*very high confidence*). Public policies supporting decision making for adaptation in rural areas exist in developed and, increasingly, developing countries, and there are also examples of private adaptations led by individuals, companies, and nongovernmental organizations (NGOs) (*high confidence*). Adaptation constraints, particularly pronounced in developing countries, result from lack of access to credit, land, water, technology, markets, information, and perceptions of the need to change. [9.4, 17.3, Tables 9-7 and 9-8]
- In Africa, most national governments are initiating governance systems for adaptation (*high confidence*). Progress on national and subnational policies and strategies has initiated the mainstreaming of adaptation into sectoral planning, but evolving institutional frameworks cannot yet effectively coordinate the range of adaptation initiatives being implemented. Disaster risk management, adjustments in technologies and infrastructure, ecosystem-based approaches, basic public health measures, and livelihood diversification are reducing vulnerability, although efforts to date tend to be isolated. [22.4]
- In Europe, adaptation policy has been developed at international (EU), national, and local government levels, with limited systematic information on current implementation or effectiveness (*high confidence*). Some adaptation planning has been integrated into coastal and water management, into environmental protection and land planning, and into disaster risk management. [23.7, Boxes 5-1 and 23-3]
- In Asia, adaptation is being facilitated in some areas through mainstreaming climate adaptation action into subnational development planning, early warning systems, integrated water resources management, agroforestry, and coastal reforestation of mangroves (*high confidence*). [24.4 to 24.6, 24.9, Box CC-TC]
- In Australasia, planning for sea level rise, and in southern Australia for reduced water availability, is becoming adopted widely. Planning

Table TS.2 | Illustrative examples of adaptation experience, as well as approaches to reducing vulnerability and enhancing resilience. Adaptation actions can be influenced by climate variability, extremes, and change, and by exposure and vulnerability at the scale of risk management. Many examples and case studies demonstrate complexity at the level of communities or specific regions within a country. It is at this spatial scale that complex interactions between vulnerability, exposure, and climate change come to the fore. [Table 21-4]

Early warning systems for heat	
Exposure and vulnerability	Factors affecting exposure and vulnerability include age, preexisting health status, level of outdoor activity, socioeconomic factors including poverty and social isolation, access to and use of cooling, physiological and behavioral adaptation of the population, urban heat island effects, and urban infrastructure. [8.2.3, 8.2.4, 11.3.3, 11.3.4, 11.4.1, 11.7, 13.2.1, 19.3.2, 23.5.1, 25.3, 25.8.1, SREX Table SPM.1]
Climate information at the global scale	**Observed:** • *Very likely* decrease in the number of cold days and nights and increase in the number of warm days and nights, on the global scale between 1951 and 2010. [WGI AR5 2.6.1] • *Medium confidence* that the length and frequency of warm spells, including heat waves, has increased globally since 1950. [WGI AR5 2.6.1] **Projected:** *Virtually certain* that, in most places, there will be more hot and fewer cold temperature extremes as global mean temperatures increase, for events defined as extremes on both daily and seasonal time scales. [WGI AR5 12.4.3]
Climate information at the regional scale	**Observed:** • *Likely* that heat wave frequency has increased since 1950 in large parts of Europe, Asia, and Australia. [WGI AR5 2.6.1] • *Medium confidence* in overall increase in heat waves and warm spells in North America since 1960. Insufficient evidence for assessment or spatially varying trends in heat waves or warm spells for South America and most of Africa. [SREX Table 3-2; WGI AR5 2.6.1] **Projected:** • *Likely* that, by the end of the 21st century under Representative Concentration Pathway 8.5 (RCP8.5) in most land regions, a current 20-year high-temperature event will at least double its frequency and in many regions occur every 2 years or annually, while a current 20-year low-temperature event will become exceedingly rare. [WGI AR5 12.4.3] • *Very likely* more frequent and/or longer heat waves or warm spells over most land areas. [WGI AR5 12.4.3]
Description	Heat-health early warning systems are instruments to prevent negative health impacts during heat waves. Weather forecasts are used to predict situations associated with increased mortality or morbidity. Components of effective heat wave and health warning systems include identifying weather situations that adversely affect human health, monitoring weather forecasts, communicating heat wave and prevention responses, targeting notifications to vulnerable populations, and evaluating and revising the system to increase effectiveness in a changing climate. Warning systems for heat waves have been planned and implemented broadly, for example in Europe, the United States, Asia, and Australia. [11.7.3, 24.4.6, 25.8.1, 26.6, Box 25-6]
Broader context	• Heat-health warning systems can be combined with other elements of a health protection plan, for example building capacity to support communities most at risk, supporting and funding health services, and distributing public health information. • In Africa, Asia, and elsewhere, early warning systems have been used to provide warning of and reduce a variety of risks related to famine and food insecurity; flooding and other weather-related hazards; exposure to air pollution from fire; and vector-borne and food-borne disease outbreaks. [7.5.1, 11.7, 15.4.2, 22.4.5, 24.4.6, 25.8.1, 26.6.3, Box 25-6]
Mangrove restoration to reduce flood risks and protect shorelines from storm surge	
Exposure and vulnerability	Loss of mangroves increases exposure of coastlines to storm surge, coastal erosion, saline intrusion, and tropical cyclones. Exposed infrastructure, livelihoods, and people are vulnerable to associated damage. Areas with development in the coastal zone, such as on small islands, can be particularly vulnerable. [5.4.3, 5.5.6, 29.7.2, Box CC-EA]
Climate information at the global scale	**Observed:** • *Likely* increase in the magnitude of extreme high sea level events since 1970, mostly explained by rising mean sea level. [WGI AR5 3.7.5] • *Low confidence* in long-term (centennial) changes in tropical cyclone activity, after accounting for past changes in observing capabilities. [WGI AR5 2.6.3] **Projected:** • *Very likely* significant increase in the occurrence of future sea level extremes by 2050 and 2100. [WGI AR5 13.7.2] • In the 21st century, *likely* that the global frequency of tropical cyclones will either decrease or remain essentially unchanged. *Likely* increase in both global mean tropical cyclone maximum wind speed and rainfall rates. [WGI AR5 14.6]
Climate information at the regional scale	**Observed:** Change in sea level relative to the land (relative sea level) can be significantly different from the global mean sea level change because of changes in the distribution of water in the ocean and vertical movement of the land. [WGI AR5 3.7.3] **Projected:** • *Low confidence* in region-specific projections of storminess and associated storm surges. [WGI AR5 13.7.2] • Projections of regional changes in sea level reach values of up to 30% above the global mean value in the Southern Ocean and around North America, and between 10% to 20% above the global mean value in equatorial regions. [WGI AR5 13.6.5] • *More likely than not* substantial increase in the frequency of the most intense tropical cyclones in the western North Pacific and North Atlantic. [WGI AR5 14.6]
Description	Mangrove restoration and rehabilitation has occurred in a number of locations (e.g., Vietnam, Djibouti, and Brazil) to reduce coastal flooding risks and protect shorelines from storm surge. Restored mangroves have been shown to attenuate wave height and thus reduce wave damage and erosion. They protect aquaculture industry from storm damage and reduce saltwater intrusion. [2.4.3, 5.5.4, 8.3.3, 22.4.5, 27.3.3]
Broader context	• Considered a low-regrets option benefiting sustainable development, livelihood improvement, and human well-being through improvements for food security and reduced risks from flooding, saline intrusion, wave damage, and erosion. Restoration and rehabilitation of mangroves, as well as of wetlands or deltas, is ecosystem-based adaptation that enhances ecosystem services. • Synergies with mitigation given that mangrove forests represent large stores of carbon. • Well-integrated ecosystem-based adaptation can be more cost effective and sustainable than non-integrated physical engineering approaches. [5.5, 8.4.2, 14.3.1, 24.6, 29.3.1, 29.7.2, 30.6.1, 30.6.2, Table 5-4, Box CC-EA]

Continued next page →

Table TS.2 (continued)

Community-based adaptation and traditional practices in small island contexts	
Exposure and vulnerability	With small land area, often low elevation coasts, and concentration of human communities and infrastructure in coastal zones, small islands are particularly vulnerable to rising sea levels and impacts such as inundation, saltwater intrusion, and shoreline change. [29.3.1, 29.3.3, 29.6.1, 29.6.2, 29.7.2]
Climate information at the global scale	**Observed:** • *Likely* increase in the magnitude of extreme high sea level events since 1970, mostly explained by rising mean sea level. [WGI AR5 3.7.5] • *Low confidence* in long-term (centennial) changes in tropical cyclone activity, after accounting for past changes in observing capabilities. [WGI AR5 2.6.3] • Since 1950 the number of heavy precipitation events over land has *likely* increased in more regions than it has decreased. [WGI AR5 2.6.2] **Projected:** • *Very likely* significant increase in the occurrence of future sea level extremes by 2050 and 2100. [WGI AR5 13.7.2] • In the 21st century, *likely* that the global frequency of tropical cyclones will either decrease or remain essentially unchanged. *Likely* increase in both global mean tropical cyclone maximum wind speed and rainfall rates. [WGI AR5 14.6] • Globally, for short-duration precipitation events, *likely* shift to more intense individual storms and fewer weak storms. [WGI AR5 12.4.5]
Climate information at the regional scale	**Observed:** Change in sea level relative to the land (relative sea level) can be significantly different from the global mean sea level change because of changes in the distribution of water in the ocean and vertical movement of the land. [WGI AR5 3.7.3] **Projected:** • *Low confidence* in region-specific projections of storminess and associated storm surges. [WGI AR5 13.7.2] • Projections of regional changes in sea level reach values of up to 30% above the global mean value in the Southern Ocean and around North America, and between 10% and 20% above the global mean value in equatorial regions. [WGI AR5 13.6.5] • *More likely than not* substantial increase in the frequency of the most intense tropical cyclones in the western North Pacific and North Atlantic. [WGI AR5 14.6]
Description	Traditional technologies and skills can be relevant for climate adaptation in small island contexts. In the Solomon Islands, relevant traditional practices include elevating concrete floors to keep them dry during heavy precipitation events and building low aerodynamic houses with palm leaves as roofing to avoid hazards from flying debris during cyclones, supported by perceptions that traditional construction methods are more resilient to extreme weather. In Fiji after Cyclone Ami in 2003, mutual support and risk sharing formed a central pillar for community-based adaptation, with unaffected households fishing to support those with damaged homes. Participatory consultations across stakeholders and sectors within communities and capacity building taking into account traditional practices can be vital to the success of adaptation initiatives in island communities, such as in Fiji or Samoa. [29.6.2]
Broader context	• Perceptions of self-efficacy and adaptive capacity in addressing climate stress can be important in determining resilience and identifying useful solutions. • The relevance of community-based adaptation principles to island communities, as a facilitating factor in adaptation planning and implementation, has been highlighted, for example, with focus on empowerment and learning-by-doing, while addressing local priorities and building on local knowledge and capacity. Community-based adaptation can include measures that cut across sectors and technological, social, and institutional processes, recognizing that technology by itself is only one component of successful adaptation. [5.5.4, 29.6.2]
Adaptive approaches to flood defense in Europe	
Exposure and vulnerability	Increased exposure of persons and property in flood risk areas has contributed to increased damages from flood events over recent decades. [5.4.3, 5.4.4, 5.5.5, 23.3.1, Box 5-1]
Climate information at the global scale	**Observed:** • *Likely* increase in the magnitude of extreme high sea level events since 1970, mostly explained by rising mean sea level. [WGI AR5 3.7.5] • Since 1950 the number of heavy precipitation events over land has *likely* increased in more regions than it has decreased. [WGI AR5 2.6.2] **Projected:** • *Very likely* that the time-mean rate of global mean sea level rise during the 21st century will exceed the rate observed during 1971–2010 for all RCP scenarios. [WGI AR5 13.5.1] • Globally, for short-duration precipitation events, *likely* shift to more intense individual storms and fewer weak storms. [WGI AR5 12.4.5]
Climate information at the regional scale	**Observed:** • *Likely* increase in the frequency or intensity of heavy precipitation in Europe, with some seasonal and/or regional variations. [WGI AR5 2.6.2] • Increase in heavy precipitation in winter since the 1950s in some areas of northern Europe (*medium confidence*). Increase in heavy precipitation since the 1950s in some parts of west-central Europe and European Russia, especially in winter (*medium confidence*). [SREX Table 3-2] • Increasing mean sea level with regional variations, except in the Baltic Sea where the relative sea level is decreasing due to vertical crustal motion. [5.3.2, 23.2.2] **Projected:** • Over most of the mid-latitude land masses, extreme precipitation events will *very likely* be more intense and more frequent in a warmer world. [WGI AR5 12.4.5] • Overall precipitation increase in northern Europe and decrease in southern Europe (*medium confidence*). [23.2.2] • Increased extreme precipitation in northern Europe during all seasons, particularly winter, and in central Europe except in summer (*high confidence*). [23.2.2; SREX Table 3-3]
Description	Several governments have made ambitious efforts to address flood risk and sea level rise over the coming century. In the Netherlands, government recommendations include "soft" measures preserving land from development to accommodate increased river inundation; maintaining coastal protection through beach nourishment; and ensuring necessary political-administrative, legal, and financial resources. Through a multi-stage process, the British government has also developed extensive adaptation plans to adjust and improve flood defenses to protect London from future storm surges and river flooding. Pathways have been analyzed for different adaptation options and decisions, depending on eventual sea level rise, with ongoing monitoring of the drivers of risk informing decisions. [5.5.4, 23.7.1, Box 5-1]
Broader context	• The Dutch plan is considered a paradigm shift, addressing coastal protection by "working with nature" and providing "room for river." • The British plan incorporates iterative, adaptive decisions depending on the eventual sea level rise with numerous and diverse measures possible over the next 50 to 100 years to reduce risk to acceptable levels. • In cities in Europe and elsewhere, the importance of strong political leadership or government champions in driving successful adaptation action has been noted. [5.5.3, 5.5.4, 8.4.3, 23.7.1, 23.7.2, 23.7.4, Boxes 5-1 and 26-3]

Continued next page →

Table TS.2 (continued)

Index-based insurance for agriculture in Africa	
Exposure and vulnerability	Susceptibility to food insecurity and depletion of farmers' productive assets following crop failure. Low prevalence of insurance due to absent or poorly developed insurance markets or to amount of premium payments. The most marginalized and resource-poor especially may have limited ability to afford insurance premiums. [10.7.6, 13.3.2, Box 22-1]
Climate information at the global scale	**Observed:** • *Very likely* decrease in the number of cold days and nights and increase in the number of warm days and nights, on the global scale between 1951 and 2010. [WGI AR5 2.6.1] • *Medium confidence* that the length and frequency of warm spells, including heat waves, has increased globally since 1950. [WGI AR5 2.6.1] • Since 1950 the number of heavy precipitation events over land has *likely* increased in more regions than it has decreased. [WGI AR5 2.6.2] • *Low confidence* in a global-scale observed trend in drought or dryness (lack of rainfall). [WGI AR5 2.6.2] **Projected:** • *Virtually certain* that, in most places, there will be more hot and fewer cold temperature extremes as global mean temperatures increase, for events defined as extremes on both daily and seasonal time scales. [WGI AR5 12.4.3] • Regional to global-scale projected decreases in soil moisture and increased risk of agricultural drought are *likely* in presently dry regions, and are projected with *medium confidence* by the end of this century under the RCP8.5 scenario. [WGI AR5 12.4.5] • Globally, for short-duration precipitation events, *likely* shift to more intense individual storms and fewer weak storms. [WGI AR5 12.4.5]
Climate information at the regional scale	**Observed:** • *Medium confidence* in increase in frequency of warm days and decrease in frequency of cold days and nights in southern Africa. [SREX Table 3-2] • *Medium confidence* in increase in frequency of warm nights in northern and southern Africa. [SREX Table 3-2] **Projected:** • *Likely* surface drying in southern Africa by the end of the 21st century under RCP8.5 (*high confidence*). [WGI AR5 12.4.5] • *Likely* increase in warm days and nights and decrease in cold days and nights in all regions of Africa (*high confidence*). Increase in warm days largest in summer and fall (*medium confidence*). [SREX Table 3-3] • *Likely* more frequent and/or longer heat waves and warm spells in Africa (*high confidence*). [SREX Table 3-3]
Description	A recently introduced mechanism that has been piloted in a number of rural locations, including in Malawi, Sudan, and Ethiopia, as well as in India. When physical conditions reach a particular predetermined threshold where significant losses are expected to occur—weather conditions such as excessively high or low cumulative rainfall or temperature peaks—the insurance pays out. [9.4.2, 13.3.2, 15.4.4, Box 22-1]
Broader context	• Index-based weather insurance is considered well suited to the agricultural sector in developing countries. • The mechanism allows risk to be shared across communities, with costs spread over time, while overcoming obstacles to traditional agricultural and disaster insurance markets. It can be integrated with other strategies such as microfinance and social protection programs. • Risk-based premiums can help encourage adaptive responses and foster risk awareness and risk reduction by providing financial incentives to policyholders to reduce their risk profile. • Challenges can be associated with limited availability of accurate weather data and difficulties in establishing which weather conditions cause losses. Basis risk (i.e., farmers suffer losses but no payout is triggered based on weather data) can promote distrust. There can also be difficulty in scaling up pilot schemes. • Insurance for work programs can enable cash-poor farmers to work for insurance premiums by engaging in community-identified disaster risk reduction projects. [10.7.4 to 10.7.6, 13.3.2, 15.4.4, Table 10-7, Boxes 22-1 and 25-7]

Continued next page →

for sea level rise has evolved considerably over the past 2 decades and shows a diversity of approaches, although its implementation remains piecemeal (*high confidence*). Adaptive capacity is generally high in many human systems, but implementation faces major constraints especially for transformational responses at local and community levels. [25.4, 25.10, Table 25-2, Boxes 25-1, 25-2, and 25-9]

- In North America, governments are engaging in incremental adaptation assessment and planning, particularly at the municipal level (*high confidence*). Some proactive adaptation is occurring to protect longer-term investments in energy and public infrastructure. [26.7 to 26.9]
- In Central and South America, ecosystem-based adaptation including protected areas, conservation agreements, and community management of natural areas is occurring (*high confidence*). Resilient crop varieties, climate forecasts, and integrated water resources management are being adopted within the agricultural sector in some areas. [27.3]
- In the Arctic, some communities have begun to deploy adaptive co-management strategies and communications infrastructure, combining traditional and scientific knowledge (*high confidence*). [28.2, 28.4]
- In small islands, which have diverse physical and human attributes, community-based adaptation has been shown to generate larger benefits when delivered in conjunction with other development activities (*high confidence*). [29.3, 29.6, Table 29-3, Figure 29-1]
- In both the open ocean and coastal areas, international cooperation and marine spatial planning are starting to facilitate adaptation to climate change, with constraints from challenges of spatial scale and governance issues (*high confidence*). Observed coastal adaptation includes major projects (e.g., Thames Estuary, Venice Lagoon, Delta Works) and specific practices in some countries (e.g., Netherlands, Australia, Bangladesh). [5.5, 7.3, 15.4, 30.6, Box CC-EA]

Table TS.2 presents examples of how climate extremes and change, as well as exposure and vulnerability at the scale of risk management, shape adaptation actions and approaches to reducing vulnerability and enhancing resilience.

A-3. The Decision-making Context

Climate variability and extremes have long been important in many decision-making contexts. Climate-related risks are now evolving over

Table TS.2 (continued)

Relocation of agricultural industries in Australia	
Exposure and vulnerability	Crops sensitive to changing patterns of temperature, rainfall, and water availability. [7.3, 7.5.2]
Climate information at the global scale	**Observed:** • *Very likely* decrease in the number of cold days and nights and increase in the number of warm days and nights, on the global scale between 1951 and 2010. [WGI AR5 2.6.1] • *Medium confidence* that the length and frequency of warm spells, including heat waves, has increased globally since 1950. [WGI AR5 2.6.1] • *Medium confidence* in precipitation change over global land areas since 1950. [WGI AR5 2.5.1] • Since 1950 the number of heavy precipitation events over land has *likely* increased in more regions than it has decreased. [WGI AR5 2.6.2] • *Low confidence* in a global-scale observed trend in drought or dryness (lack of rainfall). [WGI AR5 2.6.2] **Projected:** • *Virtually certain* that, in most places, there will be more hot and fewer cold temperature extremes as global mean temperatures increase, for events defined as extremes on both daily and seasonal time scales. [WGI AR5 12.4.3] • *Virtually certain* increase in global precipitation as global mean surface temperature increases. [WGI AR5 12.4.1] • Regional to global-scale projected decreases in soil moisture and increased risk of agricultural drought are *likely* in presently dry regions, and are projected with *medium confidence* by the end of this century under the RCP8.5 scenario. [WGI AR5 12.4.5] • Globally, for short-duration precipitation events, *likely* shift to more intense individual storms and fewer weak storms. [WGI AR5 12.4.5]
Climate information at the regional scale	**Observed:** • Cool extremes rarer and hot extremes more frequent and intense over Australia and New Zealand, since 1950 (*high confidence*). [Table 25-1] • *Likely* increase in heat wave frequency since 1950 in large parts of Australia. [WGI AR5 2.6.1] • Late autumn/winter decreases in precipitation in southwestern Australia since the 1970s and southeastern Australia since the mid-1990s, and annual increases in precipitation in northwestern Australia since the 1950s (*very high confidence*). [Table 25-1] • Mixed or insignificant trends in annual daily precipitation extremes, but a tendency to significant increase in annual intensity of heavy precipitation in recent decades for sub-daily events in Australia (*high confidence*). [Table 25-1] **Projected:** • Hot days and nights more frequent and cold days and nights less frequent during the 21st century in Australia and New Zealand (*high confidence*). [Table 25-1] • Annual decline in precipitation over southwestern Australia (*high confidence*) and elsewhere in southern Australia (*medium confidence*). Reductions strongest in the winter half-year (*high confidence*). [Table 25-1] • Increase in most regions in the intensity of rare daily rainfall extremes and in sub-daily extremes (*medium confidence*) in Australia and New Zealand. [Table 25-1] • Drought occurrence to increase in southern Australia (*medium confidence*). [Table 25-1] • Snow depth and snow area to decline in Australia (*very high confidence*). [Table 25-1] • Freshwater resources projected to decline in far southeastern and far southwestern Australia (*high confidence*). [25.5.2]
Description	Industries and individual farmers are relocating parts of their operations, for example for rice, wine, or peanuts in Australia, or are changing land use *in situ* in response to recent climate change or expectations of future change. For example, there has been some switching from grazing to cropping in southern Australia. Adaptive movement of crops has also occurred elsewhere. [7.5.1, 25.7.2, Table 9-7, Box 25-5]
Broader context	• Considered transformational adaptation in response to impacts of climate change. • Positive or negative implications for the wider communities in origin and destination regions. [25.7.2, Box 25-5]

time due to both climate change and development. This section builds from existing experience with decision making and risk management. It creates a foundation for understanding the report's assessment of future climate-related risks and potential responses.

Responding to climate-related risks involves decision making in a changing world, with continuing uncertainty about the severity and timing of climate-change impacts and with limits to the effectiveness of adaptation (*high confidence*). Iterative risk management is a useful framework for decision making in complex situations characterized by large potential consequences, persistent uncertainties, long timeframes, potential for learning, and multiple climatic and non-climatic influences changing over time. See Figure TS.4. Assessment of the widest possible range of potential impacts, including low-probability outcomes with large consequences, is central to understanding the benefits and trade-offs of alternative risk management actions. The complexity of adaptation actions across scales and contexts means that monitoring and learning are important components of effective adaptation. [2.1 to 2.4, 3.6, 14.1 to 14.3, 15.2 to 15.4, 16.2 to 16.4, 17.1 to 17.3, 17.5, 20.6, 22.4, 25.4, Figure 1-5]

Adaptation and mitigation choices in the near term will affect the risks of climate change throughout the 21st century (*high confidence*). Figure TS.5 illustrates projected climate futures under a low-emission mitigation scenario and a high-emission scenario [Representative Concentration Pathways (RCPs) 2.6 and 8.5], along with observed temperature and precipitation changes. The benefits of adaptation and mitigation occur over different but overlapping timeframes. Projected global temperature increase over the next few decades is similar across emission scenarios (Figure TS.5A, middle panel) (WGI AR5 Section 11.3). During this near-term era of committed climate change, risks will evolve as socioeconomic trends interact with the changing climate. Societal responses, particularly adaptations, will influence near-term outcomes. In the second half of the 21st century and beyond, global temperature increase diverges across emission scenarios (Figure TS.5A, middle and bottom panels) (WGI AR5 Section 12.4 and Table SPM.2). For this longer-term era of climate options, near-term and longer-term adaptation and mitigation, as well as development pathways, will determine the risks of climate change. [2.5, 21.2, 21.3, 21.5, Box CC-RC]

Assessment of risks in the WGII AR5 relies on diverse forms of evidence. Expert judgment is used to integrate evidence into evaluations of risks. Forms of evidence include, for example, empirical observations, experimental results, process-based understanding, statistical approaches, and simulation and descriptive models. Future

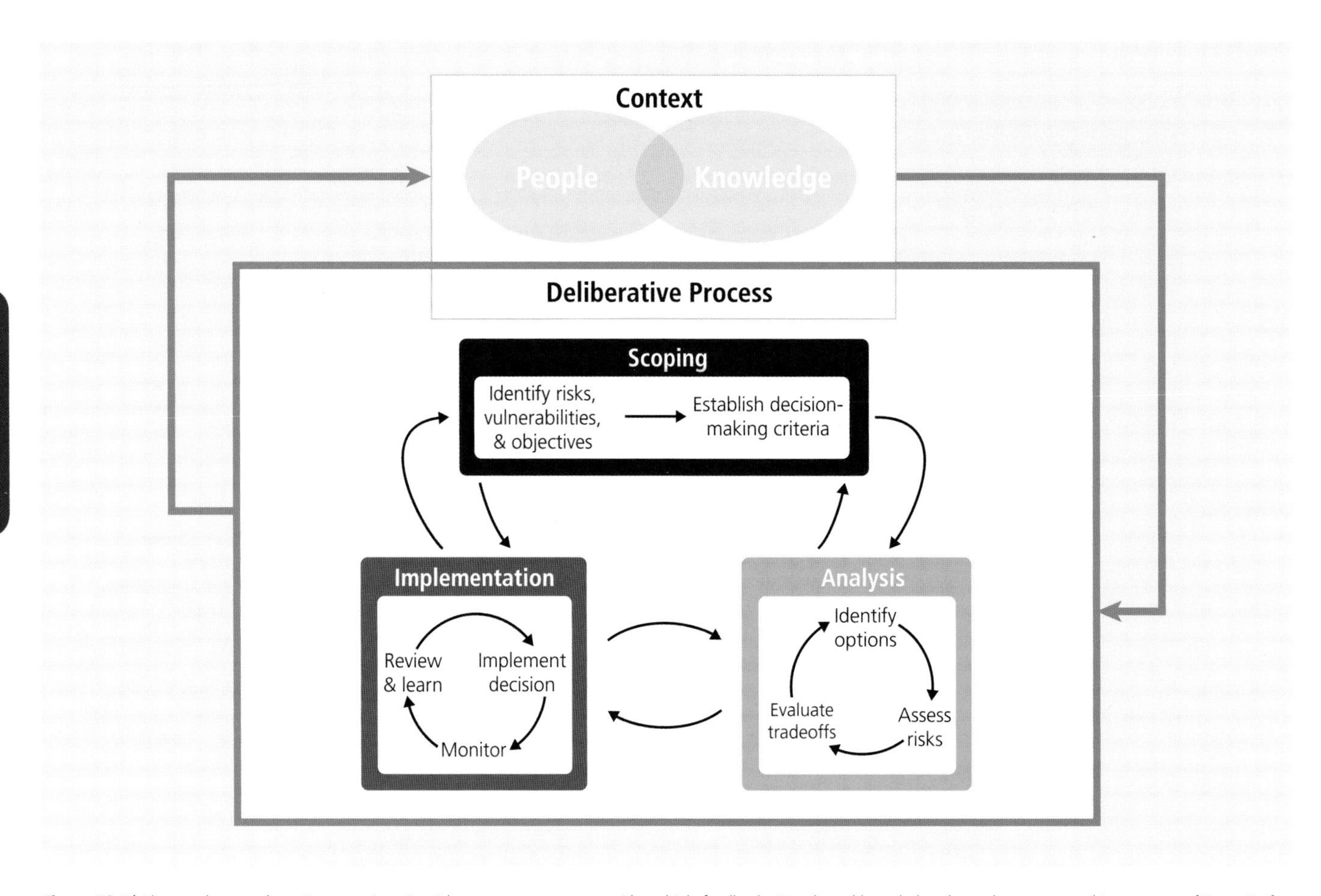

Figure TS.4 | Climate-change adaptation as an iterative risk management process with multiple feedbacks. People and knowledge shape the process and its outcomes. [Figure 2-1]

risks related to climate change vary substantially across plausible alternative development pathways, and the relative importance of development and climate change varies by sector, region, and time period (*high confidence*). Scenarios are useful tools for characterizing possible future socioeconomic pathways, climate change and its risks, and policy implications. Climate-model projections informing evaluations of risks in this report are generally based on the RCPs (Figure TS.5), as well as the older IPCC *Special Report on Emissions Scenarios* (SRES) scenarios. [1.1, 1.3, 2.2, 2.3, 19.6, 20.2, 21.3, 21.5, 26.2, Box CC-RC; WGI AR5 Box SPM.1]

Scenarios can be divided into those that explore how futures may unfold under various drivers (problem exploration) and those that test how various interventions may play out (solution exploration) (*robust evidence*, *high agreement*). Adaptation approaches address uncertainties associated with future climate and socioeconomic conditions and with the diversity of specific contexts (*medium evidence*, *high agreement*). Although many national studies identify a variety of strategies and approaches for adaptation, they can be classified into two broad categories: "top-down" and "bottom-up" approaches. The top-down approach is a scenario-impact approach, consisting of downscaled climate projections, impact assessments, and formulation of strategies and options. The bottom-up approach is a vulnerability-threshold approach, starting with the identification of vulnerabilities, sensitivities, and thresholds for specific sectors or communities. Iterative assessments of impacts and adaptation in the top-down approach and building adaptive capacity of local communities are typical strategies for responding to uncertainties. [2.2, 2.3, 15.3]

Uncertainties about future vulnerability, exposure, and responses of interlinked human and natural systems are large (*high confidence*). This motivates exploration of a wide range of socioeconomic futures in assessments of risks. Understanding future vulnerability, exposure, and response capacity of interlinked human and natural systems is challenging due to the number of interacting social, economic, and cultural factors, which have been incompletely considered to date. These factors include wealth and its distribution across society, demographics, migration, access to technology and information, employment patterns, the quality of adaptive responses, societal values, governance structures, and institutions to resolve conflicts. International dimensions such as trade and relations among states are also important for understanding the risks of climate change at regional scales. [11.3, 12.6, 21.3 to 21.5, 25.3, 25.4, 25.11, 26.2]

(A)

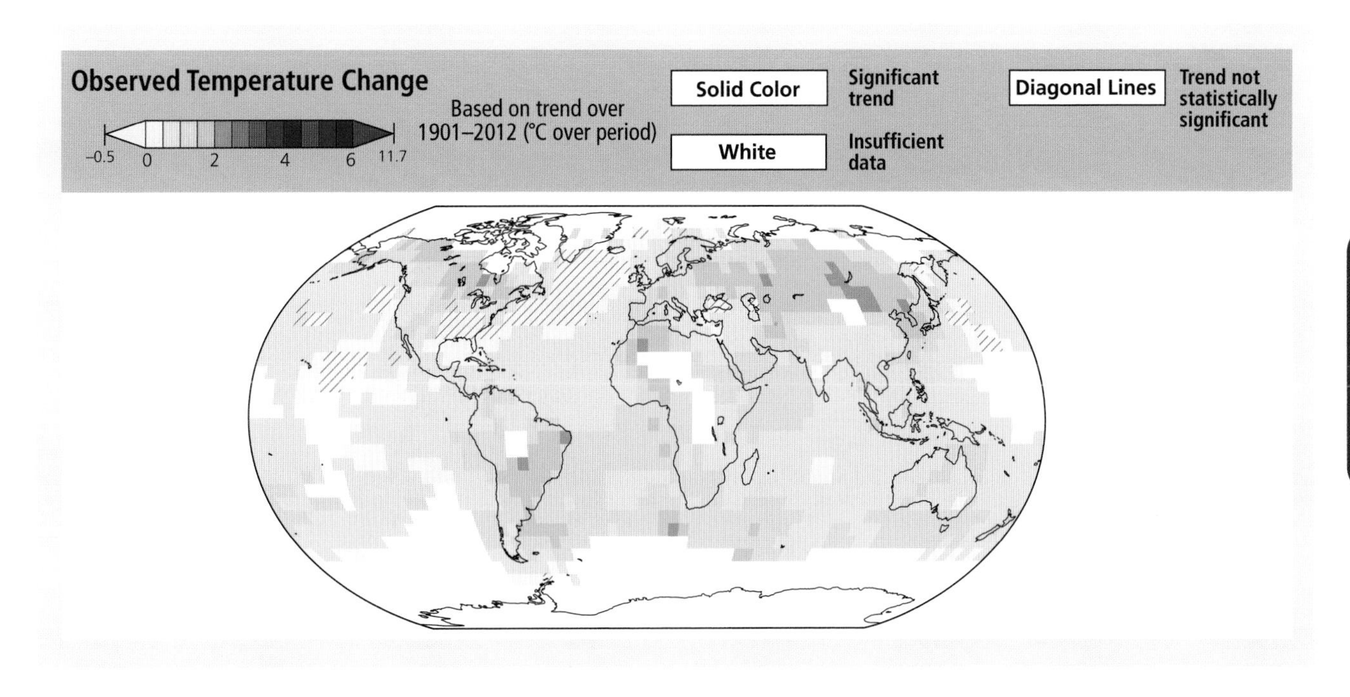

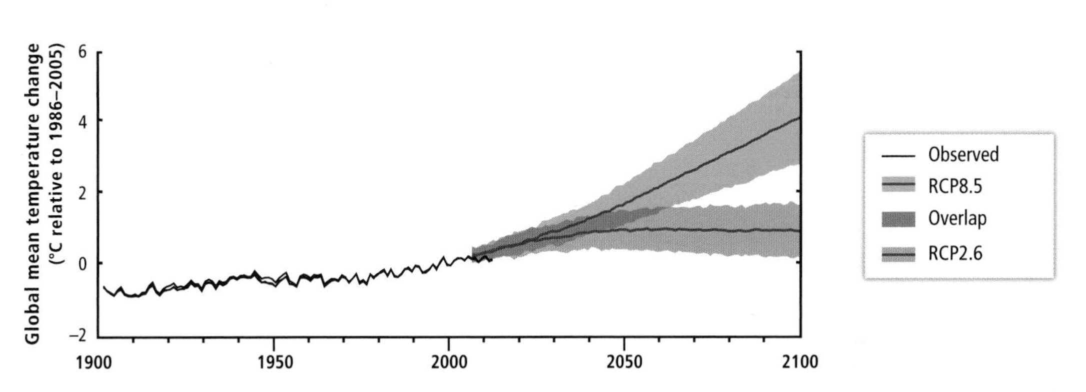

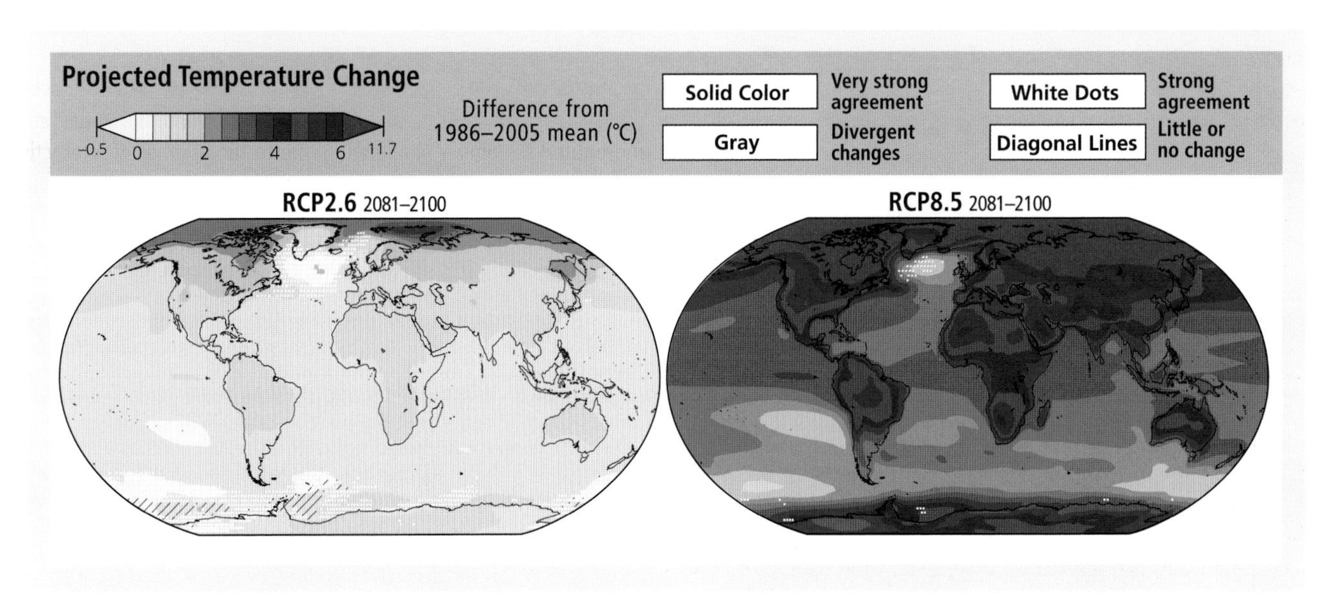

Figure TS.5

Continued next page →

Figure TS.5 (continued)

(B)

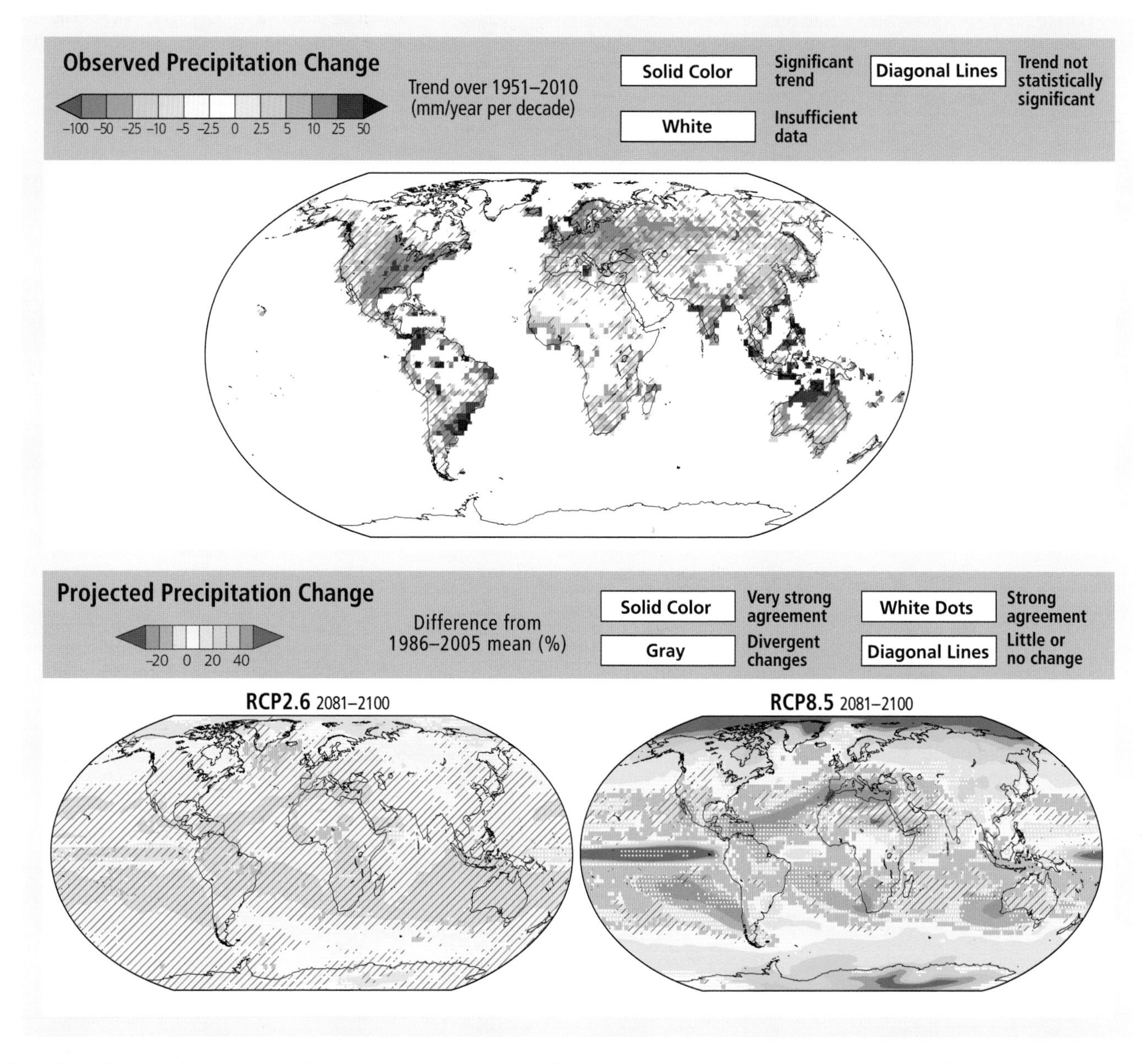

Figure TS.5 | Observed and projected changes in annual average surface temperature (A) and precipitation (B). This figure informs understanding of climate-related risks in the WGII AR5. It illustrates changes observed to date and projected changes under continued high emissions and under ambitious mitigation.

Technical details: (A, top panel) Map of observed annual mean temperature change from 1901–2012, derived from a linear trend. Observed data (range of grid-point values: –0.53 to 2.50°C over period) are from WGI AR5 Figures SPM.1 and 2.21. (B, top panel) Map of observed annual precipitation change from 1951–2010, derived from a linear trend. Observed data (range of grid-point values: –185 to 111 mm/year per decade) are from WGI AR5 Figures SPM.2 and 2.29. For observed temperature and precipitation, trends have been calculated where sufficient data permit a robust estimate (i.e., only for grid boxes with greater than 70% complete records and more than 20% data availability in the first and last 10% of the time period). Other areas are white. Solid colors indicate areas where trends are significant at the 10% level. Diagonal lines indicate areas where trends are not significant. (A, middle panel) Observed and projected future global annual mean temperature relative to 1986–2005. Observed warming from 1850–1900 to 1986–2005 is 0.61°C (5–95% confidence interval: 0.55 to 0.67°C). Black lines show temperature estimates from three datasets. Blue and red lines and shading denote the ensemble mean and ±1.64 standard deviation range, based on Coupled Model Intercomparison Project Phase 5 (CMIP5) simulations from 32 models for RCP2.6 and 39 models for RCP8.5. (A and B, bottom panel) CMIP5 multi-model mean projections of annual mean temperature changes (A) and mean percent changes in annual mean precipitation (B) for 2081–2100 under RCP2.6 and 8.5, relative to 1986–2005. Solid colors indicate areas with very strong agreement, where the multi-model mean change is greater than twice the baseline variability (natural internal variability in 20-yr means) and ≥90% of models agree on sign of change. Colors with white dots indicate areas with strong agreement, where ≥66% of models show change greater than the baseline variability and ≥66% of models agree on sign of change. Gray indicates areas with divergent changes, where ≥66% of models show change greater than the baseline variability, but <66% agree on sign of change. Colors with diagonal lines indicate areas with little or no change, where <66% of models show change greater than the baseline variability, although there may be significant change at shorter timescales such as seasons, months, or days. For temperature projections, analysis uses model data (range of grid-point values across RCP2.6 and 8.5: 0.06 to 11.71°C) from WGI AR5 Figure SPM.8. For precipitation projections, analysis uses model data (range of grid-point values: –9 to 22% for RCP2.6 and –34 to 112% for RCP8.5) from WGI AR5 Figure SPM.8, Box 12.1, and Annex I. For a full description of methods, see Box CC-RC. See also Annex I of WGI AR5. [Boxes 21-2 and CC-RC; WGI AR5 2.4 and 2.5, Figures SPM.1, SPM.2, SPM.7, SPM.8, 2.21, and 2.29]

B: FUTURE RISKS AND OPPORTUNITIES FOR ADAPTATION

This section presents future risks and more limited potential benefits across sectors and regions, examining how they are affected by the magnitude and rate of climate change and by socioeconomic choices. It also assesses opportunities for reducing impacts and managing risks through adaptation and mitigation. The section examines the distribution of risks across populations with contrasting vulnerability and adaptive capacity, across sectors where metrics for quantifying impacts may be quite different, and across regions with varying traditions and resources. The assessment features interactions across sectors and regions and among climate change and other stressors. For different sectors and regions, the section describes risks and potential benefits over the next few decades, the near-term era of committed climate change. Over this timeframe, projected global temperature increase is similar across emission scenarios. The section also provides information on risks and potential benefits in the second half of the 21st century and beyond, the longer-term era of climate options. Over this longer term, global temperature increase diverges across emission scenarios, and the assessment distinguishes potential outcomes for 2°C and 4°C global mean temperature increase above preindustrial levels. The section elucidates how and when choices matter in reducing future risks, highlighting the differing timeframes for mitigation and adaptation benefits.

B-1. Key Risks across Sectors and Regions

Key risks are potentially severe impacts relevant to Article 2 of the UN Framework Convention on Climate Change, which refers to "dangerous anthropogenic interference with the climate system." Risks are considered key due to high hazard or high vulnerability of societies and systems exposed, or both. Identification of key risks was based on expert judgment using the following specific criteria: large magnitude, high probability, or irreversibility of impacts; timing of impacts; persistent vulnerability or exposure contributing to risks; or limited potential to reduce risks through adaptation or mitigation. Key risks are integrated into five complementary and overarching reasons for concern (RFCs) in Box TS.5.

The key risks that follow, all of which are identified with *high confidence*, span sectors and regions. Each of these key risks contributes to one or more RFCs. Roman numerals correspond to entries in Table TS.3, which further illustrates relevant examples and interactions. [19.2 to 19.4, 19.6, Table 19-4, Boxes 19-2 and CC-KR]

i) Risk of death, injury, ill-health, or disrupted livelihoods in low-lying coastal zones and small island developing states and other small islands, due to storm surges, coastal flooding, and sea level rise. See RFCs 1 to 5. [5.4, 8.2, 13.2, 19.2 to 19.4, 19.6, 19.7, 24.4, 24.5, 26.7, 26.8, 29.3, 30.3, Tables 19-4 and 26-1, Figure 26-2, Boxes 25-1, 25-7, and CC-KR]

Table TS.3 | A selection of the hazards, key vulnerabilities, key risks, and emergent risks identified in chapters of this report. The examples underscore the complexity of risks determined by various interacting climate-related hazards, non-climatic stressors, and multifaceted vulnerabilities (see also Figure TS.1). Vulnerabilities identified as key arise when exposure to hazards combines with social, institutional, economic, or environmental vulnerability, as indicated by icons in the table. Emergent risks arise from complex system interactions. Roman numerals correspond with key risks listed in Section B-1. [19.6, Table 19-4]

No.	Hazard	Key vulnerabilities	Key risks	Emergent risks
i	Sea level rise and coastal flooding including storm surges [5.4.3, 8.1.4, 8.2.3, 8.2.4, 13.1.4, 13.2.2, 24.4, 24.5, 26.7, 26.8, 29.3, 30.3.1, Boxes 25-1 and 25-7; WGI AR5 3.7, 13.5, Table 13-5]	High exposure of people, economic activity, and infrastructure in low-lying coastal zones and Small Island Developing States (SIDS) and other small islands Urban population unprotected due to substandard housing and inadequate insurance. Marginalized rural population with multidimensional poverty and limited alternative livelihoods Insufficient local governmental attention to disaster risk reduction	Death, injury, and disruption to livelihoods, food supplies, and drinking water Loss of common-pool resources, sense of place, and identity, especially among indigenous populations in rural coastal zones	Interaction of rapid urbanization, sea level rise, increasing economic activity, disappearance of natural resources, and limits of insurance; burden of risk management shifted from the state to those at risk leading to greater inequality
ii	Extreme precipitation and inland flooding [3.2.7, 3.4.8, 8.2.3, 8.2.4, 13.2.1, 25.10, 26.3, 26.7, 26.8, 27.3.5, Box 25-8; WGI AR5 11.3.2]	Large numbers of people exposed in urban areas to flood events, particularly in low-income informal settlements Overwhelmed, aging, poorly maintained, and inadequate urban drainage infrastructure and limited ability to cope and adapt due to marginalization, high poverty, and culturally imposed gender roles Inadequate governmental attention to disaster risk reduction	Death, injury, and disruption of human security, especially among children, elderly, and disabled persons	Interaction of increasing frequency of intense precipitation, urbanization, and limits of insurance; burden of risk management shifted from the state to those at risk leading to greater inequality, eroded assets due to infrastructure damage, abandonment of urban districts, and the creation of high risk/high poverty spatial traps
iii	Novel hazards yielding systemic risks [8.1.4, 8.2.4, 10.2, 10.3, 12.6, 23.9, 25.10, 26.7, 26.8; WGI AR5 11.3.2]	Populations and infrastructure exposed and lacking historical experience with these hazards Overly hazard-specific management planning and infrastructure design, and/or low forecasting capability	Failure of systems coupled to electric power system, e.g., drainage systems reliant on electric pumps or emergency services reliant on telecommunications. Collapse of health and emergency services in extreme events	Interactions due to dependence on coupled systems lead to magnification of impacts of extreme events. Reduced social cohesion due to loss of faith in management institutions undermines preparation and capacity for response.
iv	Increasing frequency and intensity of extreme heat, including urban heat island effect [8.2.3, 11.3, 11.4.1, 13.2, 23.5.1, 24.4.6, 25.8.1, 26.6, 26.8, Box CC-HS; WGI AR5 11.3.2]	Increasing urban population of the elderly, the very young, expectant mothers, and people with chronic health problems in settlements subject to higher temperatures Inability of local organizations that provide health, emergency, and social services to adapt to new risk levels for vulnerable groups	Increased mortality and morbidity during periods of extreme heat	Interaction of demographic shifts with changes in regional temperature extremes, local heat island, and air pollution Overloading of health and emergency services. Higher mortality, morbidity, and productivity loss among manual workers in hot climates

Table TS.3 (continued)

No.	Hazard	Key vulnerabilities	Key risks	Emergent risks
v	Warming, drought, and precipitation variability [7.3 to 7.5, 11.3, 11.6.1, 13.2, 19.3.2, 19.4.1, 22.3.4, 24.4, 26.8, 27.3.4; WGI AR5 11.3.2]	Poorer populations in urban and rural settings are susceptible to resulting food insecurity; includes particularly farmers who are net food buyers and people in low-income, agriculturally dependent economies that are net food importers. Limited ability to cope among the elderly and female-headed households	Risk of harm and loss of life due to reversal of progress in reducing malnutrition	Interactions of climate changes, population growth, reduced productivity, biofuel crop cultivation, and food prices with persistent inequality, and ongoing food insecurity for the poor increase malnutrition, giving rise to larger burden of disease. Exhaustion of social networks reduces coping capacity.
vi	Drought [3.2.7, 3.4.8, 3.5.1, 8.2.3, 8.2.4, 9.3.3, 9.3.5, 13.2.1, 19.3.2, 24.4, 25.7, Box 25-5; WGI AR5 12.4.1, 12.4.5]	Urban populations with inadequate water services. Existing water shortages (and irregular supplies), and constraints on increasing supplies Lack of capacity and resilience in water management regimes including rural–urban linkages	Insufficient water supply for people and industry yielding severe harm and economic impacts	Interaction of urbanization, infrastructure insufficiency, groundwater depletion
		Poorly endowed farmers in drylands or pastoralists with insufficient access to drinking and irrigation water Limited ability to compensate for losses in water-dependent farming and pastoral systems, and conflict over natural resources Lack of capacity and resilience in water management regimes, inappropriate land policy, and misperception and undermining of pastoral livelihoods	Loss of agricultural productivity and/or income of rural people. Destruction of livelihoods particularly for those depending on water-intensive agriculture. Risk of food insecurity	Interactions across human vulnerabilities: deteriorating livelihoods, poverty traps, heightened food insecurity, decreased land productivity, rural outmigration, and increase in new urban poor in developing countries. Potential tipping point in rain-fed farming system and/or pastoralism
vii	Rising ocean temperature, ocean acidification, and loss of Arctic sea ice [5.4.2, 6.3.1, 6.3.2, 7.4.2, 9.3.5, 22.3.2, 24.4, 25.6, 27.3.3, 28.2, 28.3, 29.3.1, 30.5, 30.6, Boxes CC-OA and CC-CR; WGI AR5 11.3.3]	High susceptibility of warm-water coral reefs and respective ecosystem services for coastal communities; high susceptibility of polar systems, e.g., to invasive species Susceptibility of coastal and SIDS fishing communities depending on these ecosystem services; and of Arctic settlements and culture	Loss of coral cover, Arctic species, and associated ecosystems with reduction of biodiversity and potential losses of important ecosystem services. Risk of loss of endemic species, mixing of ecosystem types, and increased dominance of invasive organisms	Interactions of stressors such as acidification and warming on calcareous organisms enhancing risk
viii	Rising land temperatures, and changes in precipitation patterns and in frequency and intensity of extreme heat [4.3.4, 19.3.2, 22.4.5, 27.3, Boxes 23-1 and CC-WE; WGI AR5 11.3.2]	Susceptibility of human systems, agro-ecosystems, and natural ecosystems to (1) loss of regulation of pests and diseases, fire, landslide, erosion, flooding, avalanche, water quality, and local climate; (2) loss of provision of food, livestock, fiber, and bioenergy; (3) loss of recreation, tourism, aesthetic and heritage values, and biodiversity	Reduction of biodiversity and potential losses of important ecosystem services. Risk of loss of endemic species, mixing of ecosystem types, and increased dominance of invasive organisms	Interaction of social-ecological systems with loss of ecosystem services on which they depend

Social vulnerability

Economic vulnerability

Environmental vulnerability

Institutional vulnerability

Exposure

ii) Risk of severe ill-health and disrupted livelihoods for large urban populations due to inland flooding in some regions. See RFCs 2 and 3. [3.4, 3.5, 8.2, 13.2, 19.6, 25.10, 26.3, 26.8, 27.3, Tables 19-4 and 26-1, Boxes 25-8 and CC-KR]

iii) Systemic risks due to extreme weather events leading to breakdown of infrastructure networks and critical services such as electricity, water supply, and health and emergency services. See RFCs 2 to 4. [5.4, 8.1, 8.2, 9.3, 10.2, 10.3, 12.6, 19.6, 23.9, 25.10, 26.7, 26.8, 28.3, Table 19-4, Boxes CC-KR and CC-HS]

iv) Risk of mortality and morbidity during periods of extreme heat, particularly for vulnerable urban populations and those working outdoors in urban or rural areas. See RFCs 2 and 3. [8.1, 8.2, 11.3, 11.4, 11.6, 13.2, 19.3, 19.6, 23.5, 24.4, 25.8, 26.6, 26.8, Tables 19-4 and 26-1, Boxes CC-KR and CC-HS]

v) Risk of food insecurity and the breakdown of food systems linked to warming, drought, flooding, and precipitation variability and extremes, particularly for poorer populations in urban and rural settings. See RFCs 2 to 4. [3.5, 7.4, 7.5, 8.2, 8.3, 9.3, 11.3, 11.6, 13.2, 19.3, 19.4, 19.6, 22.3, 24.4, 25.5, 25.7, 26.5, 26.8, 27.3, 28.2, 28.4, Table 19-4, Box CC-KR]

vi) Risk of loss of rural livelihoods and income due to insufficient access to drinking and irrigation water and reduced agricultural productivity, particularly for farmers and pastoralists with minimal capital in semi-arid regions. See RFCs 2 and 3. [3.4, 3.5, 9.3, 12.2, 13.2, 19.3, 19.6, 24.4, 25.7, 26.8, Table 19-4, Boxes 25-5 and CC-KR]

vii) Risk of loss of marine and coastal ecosystems, biodiversity, and the ecosystem goods, functions, and services they provide for coastal livelihoods, especially for fishing communities in the tropics and the

Box TS.5 | Human Interference with the Climate System

Human influence on the climate system is clear (WGI AR5 SPM Section D.3; WGI AR5 Sections 2.2, 6.3, 10.3 to 10.6, 10.9). Yet determining whether such influence constitutes "dangerous anthropogenic interference" in the words of Article 2 of the UNFCCC involves both risk assessment and value judgments. Scientific assessment can characterize risks based on the likelihood, magnitude, and scope of potential consequences of climate change. Science can also evaluate risks varying spatially and temporally across alternative development pathways, which affect vulnerability, exposure, and level of climate change. Interpreting the potential danger of risks, however, also requires value judgments by people with differing goals and worldviews. Judgments about the risks of climate change depend on the relative importance ascribed to economic versus ecosystem assets, to the present versus the future, and to the distribution versus aggregation of impacts. From some perspectives, isolated or infrequent impacts from climate change may not rise to the level of dangerous anthropogenic interference, but accumulation of the same kinds of impacts could, as they become more widespread, more frequent, or more severe. The rate of climate change can also influence risks. This report assesses risks across contexts and through time, providing a basis for judgments about the level of climate change at which risks become dangerous.

Five integrative reasons for concern (RFCs) provide a framework for summarizing key risks across sectors and regions. First identified in the IPCC Third Assessment Report, the RFCs illustrate the implications of warming and of adaptation limits for people, economies, and ecosystems. They provide one starting point for evaluating dangerous anthropogenic interference with the climate system. Risks for each RFC, updated based on assessment of the literature and expert judgments, are presented below and in Box TS.5 Figure 1. All temperatures below are given as global average temperature change relative to 1986–2005 ("recent").[1] [18.6, 19.6]

1) **Unique and threatened systems**: Some unique and threatened systems, including ecosystems and cultures, are already at risk from climate change (*high confidence*). The number of such systems at risk of severe consequences is higher with additional warming of around 1°C. Many species and systems with limited adaptive capacity are subject to very high risks with additional warming of 2°C, particularly Arctic-sea-ice and coral-reef systems.
2) **Extreme weather events**: Climate-change-related risks from extreme events, such as heat waves, extreme precipitation, and coastal flooding, are already moderate (*high confidence*) and high with 1°C additional warming (*medium confidence*). Risks associated with some types of extreme events (e.g., extreme heat) increase further at higher temperatures (*high confidence*).
3) **Distribution of impacts**: Risks are unevenly distributed and are generally greater for disadvantaged people and communities in countries at all levels of development. Risks are already moderate because of regionally differentiated climate-change impacts on crop production in particular (*medium* to *high confidence*). Based on projected decreases in regional crop yields and water availability, risks of unevenly distributed impacts are high for additional warming above 2°C (*medium confidence*).
4) **Global aggregate impacts**: Risks of global aggregate impacts are moderate for additional warming between 1–2°C, reflecting impacts to both Earth's biodiversity and the overall global economy (*medium confidence*). Extensive biodiversity loss with associated loss of ecosystem goods and services results in high risks around 3°C additional warming (*high confidence*). Aggregate economic damages accelerate with increasing temperature (*limited evidence*, *high agreement*), but few quantitative estimates have been completed for additional warming around 3°C or above.
5) **Large-scale singular events**: With increasing warming, some physical systems or ecosystems may be at risk of abrupt and irreversible changes. Risks associated with such tipping points become moderate between 0–1°C additional warming, due to early warning signs that both warm-water coral reef and Arctic ecosystems are already experiencing irreversible regime shifts (*medium confidence*). Risks increase disproportionately as temperature increases between 1–2°C additional warming and become high above 3°C, due to the potential for a large and irreversible sea level rise from ice sheet loss. For sustained warming greater than some threshold,[2] near-complete loss of the Greenland ice sheet would occur over a millennium or more, contributing up to 7 m of global mean sea level rise.

Continued next page →

[1] Observed warming from 1850–1900 to 1986–2005 is 0.61°C (5–95% confidence interval: 0.55 to 0.67°C). [WGI AR5 2.4]

[2] Current estimates indicate that this threshold is greater than about 1°C (*low confidence*) but less than about 4°C (*medium confidence*) sustained global mean warming above preindustrial levels. [WGI AR5 SPM, 5.8, 13.4, 13.5]

TS

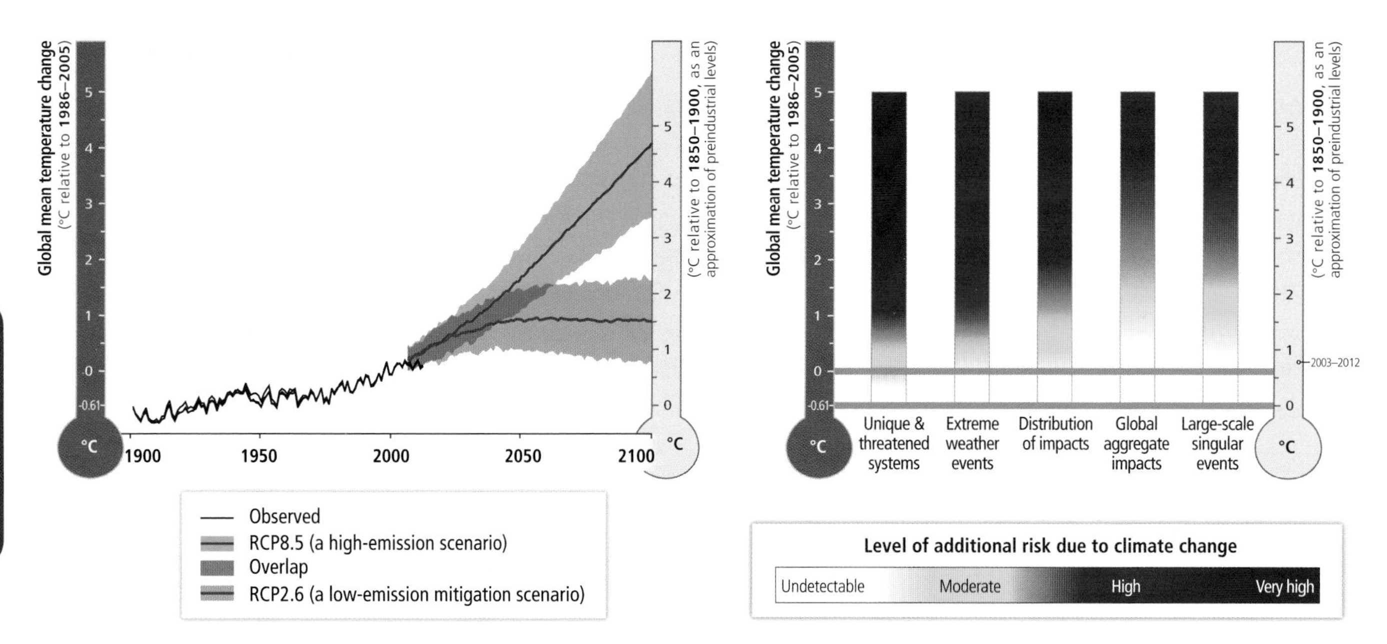

Box TS.5 Figure 1 | A global perspective on climate-related risks. Risks associated with reasons for concern are shown at right for increasing levels of climate change. The color shading indicates the additional risk due to climate change when a temperature level is reached and then sustained or exceeded. Undetectable risk (white) indicates no associated impacts are detectable and attributable to climate change. Moderate risk (yellow) indicates that associated impacts are both detectable and attributable to climate change with at least *medium confidence*, also accounting for the other specific criteria for key risks. High risk (red) indicates severe and widespread impacts, also accounting for the other specific criteria for key risks. Purple, introduced in this assessment, shows that very high risk is indicated by all specific criteria for key risks. [Figure 19-4] For reference, past and projected global annual average surface temperature is shown at left, as in Figure TS.5. [Figure RC-1, Box CC-RC; WGI AR5 Figures SPM.1 and SPM.7] Based on the longest global surface temperature dataset available, the observed change between the average of the period 1850–1900 and of the AR5 reference period (1986–2005) is 0.61°C (5–95% confidence interval: 0.55 to 0.67°C) [WGI AR5 SPM, 2.4], which is used here as an approximation of the change in global mean surface temperature since preindustrial times, referred to as the period before 1750. [WGI and WGII AR5 glossaries]

Arctic. See RFCs 1, 2, and 4. [5.4, 6.3, 7.4, 9.3, 19.5, 19.6, 22.3, 25.6, 27.3, 28.2, 28.3, 29.3, 30.5 to 30.7, Table 19-4, Boxes CC-OA, CC-CR, CC-KR, and CC-HS]

viii) Risk of loss of terrestrial and inland water ecosystems, biodiversity, and the ecosystem goods, functions, and services they provide for livelihoods. See RFCs 1, 3, and 4. [4.3, 9.3, 19.3 to 19.6, 22.3, 25.6, 27.3, 28.2, 28.3, Table 19-4, Boxes CC-KR and CC-WE]

Many key risks constitute particular challenges for the least developed countries and vulnerable communities, given their limited ability to cope.

Increasing magnitudes of warming increase the likelihood of severe, pervasive, and irreversible impacts. Some risks of climate change are considerable at 1°C or 2°C above preindustrial levels (as shown in Box TS.5). Global climate change risks are high to very high with global mean temperature increase of 4°C or more above preindustrial levels in all reasons for concern (Box TS.5), and include severe and widespread impacts on unique and threatened systems, substantial species extinction, large risks to global and regional food security, and the combination of high temperature and humidity compromising normal human activities, including growing food or working outdoors in some areas for parts of the year (*high confidence*). See Box TS.6. The precise levels of climate change sufficient to trigger tipping points (thresholds for abrupt and irreversible change) remain uncertain, but the risk associated with crossing multiple tipping points in the earth system or in interlinked human and natural systems increases with rising temperature (*medium confidence*). [4.2, 4.3, 11.8, 19.5, 19.7, 26.5, Box CC-HS]

The overall risks of climate change impacts can be reduced by limiting the rate and magnitude of climate change. Risks are reduced substantially under the assessed scenario with the lowest temperature projections (RCP2.6 – low emissions) compared to the highest temperature projections (RCP8.5 – high emissions), particularly in the second half of the 21st century (*very high confidence*). Examples include reduced risk of negative agricultural yield impacts; of water scarcity; of major challenges to urban settlements and infrastructure from sea level rise; and of adverse impacts from heat extremes, floods, and droughts in areas where increased occurrence of these extremes is projected. Reducing climate change can also reduce the scale of adaptation that might be required. Under all assessed scenarios for adaptation and mitigation, some risk from adverse impacts remains (*very high confidence*). Because mitigation reduces the rate as well as the magnitude of warming, it also increases the time available for adaptation to a particular level of climate change, potentially by several decades, but adaptation cannot generally overcome all climate change effects. In addition to biophysical limits to adaptation for example under high temperatures, some adaptation options will be too costly or resource intensive or will be cost ineffective until climate change effects grow to merit investment costs (*high confidence*). Some mitigation or adaptation options also pose risks. [3.4, 3.5, 4.2, 4.4, 16.3, 16.6, 17.2, 19.7, 20.3, 22.4, 22.5, 25.10, Tables 3-2, 8-3, and 8-6, Boxes 16-3 and 25-1]

B-2. Sectoral Risks and Potential for Adaptation

For the near-term era of committed climate change (the next few decades) and the longer-term era of climate options (the second half

Box TS.6 | Consequences of Large Temperature Increase

This box provides a selection of salient climate change impacts projected for large temperature rise. Warming levels described here (e.g., 4°C warming) refer to global mean temperature increase above preindustrial levels, unless otherwise indicated.

With 4°C warming, climate change is projected to become an increasingly important driver of impacts on ecosystems, becoming comparable with land-use change. [4.2, 19.5] A number of studies project large increases in water stress, groundwater supplies, and drought in a number of regions with greater than 4°C warming, and decreases in others, generally placing already arid regions at greater water stress. [19.5]

Risks of large-scale singular events such as ice sheet disintegration, methane release from clathrates, and onset of long-term droughts in areas such as southwest North America [19.6, Box 26-1; WGI AR5 12.4, 12.5, 13.4], as well as regime shifts in ecosystems and substantial species loss [4.3, 19.6], are higher with increased warming. Sustained warming greater than some threshold would lead to the near-complete loss of the Greenland ice sheet over a millennium or more, causing a global mean sea level rise of up to 7 m (*high confidence*); current estimates indicate that the threshold is greater than about 1°C (*low confidence*) but less than about 4°C (*medium confidence*) global mean warming. [WGI AR5 SPM, 5.8, 13.4, 13.5] Abrupt and irreversible ice loss from a potential instability of marine-based areas of the Antarctic ice sheet in response to climate forcing is possible, but current evidence and understanding is insufficient to make a quantitative assessment. [19.6; WGI AR5 SPM, 5.8, 13.4, 13.5] Sea level rise of 0.45 to 0.82 m (mean 0.63 m) is *likely* by 2081–2100 under RCP8.5 (*medium confidence*) [WGI AR5 Tables SPM.2 and 13.5], with sea level continuing to rise beyond 2100.

The Atlantic Meridional Overturning Circulation (AMOC) will *very likely* weaken over the 21st century, with a best estimate of 34% loss (range 12 to 54%) under RCP8.5. [WGI AR5 SPM, 12.4] The release of carbon dioxide (CO_2) or methane (CH_4) to the atmosphere from thawing permafrost carbon stocks over the 21st century is assessed to be in the range of 50 to 250 GtC for Representative Concentration Pathway 8.5 (RCP8.5) (*low confidence*). [WGI AR5 SPM, 6.4] A nearly ice-free Arctic Ocean in September before mid-century is *likely* under RCP8.5 (*medium confidence*). [WGI AR5 SPM, 11.3, 12.4, 12.5]

By 2100 for the high-emission scenario RCP8.5, the combination of high temperature and humidity in some areas for parts of the year is projected to compromise normal human activities, including growing food or working outdoors (*high confidence*). [11.8] Global temperature increases of ~4°C or more above late-20th-century levels, combined with increasing food demand, would pose large risks to food security globally and regionally (*high confidence*). [7.4, 7.5, Table 7-3, Figures 7-1, 7-4, and 7-7, Box 7-1]

Under 4°C warming, some models project large increases in fire risk in parts of the world. [4.3, Figure 4-6] 4°C warming implies a substantial increase in extinction risk for terrestrial and freshwater species, although there is *low agreement* concerning the fraction of species at risk. [4.3] Widespread coral reef mortality is expected with significant impacts on coral reef ecosystems (*high confidence*). [5.4, Box CC-CR] Assessments of potential ecological impacts at and above 4°C warming imply a high risk of extensive loss of biodiversity with concomitant loss of ecosystem services (*high confidence*). [4.3, 19.3, 19.5, Box 25-6]

Projected large increases in exposure to water stress, fluvial and coastal flooding, negative impacts on crop yields, and disruption of ecosystem function and services would represent large, potentially compounding impacts of climate change on society generally and on the global economy. [19.4 to 19.6]

of the 21st century and beyond), climate change will amplify existing climate-related risks and create new risks for natural and human systems, dependent on the magnitude and rate of climate change and on the vulnerability and exposure of interlinked human and natural systems. Some of these risks will be limited to a particular sector or region, and others will have cascading effects. To a lesser extent, climate change will also have some potential benefits. A selection of key sectoral risks identified with *medium* to *high confidence* is presented in Table TS.4.

Table TS.4 | Key sectoral risks from climate change and the potential for reducing risks through adaptation and mitigation. Key risks have been identified based on assessment of the relevant scientific, technical, and socioeconomic literature detailed in supporting chapter sections. Identification of key risks was based on expert judgment using the following specific criteria: large magnitude, high probability, or irreversibility of impacts; timing of impacts; persistent vulnerability or exposure contributing to risks; or limited potential to reduce risks through adaptation or mitigation. Each key risk is characterized as very low to very high for three timeframes: the present, near term (here, assessed over 2030–2040), and longer term (here, assessed over 2080–2100). The risk levels integrate probability and consequence over the widest possible range of potential outcomes, based on available literature. These potential outcomes result from the interaction of climate-related hazards, vulnerability, and exposure. Each risk level reflects total risk from climatic and non-climatic factors. For the near-term era of committed climate change, projected levels of global mean temperature increase do not diverge substantially for different emission scenarios. For the longer-term era of climate options, risk levels are presented for two scenarios of global mean temperature increase (2°C and 4°C above preindustrial levels). These scenarios illustrate the potential for mitigation and adaptation to reduce the risks related to climate change. For the present, risk levels were estimated for current adaptation and a hypothetical highly adapted state, identifying where current adaptation deficits exist. For the two future timeframes, risk levels were estimated for a continuation of current adaptation and for a highly adapted state, representing the potential for and limits to adaptation. Climate-related drivers of impacts are indicated by icons. Risk levels are not necessarily comparable because the assessment considers potential impacts and adaptation in different physical, biological, and human systems across diverse contexts. This assessment of risks acknowledges the importance of differences in values and objectives in interpretation of the assessed risk levels.

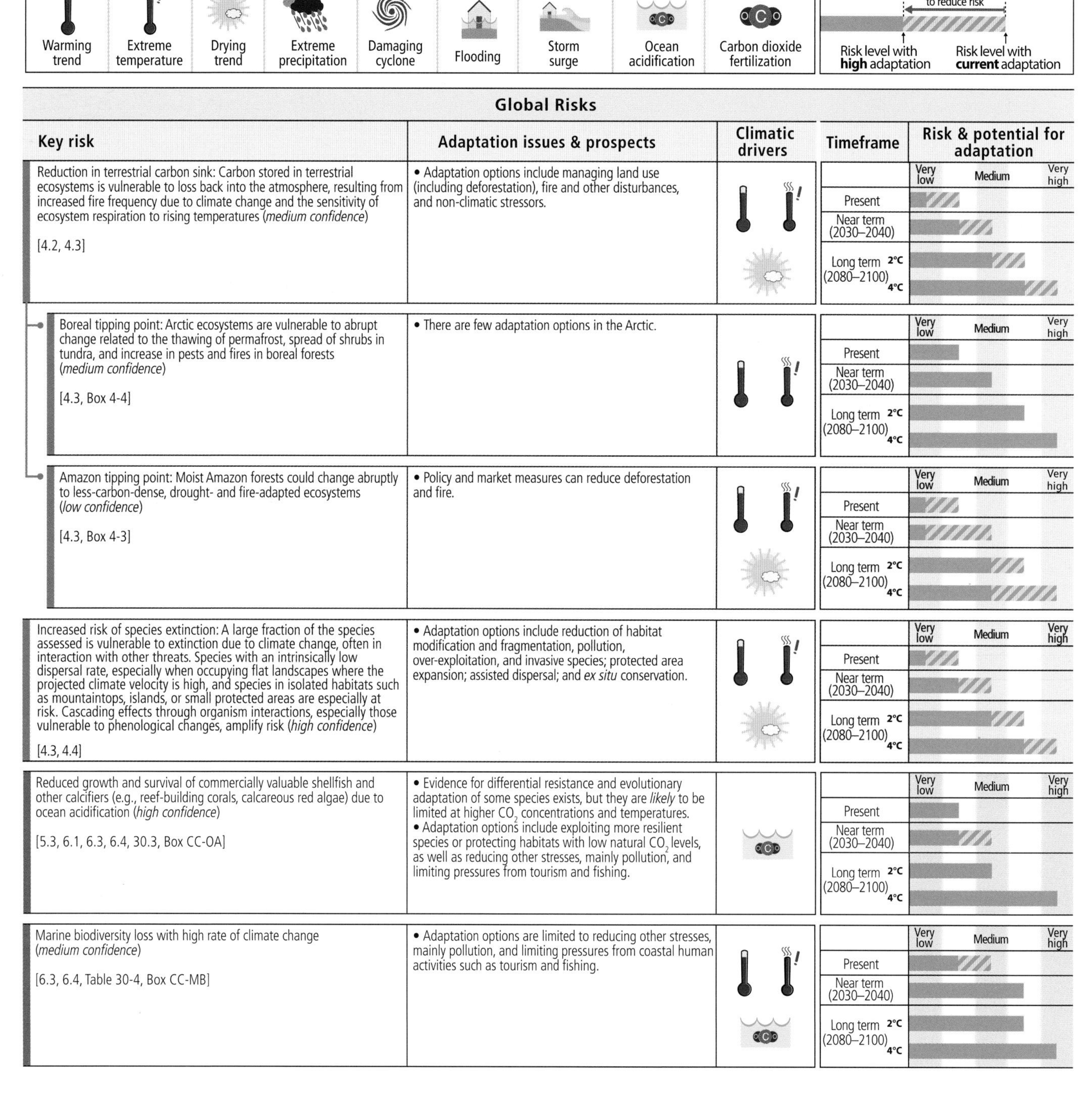

Climate-related drivers of impacts								
Warming trend	Extreme temperature	Drying trend	Extreme precipitation	Damaging cyclone	Flooding	Storm surge	Ocean acidification	Carbon dioxide fertilization

Level of risk & potential for adaptation
Potential for additional adaptation to reduce risk
Risk level with **high** adaptation — Risk level with **current** adaptation

Global Risks

Key risk	Adaptation issues & prospects	Climatic drivers	Timeframe	Risk & potential for adaptation (Very low – Medium – Very high)
Reduction in terrestrial carbon sink: Carbon stored in terrestrial ecosystems is vulnerable to loss back into the atmosphere, resulting from increased fire frequency due to climate change and the sensitivity of ecosystem respiration to rising temperatures (*medium confidence*) [4.2, 4.3]	• Adaptation options include managing land use (including deforestation), fire and other disturbances, and non-climatic stressors.	Warming trend, Extreme temperature, Drying trend	Present Near term (2030–2040) Long term (2080–2100) 2°C / 4°C	
Boreal tipping point: Arctic ecosystems are vulnerable to abrupt change related to the thawing of permafrost, spread of shrubs in tundra, and increase in pests and fires in boreal forests (*medium confidence*) [4.3, Box 4-4]	• There are few adaptation options in the Arctic.	Warming trend, Extreme temperature	Present Near term (2030–2040) Long term (2080–2100) 2°C / 4°C	
Amazon tipping point: Moist Amazon forests could change abruptly to less-carbon-dense, drought- and fire-adapted ecosystems (*low confidence*) [4.3, Box 4-3]	• Policy and market measures can reduce deforestation and fire.	Warming trend, Extreme temperature, Drying trend	Present Near term (2030–2040) Long term (2080–2100) 2°C / 4°C	
Increased risk of species extinction: A large fraction of the species assessed is vulnerable to extinction due to climate change, often in interaction with other threats. Species with an intrinsically low dispersal rate, especially when occupying flat landscapes where the projected climate velocity is high, and species in isolated habitats such as mountaintops, islands, or small protected areas are especially at risk. Cascading effects through organism interactions, especially those vulnerable to phenological changes, amplify risk (*high confidence*) [4.3, 4.4]	• Adaptation options include reduction of habitat modification and fragmentation, pollution, over-exploitation, and invasive species; protected area expansion; assisted dispersal; and *ex situ* conservation.	Warming trend, Extreme temperature, Drying trend	Present Near term (2030–2040) Long term (2080–2100) 2°C / 4°C	
Reduced growth and survival of commercially valuable shellfish and other calcifiers (e.g., reef-building corals, calcareous red algae) due to ocean acidification (*high confidence*) [5.3, 6.1, 6.3, 6.4, 30.3, Box CC-OA]	• Evidence for differential resistance and evolutionary adaptation of some species exists, but they are *likely* to be limited at higher CO_2 concentrations and temperatures. • Adaptation options include exploiting more resilient species or protecting habitats with low natural CO_2 levels, as well as reducing other stresses, mainly pollution, and limiting pressures from tourism and fishing.	Ocean acidification	Present Near term (2030–2040) Long term (2080–2100) 2°C / 4°C	
Marine biodiversity loss with high rate of climate change (*medium confidence*) [6.3, 6.4, Table 30-4, Box CC-MB]	• Adaptation options are limited to reducing other stresses, mainly pollution, and limiting pressures from coastal human activities such as tourism and fishing.	Warming trend, Extreme temperature, Ocean acidification	Present Near term (2030–2040) Long term (2080–2100) 2°C / 4°C	

Continued next page →

Table TS.4 (continued)

Global Risks				
Key risk	**Adaptation issues & prospects**	**Climatic drivers**	**Timeframe**	**Risk & potential for adaptation**
Negative impacts on average crop yields and increases in yield variability due to climate change (*high confidence*) [7.2 to 7.5, Figure 7-5, Box 7-1]	• Projected impacts vary across crops and regions and adaptation scenarios, with about 10% of projections for the period 2030–2049 showing yield gains of more than 10%, and about 10% of projections showing yield losses of more than 25%, compared to the late 20th century. After 2050 the risk of more severe yield impacts increases and depends on the level of warming.		Present Near term (2030–2040) Long term (2080–2100) 2°C / 4°C	Very low – Medium – Very high
Urban risks associated with water supply systems (*high confidence*) [8.2, 8.3]	• Adaptation options include changes to network infrastructure as well as demand-side management to ensure sufficient water supplies and quality, increased capacities to manage reduced freshwater availability, and flood risk reduction.		Present Near term (2030–2040) Long term (2080–2100) 2°C / 4°C	Very low – Medium – Very high
Urban risks associated with energy systems (*high confidence*) [8.2, 8.4]	• Most urban centers are energy intensive, with energy-related climate policies focused only on mitigation measures. A few cities have adaptation initiatives underway for critical energy systems. There is potential for non-adapted, centralized energy systems to magnify impacts, leading to national and transboundary consequences from localized extreme events.		Present Near term (2030–2040) Long term (2080–2100) 2°C / 4°C	Very low – Medium – Very high
Urban risks associated with housing (*high confidence*) [8.3]	• Poor quality, inappropriately located housing is often most vulnerable to extreme events. Adaptation options include enforcement of building regulations and upgrading. Some city studies show the potential to adapt housing and promote mitigation, adaptation, and development goals simultaneously. Rapidly growing cities, or those rebuilding after a disaster, especially have opportunities to increase resilience, but this is rarely realized. Without adaptation, risks of economic losses from extreme events are substantial in cities with high-value infrastructure and housing assets, with broader economic effects possible.		Present Near term (2030–2040) Long term (2080–2100) 2°C / 4°C	Very low – Medium – Very high
Displacement associated with extreme events (*high confidence*) [12.4]	• Adaptation to extreme events is well understood, but poorly implemented even under present climate conditions. Displacement and involuntary migration are often temporary. With increasing climate risks, displacement is more likely to involve permanent migration.		Present Near term (2030–2040) Long term (2080–2100) 2°C / 4°C	Very low – Medium – Very high
Violent conflict arising from deterioration in resource-dependent livelihoods such as agriculture and pastoralism (*high confidence*) [12.5]	Adaptation options: • Buffering rural incomes against climate shocks, for example through livelihood diversification, income transfers, and social safety net provision • Early warning mechanisms to promote effective risk reduction • Well-established strategies for managing violent conflict that are effective but require significant resources, investment, and political will		Present Near term (2030–2040) Long term (2080–2100) 2°C / 4°C	Very low – Medium – Very high
Declining work productivity, increasing morbidity (e.g., dehydration, heat stroke, and heat exhaustion), and mortality from exposure to heat waves. Particularly at risk are agricultural and construction workers as well as children, homeless people, the elderly, and women who have to walk long hours to collect water (*high confidence*) [13.2, Box 13-1]	• Adaptation options are limited for people who are dependent on agriculture and cannot afford agricultural machinery. • Adaptation options are limited in the construction sector where many poor people work under insecure arrangements. • Adaptation limits may be exceeded in certain areas in a +4°C world.		Present Near term (2030–2040) Long term (2080–2100) 2°C / 4°C	Very low – Medium – Very high
Reduced access to water for rural and urban poor people due to water scarcity and increasing competition for water (*high confidence*) [13.2, Box 13-1]	• Adaptation through reducing water use is not an option for the many people already lacking adequate access to safe water. Access to water is subject to various forms of discrimination, for instance due to gender and location. Poor and marginalized water users are unable to compete with water extraction by industries, large-scale agriculture, and other powerful users.		Present Near term (2030–2040) Long term (2080–2100) 2°C / 4°C	Very low – Medium – Very high

For extended summary of sectoral risks and the more limited potential benefits, see introductory overviews for each sector below and also Chapters 3 to 13.

Freshwater Resources

Freshwater-related risks of climate change increase significantly with increasing greenhouse gas concentrations (*robust evidence, high agreement*). The fraction of global population experiencing water scarcity and the fraction affected by major river floods increase with the level of warming in the 21st century. See, for example, Figure TS.6. [3.4, 3.5, 26.3, Table 3-2, Box 25-8]

Climate change over the 21st century is projected to reduce renewable surface water and groundwater resources significantly in most dry subtropical regions (*robust evidence, high agreement*), intensifying competition for water among sectors (*limited evidence, medium agreement*). In presently dry regions, drought frequency will *likely* increase by the end of the 21st century under RCP8.5 (*medium confidence*). In contrast, water resources are projected to increase at high latitudes (*robust evidence, high agreement*). Climate change is projected to reduce raw water quality and pose risks to drinking water quality even with conventional treatment, due to interacting factors: increased temperature; increased sediment, nutrient, and pollutant loadings from heavy rainfall; increased concentration of pollutants during droughts; and disruption of treatment facilities during floods (*medium evidence, high agreement*). [3.2, 3.4, 3.5, 22.3, 23.9, 25.5, 26.3, Tables 3-2 and 23-3, Boxes CC-RF and CC-WE; WGI AR5 12.4]

Adaptive water management techniques, including scenario planning, learning-based approaches, and flexible and low-regret solutions, can help create resilience to uncertain hydrological changes and impacts due to climate change (*limited evidence, high agreement*). Barriers to progress include lack of human and institutional capacity, financial resources, awareness, and communication. [3.6, Box 25-2]

(A)

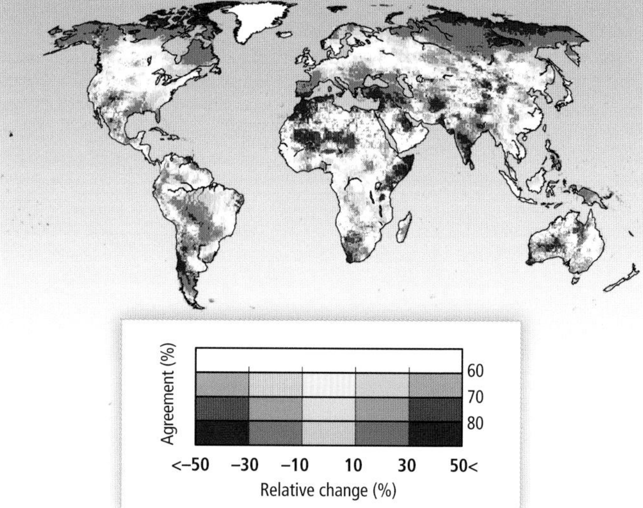

(B)

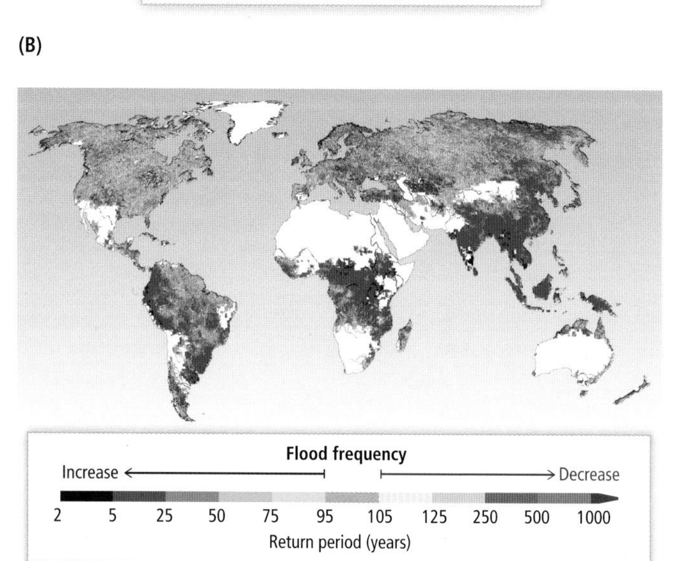

(C)

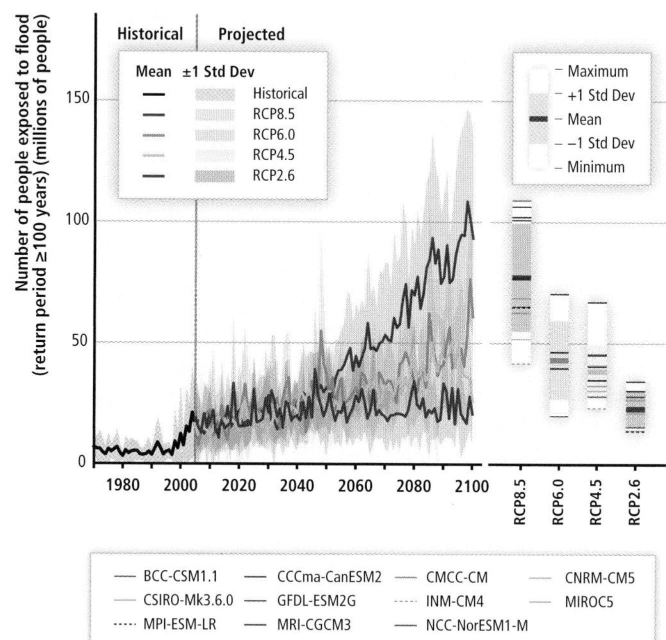

Figure TS.6 | (A) Percentage change of mean annual streamflow for a global mean temperature rise of 2°C above 1980–2010. Color hues show the multi-model mean change across 5 General Circulation Models (GCMs) and 11 Global Hydrological Models (GHMs), and saturation shows the agreement on the sign of change across all 55 GHM–GCM combinations (percentage of model runs agreeing on the sign of change). (B and C) Projected change in river flood return period and exposure, based on one hydrological model driven by 11 GCMs and on global population in 2005. (B) In the 2080s under RCP8.5, multi-model median return period (years) for the 20th-century 100-year flood. (C) Global exposure to the 20th-century 100-year flood in millions of people. Left: Ensemble means of historical (black line) and future simulations (colored lines) for each scenario. Shading denotes ±1 standard deviation. Right: Maximum and minimum (extent of white), mean (thick colored lines), ±1 standard deviation (extent of shading), and projections of each GCM (thin colored lines) averaged over the 21st century. [Figures 3-4 and 3-6]

Terrestrial and Freshwater Ecosystems

Climate change is projected to be a powerful stressor on terrestrial and freshwater ecosystems in the second half of the 21st century, especially under high-warming scenarios such as RCP6.0 and 8.5 (*high confidence*). Through to 2040 globally, direct human impacts such as land-use change, pollution, and water resource development will continue to dominate threats to most freshwater ecosystems (*high confidence*) and most terrestrial ecosystems (*medium confidence*). Many species will be unable to track suitable climates under mid- and high-range rates of climate change (i.e., RCP4.5, 6.0, and 8.5) during the 21st century (*medium confidence*). Lower rates of change (i.e., RCP2.6) will pose fewer problems. See Figure TS.7. Some species will adapt to new climates. Those that cannot adapt sufficiently fast will decrease in abundance or go extinct in part or all of their ranges. Increased tree mortality and associated forest dieback is projected to occur in many regions over the 21st century, due to increased temperatures and drought (*medium confidence*). Forest dieback poses risks for carbon storage, biodiversity, wood production, water quality, amenity, and economic activity. Management actions, such as maintenance of genetic diversity, assisted species migration and dispersal, manipulation of disturbance regimes (e.g., fires, floods), and reduction of other stressors, can reduce, but not eliminate, risks of impacts to terrestrial and freshwater ecosystems due to climate change, as well as increase the inherent capacity of ecosystems and their species to adapt to a changing climate (*high confidence*). [4.3, 4.4, 25.6, 26.4, Boxes 4-2, 4-3, and CC-RF]

A large fraction of both terrestrial and freshwater species faces increased extinction risk under projected climate change during and beyond the 21st century, especially as climate change interacts with other stressors, such as habitat modification, over-exploitation, pollution, and invasive species (*high confidence*). Extinction risk is increased under all RCP scenarios, with risk increasing with both magnitude and rate of climate change. Models project that the risk of species extinctions will increase in the future due to climate change, but there is *low agreement* concerning the fraction of species at increased risk, the regional and taxonomic distribution of such extinctions, and the timeframe over which extinctions could occur. Some aspects leading to uncertainty in the quantitative projections of extinction risks were not taken into account in previous models; as more realistic details are included, it has been shown that the extinction risks may be either under- or overestimated when based on simpler models. [4.3, 25.6]

Within this century, magnitudes and rates of climate change associated with medium- to high-emission scenarios (RCP4.5, 6.0, and 8.5) pose high risk of abrupt and irreversible regional-scale change in the composition, structure, and function of

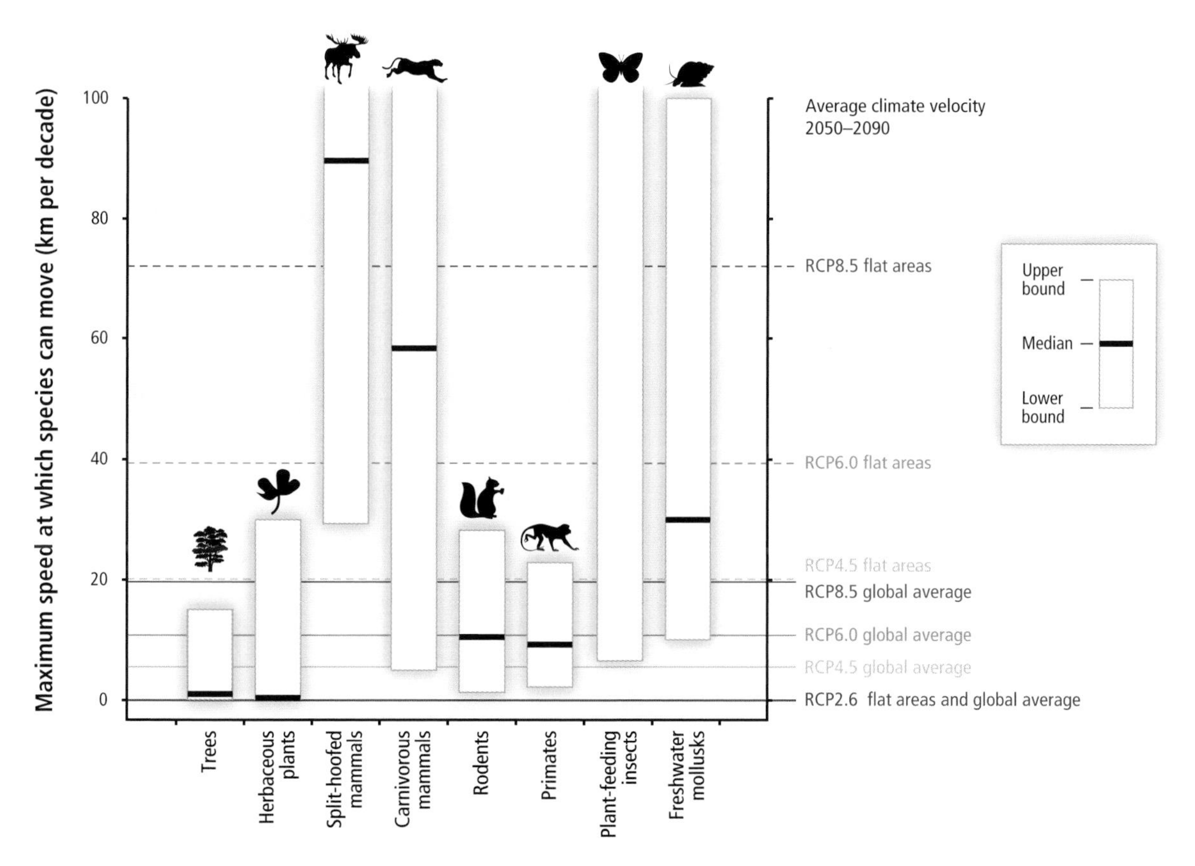

Figure TS.7 | Maximum speeds at which species can move across landscapes (based on observations and models; vertical axis on left), compared with speeds at which temperatures are projected to move across landscapes (climate velocities for temperature; vertical axis on right). Human interventions, such as transport or habitat fragmentation, can greatly increase or decrease speeds of movement. White boxes with black bars indicate ranges and medians of maximum movement speeds for trees, plants, mammals, plant-feeding insects (median not estimated), and freshwater mollusks. For RCP2.6, 4.5, 6.0, and 8.5 for 2050–2090, horizontal lines show climate velocity for the global-land-area average and for large flat regions. Species with maximum speeds below each line are expected to be unable to track warming in the absence of human intervention. [Figure 4-5]

terrestrial and freshwater ecosystems, including wetlands (*medium confidence*). Examples that could lead to substantial impact on climate are the boreal–tundra Arctic system (*medium confidence*) and the Amazon forest (*low confidence*). For the boreal–tundra system, continued climate change will transform the species composition, land cover, drainage, and permafrost extent of the boreal–tundra system, leading to decreased albedo and the release of greenhouse gases (*medium confidence*), with adaptation measures unable to prevent substantial change (*high confidence*). Increased severe drought together with land-use change and forest fire would cause much of the Amazon forest to transform to less-dense drought- and fire-adapted ecosystems, increasing risk for biodiversity while decreasing net carbon uptake from the atmosphere (*low confidence*). Large reductions in deforestation, as well as wider application of effective wildfire management, will lower the risk of abrupt change in the Amazon, as well as potential negative impacts of that change (*medium confidence*). [4.2, 4.3, Figure 4-8, Boxes 4-3 and 4-4]

The natural carbon sink provided by terrestrial ecosystems is partially offset at the decadal timescale by carbon released through the conversion of natural ecosystems (principally forests) to farm and grazing land and through ecosystem degradation (*high confidence*). Carbon stored in the terrestrial biosphere (e.g., in peatlands, permafrost, and forests) is susceptible to loss to the atmosphere as a result of climate change, deforestation, and ecosystem degradation. [4.2, 4.3, Box 4-3]

Coastal Systems and Low-lying Areas

Due to sea level rise projected throughout the 21st century and beyond, coastal systems and low-lying areas will increasingly experience adverse impacts such as submergence, coastal flooding, and coastal erosion (*very high confidence*). The population and assets projected to be exposed to coastal risks as well as human pressures on coastal ecosystems will increase significantly in the coming decades due to population growth, economic development, and urbanization (*high confidence*). The relative costs of coastal adaptation vary strongly among and within regions and countries for the 21st century. Some low-lying developing countries and small island states are expected to face very high impacts that, in some cases, could have associated damage and adaptation costs of several percentage points of GDP. [5.3 to 5.5, 8.2, 22.3, 24.4, 25.6, 26.3, 26.8, Table 26-1, Box 25-1]

Marine Systems

By mid 21st century, spatial shifts of marine species will cause species richness and fisheries catch potential to increase, on average, at mid and high latitudes (*high confidence*) and to decrease at tropical latitudes (*medium confidence*), resulting in global redistribution of catch potential for fishes and invertebrates, with implications for food security (*medium confidence*). Spatial shifts of marine species due to projected warming will cause high-latitude invasions and high local-extinction rates in the tropics and semi-enclosed seas (*medium confidence*). Animal displacements will cause a 30 to 70% increase in the fisheries yield of some high-latitude regions by 2055 (relative to 2005), a redistribution at mid latitudes, and a drop of 40 to 60% in some of the tropics and the Antarctic, for 2°C warming above preindustrial levels (*medium confidence* for direction of fisheries' yield trends, *low confidence* for the precise magnitudes of yield change). See Figure TS.8A. The progressive expansion of oxygen minimum zones and anoxic "dead zones" is projected to further constrain the habitat of fishes and other O_2-dependent organisms (*medium confidence*). Open-ocean net primary production is projected to redistribute and, by 2100, fall globally under all RCP scenarios. [6.3 to 6.5, 7.4, 25.6, 28.3, 30.4 to 30.6, Boxes CC-MB and CC-PP]

Due to projected climate change by the mid 21st century and beyond, global marine-species redistribution and marine-biodiversity reduction in sensitive regions will challenge the sustained provision of fisheries productivity and other ecosystem goods and services (*high confidence*). Socioeconomic vulnerability is highest in developing tropical countries, leading to risks from reduced supplies, income, and employment from marine fisheries. [6.4, 6.5]

For medium- to high-emission scenarios (RCP4.5, 6.0, and 8.5), ocean acidification poses substantial risks to marine ecosystems, especially polar ecosystems and coral reefs, associated with impacts on the physiology, behavior, and population dynamics of individual species from phytoplankton to animals (*medium* to *high confidence*). See Box TS.7. Highly calcified mollusks, echinoderms, and reef-building corals are more sensitive than crustaceans (*high confidence*) and fishes (*low confidence*), with potentially detrimental consequences for fisheries and livelihoods (Figure TS.8B). Ocean acidification acts together with other global changes (e.g., warming, decreasing oxygen levels) and with local changes (e.g., pollution, eutrophication) (*high confidence*). Simultaneous drivers, such as warming and ocean acidification, can lead to interactive, complex, and amplified impacts for species and ecosystems. [5.4, 6.3 to 6.5, 22.3, 25.6, 28.3, 30.5, Boxes CC-CR and CC-OA]

Climate change adds to the threats of over-fishing and other non-climatic stressors, thus complicating marine management regimes (*high confidence*). In the short term, strategies including climate forecasting and early warning systems can reduce risks from ocean warming and acidification for some fisheries and aquaculture industries. Fisheries and aquaculture industries with high-technology

⟶

Figure TS.8 | Climate change risks for fisheries. (A) Projected global redistribution of maximum catch potential of ~1000 exploited fish and invertebrate species. Projections compare the 10-year averages 2001–2010 and 2051–2060 using SRES A1B, without analysis of potential impacts of overfishing or ocean acidification. (B) Marine mollusk and crustacean fisheries (present-day estimated annual catch rates ≥0.005 tonnes km^{-2}) and known locations of cold- and warm-water corals, depicted on a global map showing the projected distribution of ocean acidification under RCP8.5 (pH change from 1986–2005 to 2081–2100). [WGI AR5 Figure SPM.8] The bottom panel compares sensitivity to ocean acidification across mollusks, crustaceans, and corals, vulnerable animal phyla with socioeconomic relevance (e.g., for coastal protection and fisheries). The number of species analyzed across studies is given for each category of elevated CO_2. For 2100, RCP scenarios falling within each CO_2 partial pressure (pCO_2) category are as follows: RCP4.5 for 500–650 μatm (approximately equivalent to ppm in the atmosphere), RCP6.0 for 651–850 μatm, and RCP8.5 for 851–1370 μatm. By 2150, RCP8.5 falls within the 1371–2900 μatm category. The control category corresponds to 380 μatm. [6.1, 6.3, 30.5, Figures 6-10 and 6-14; WGI AR5 Box SPM.1]

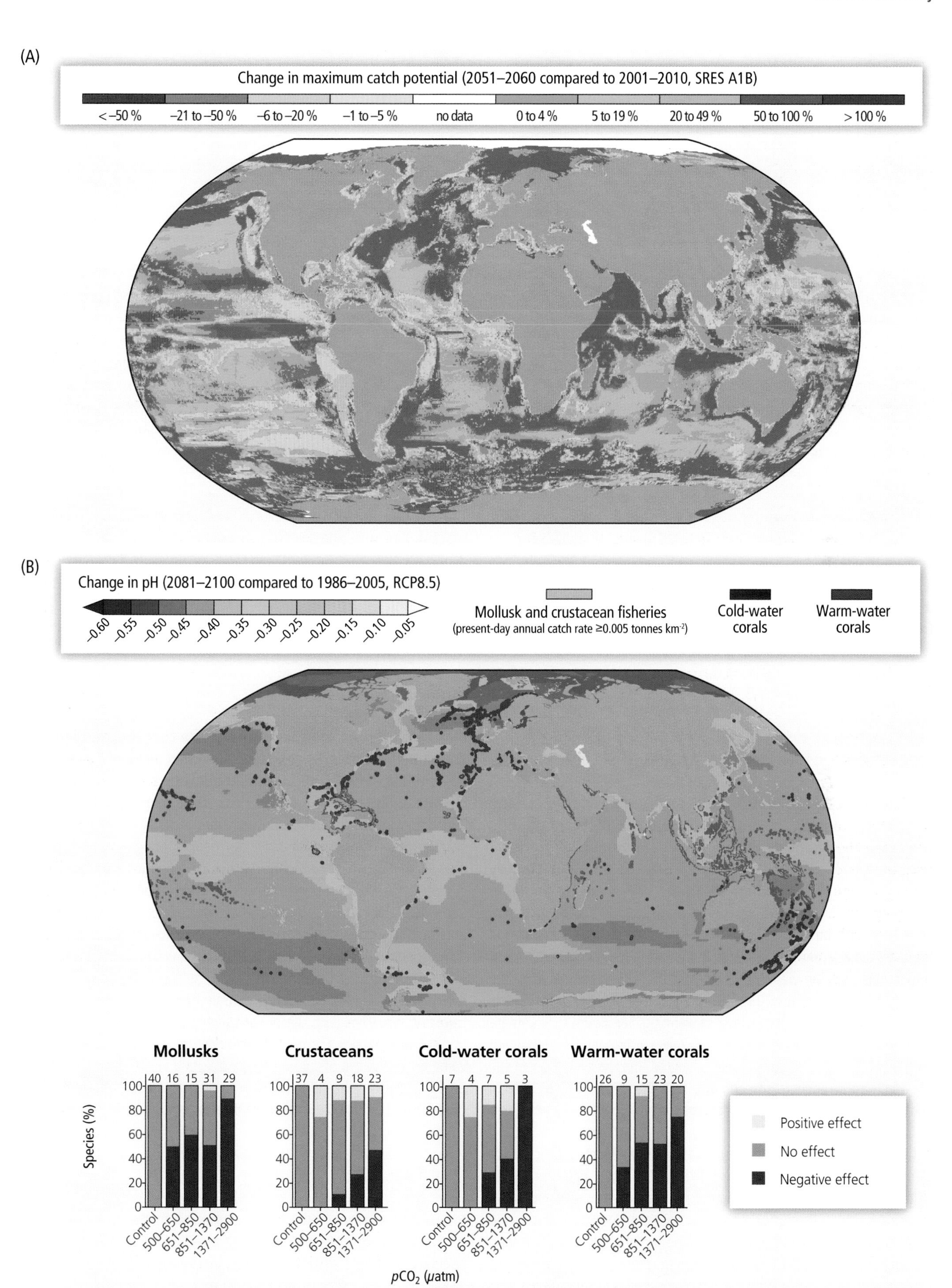

(A)
Change in maximum catch potential (2051–2060 compared to 2001–2010, SRES A1B)
<–50 %
–21 to –50 %
–6 to –20 %
–1 to –5 %
no data
0 to 4 %
5 to 19 %
20 to 49 %
50 to 100 %
>100 %
(B)
Change in pH (2081–2100 compared to 1986–2005, RCP8.5)
–0.60
–0.55
–0.50
–0.45
–0.40
–0.35
–0.30
–0.25
–0.20
–0.15
–0.10
–0.05
Mollusk and crustacean fisheries
(present-day annual catch rate ≥0.005 tonnes km-2)
Cold-water corals
Warm-water corals
Mollusks
Crustaceans
Cold-water corals
Warm-water corals
40 16 15 31 29
37 4 9 18 23
7 4 7 5 3
26 9 15 23 20
Species (%)
Control
500–650
651–850
851–1370
1371–2900
pCO2 (μatm)
Positive effect
No effect
Negative effect
TS

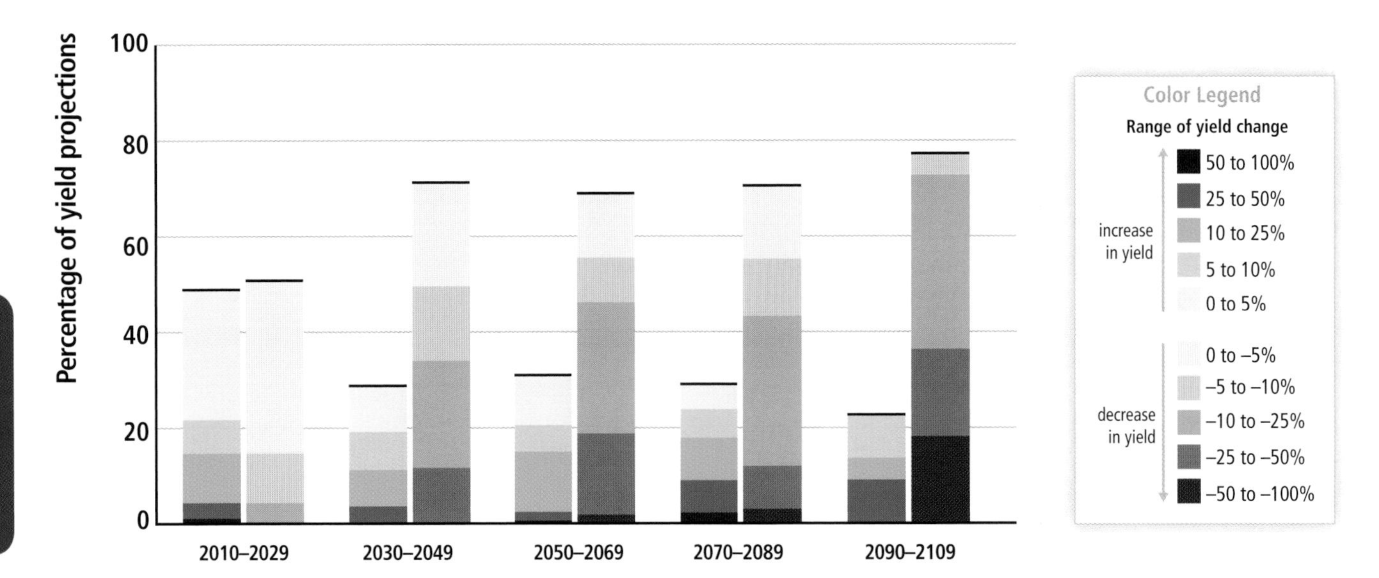

Figure TS.9 | Summary of projected changes in crop yields, due to climate change over the 21st century. The figure includes projections for different emission scenarios, for tropical and temperate regions, and for adaptation and no-adaptation cases combined. Relatively few studies have considered impacts on cropping systems for scenarios where global mean temperatures increase by 4°C or more. For five timeframes in the near term and long term, data (n=1090) are plotted in the 20-year period on the horizontal axis that includes the midpoint of each future projection period. Changes in crop yields are relative to late-20th-century levels. Data for each timeframe sum to 100%. [Figure 7-5]

and/or large investments, as well as marine shipping and oil and gas industries, have high capacities for adaptation due to greater development of environmental monitoring, modeling, and resource assessments. For smaller-scale fisheries and developing countries, building social resilience, alternative livelihoods, and occupational flexibility represent important strategies for reducing the vulnerability of ocean-dependent human communities. [6.4, 7.3, 7.4, 25.6, 29.4, 30.6, 30.7]

Food Security and Food Production Systems

For the major crops (wheat, rice, and maize) in tropical and temperate regions, climate change without adaptation is projected to negatively impact aggregate production for local temperature increases of 2°C or more above late-20th-century levels, although individual locations may benefit (*medium confidence*). Projected impacts vary across crops and regions and adaptation scenarios, with about 10% of projections for the period 2030–2049 showing yield gains of more than 10%, and about 10% of projections showing yield losses of more than 25%, compared to the late 20th century. After 2050 the risk of more severe yield impacts increases and depends on the level of warming. See Figure TS.9. Climate change is projected to progressively increase inter-annual variability of crop yields in many regions. These projected impacts will occur in the context of rapidly rising crop demand. [7.4, 7.5, 22.3, 24.4, 25.7, 26.5, Table 7-2, Figures 7-4, 7-5, 7-6, 7-7, and 7-8]

All aspects of food security are potentially affected by climate change, including food access, utilization, and price stability (*high confidence*). Redistribution of marine fisheries catch potential towards higher latitudes poses risk of reduced supplies, income, and employment in tropical countries, with potential implications for food security (*medium confidence*). Global temperature increases of ~4°C or more above late-20th-century levels, combined with increasing food demand, would pose large risks to food security globally and regionally (*high confidence*). Risks to food security are generally greater in low-latitude areas. [6.3 to 6.5, 7.4, 7.5, 9.3, 22.3, 24.4, 25.7, 26.5, Table 7-3, Figures 7-1, 7-4, and 7-7, Box 7-1]

Urban Areas

Many global risks of climate change are concentrated in urban areas (*medium confidence*). Steps that build resilience and enable sustainable development can accelerate successful climate-change adaptation globally. Heat stress, extreme precipitation, inland and coastal flooding, landslides, air pollution, drought, and water scarcity pose risks in urban areas for people, assets, economies, and ecosystems (*very high confidence*). Risks are amplified for those lacking essential infrastructure and services or living in poor-quality housing and exposed areas. Reducing basic service deficits, improving housing, and building resilient infrastructure systems could significantly reduce vulnerability and exposure in urban areas. Urban adaptation benefits from effective multi-level urban risk governance, alignment of policies and incentives, strengthened local government and community adaptation capacity, synergies with the private sector, and appropriate financing and institutional development (*medium confidence*). Increased capacity, voice, and influence of low-income groups and vulnerable communities and their partnerships with local governments also benefit adaptation. [3.5, 8.2 to 8.4, 22.3, 24.4, 24.5, 26.8, Table 8-2, Boxes 25-9 and CC-HS]

Rural Areas

Major future rural impacts are expected in the near term and beyond through impacts on water availability and supply, food security, and agricultural incomes, including shifts in production areas of food and non-food crops across the world (*high

***confidence*)**. These impacts are expected to disproportionately affect the welfare of the poor in rural areas, such as female-headed households and those with limited access to land, modern agricultural inputs, infrastructure, and education. Climate change will increase international agricultural trade volumes in both physical and value terms (*limited evidence, medium agreement*). Importing food can help countries adjust to climate change-induced domestic productivity shocks while short-term food deficits in developing countries with low income may have to be met through food aid. Further adaptations for agriculture, water, forestry, and biodiversity can occur through policies taking account of rural decision-making contexts. Trade reform and investment can improve market access for small-scale farms (*medium confidence*). Valuation of non-marketed ecosystem services and limitations of economic valuation models that aggregate across contexts pose challenges for valuing rural impacts. [9.3, 25.9, 26.8, 28.2, 28.4, Box 25-5]

Key Economic Sectors and Services

For most economic sectors, the impacts of drivers such as changes in population, age structure, income, technology, relative prices, lifestyle, regulation, and governance are projected to be large relative to the impacts of climate change (*medium evidence, high agreement*). Climate change is projected to reduce energy demand for heating and increase energy demand for cooling in the residential and commercial sectors (*robust evidence, high agreement*). Climate change is projected to affect energy sources and technologies differently, depending on resources (e.g., water flow, wind, insolation), technological processes (e.g., cooling), or locations (e.g., coastal regions, floodplains) involved. More severe and/or frequent extreme weather events and/or hazard types are projected to increase losses and loss variability in various regions and challenge insurance systems to offer affordable coverage while raising more risk-based capital, particularly in developing countries. Large-scale public-private risk reduction initiatives and economic diversification are examples of adaptation actions. [3.5, 10.2, 10.7, 10.10, 17.4, 17.5, 25.7, 26.7 to 26.9, Box 25-7]

Climate change may influence the integrity and reliability of pipelines and electricity grids (*medium evidence, medium agreement*). Climate change may require changes in design standards for the construction and operation of pipelines and of power transmission and distribution lines. Adopting existing technology from other geographical and climatic conditions may reduce the cost of adapting new infrastructure as well as the cost of retrofitting existing pipelines and grids. Climate change may negatively affect transport infrastructure (*limited evidence, high agreement*). All infrastructure is vulnerable to freeze-thaw cycles; paved roads are particularly vulnerable to temperature extremes, unpaved roads and bridges to precipitation extremes. Transport infrastructure on ice or permafrost is especially vulnerable. [10.2, 10.4, 25.7, 26.7]

Climate change will affect tourism resorts, particularly ski resorts, beach resorts, and nature resorts (*robust evidence, high agreement*), and tourists may spend their holidays at higher altitudes and latitudes (*medium evidence, high agreement*). The economic implications of climate-change-induced changes in tourism demand and supply entail gains for countries closer to the poles and countries with higher elevations and losses for other countries. [10.6, 25.7]

Global economic impacts from climate change are difficult to estimate. Economic impact estimates completed over the past 20 years vary in their coverage of subsets of economic sectors and depend on a large number of assumptions, many of which are disputable, and many estimates do not account for catastrophic changes, tipping points, and many other factors. With these recognized limitations, the incomplete estimates of global annual economic losses for additional temperature increases of ~2°C are between 0.2 and 2.0% of income (±1 standard deviation around the mean) (*medium evidence, medium agreement*). Losses are *more likely than not* to be greater, rather than smaller, than this range (*limited evidence, high agreement*). Additionally, there are large differences between and within countries. Losses accelerate with greater warming (*limited evidence, high agreement*), but few quantitative estimates have been completed for additional warming around 3°C or above. Estimates of the incremental economic impact of emitting carbon dioxide lie between a few dollars and several hundreds of dollars per tonne of carbon[3] (*robust evidence, medium agreement*). Estimates vary strongly with the assumed damage function and discount rate. [10.9]

Human Health

Until mid-century, projected climate change will impact human health mainly by exacerbating health problems that already exist (*very high confidence*). Throughout the 21st century, climate change is expected to lead to increases in ill-health in many regions and especially in developing countries with low income, as compared to a baseline without climate change (*high confidence*). Examples include greater likelihood of injury, disease, and death due to more intense heat waves and fires (*very high confidence*); increased likelihood of under-nutrition resulting from diminished food production in poor regions (*high confidence*); risks from lost work capacity and reduced labor productivity in vulnerable populations; and increased risks from food- and water-borne diseases (*very high confidence*) and vector-borne diseases (*medium confidence*). Impacts on health will be reduced, but not eliminated, in populations that benefit from rapid social and economic development, particularly among the poorest and least healthy groups (*high confidence*). Climate change will increase demands for health care services and facilities, including public health programs, disease prevention activities, health care personnel, infrastructure, and supplies for treatment (*medium evidence, high agreement*). Positive effects are expected to include modest reductions in cold-related mortality and morbidity in some areas due to fewer cold extremes (*low confidence*), geographical shifts in food production (*medium confidence*), and reduced capacity of vectors to transmit some diseases. But globally over the 21st century, the magnitude and severity of negative impacts are projected to increasingly outweigh positive impacts (*high confidence*). The most effective vulnerability reduction measures for health in the near term are programs that implement and improve basic public health measures such as provision of clean water and sanitation, secure essential health care including vaccination and child health services,

[3] 1 tonne of carbon = 3.667 tonne of CO_2

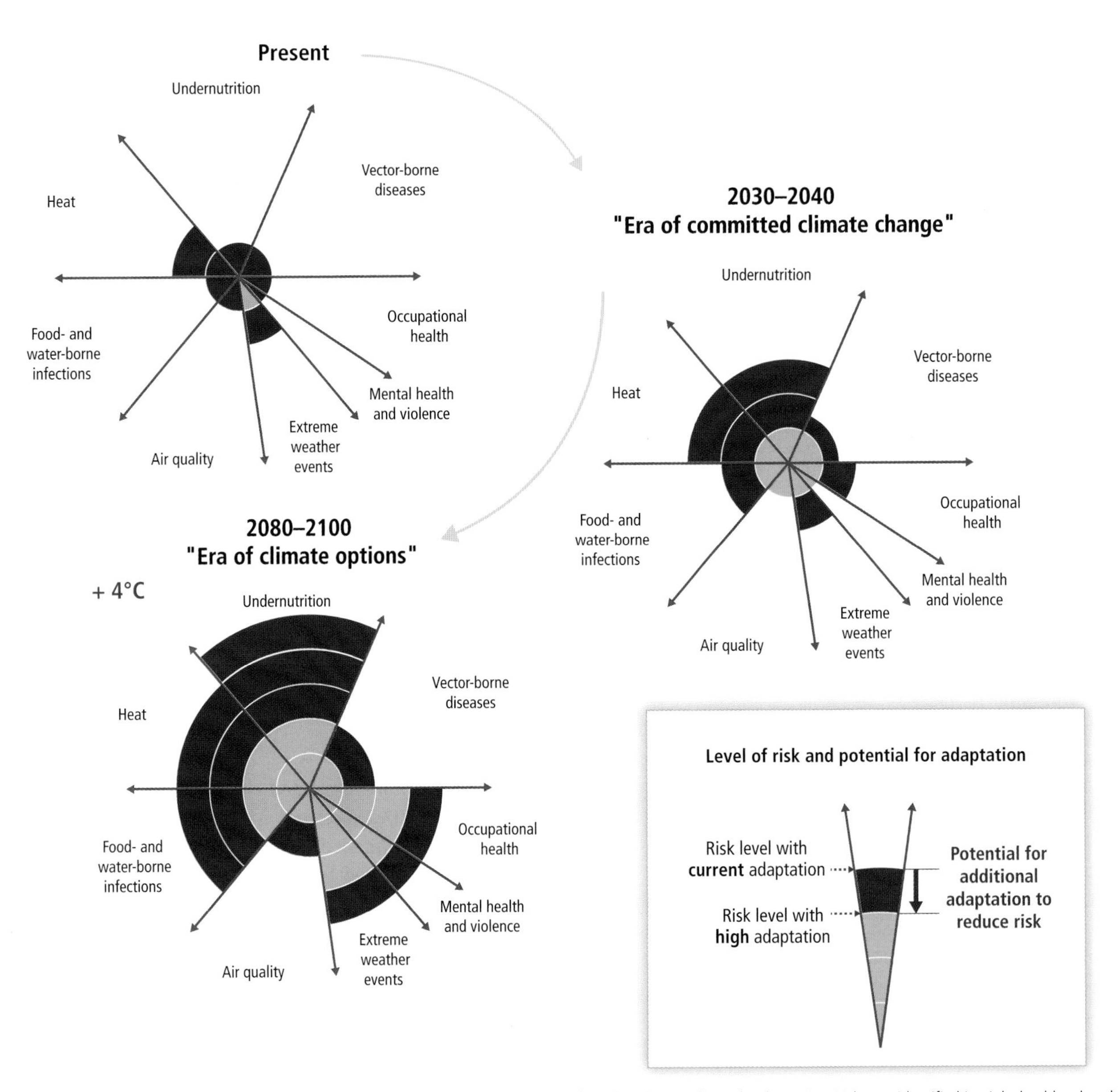

Figure TS.10 | Conceptual presentation of health risks from climate change and the potential for risk reduction through adaptation. Risks are identified in eight health-related categories based on assessment of the literature and expert judgments by authors of Chapter 11. The width of the slices indicates in a qualitative way relative importance in terms of burden of ill-health globally at present. Risk levels are assessed for the present and for the near-term era of committed climate change (here, for 2030–2040). For some categories, for example, vector-borne diseases, heat/cold stress, and agricultural production and undernutrition, there may be benefits to health in some areas, but the net impact is expected to be negative. Risk levels are also presented for the longer-term era of climate options (here, for 2080–2100) for global mean temperature increase of 4°C above preindustrial levels. For each timeframe, risk levels are estimated for the current state of adaptation and for a hypothetical highly adapted state, indicated by different colors. [Figure 11-6]

increase capacity for disaster preparedness and response, and alleviate poverty (*very high confidence*). By 2100 for the high-emission scenario RCP8.5, the combination of high temperature and humidity in some areas for parts of the year is projected to compromise normal human activities, including growing food or working outdoors (*high confidence*). See Figure TS.10. [8.2, 11.3 to 11.8, 19.3, 22.3, 25.8, 26.6, Figure 25-5, Box CC-HS]

Human Security

Human security will be progressively threatened as the climate changes (*robust evidence, high agreement*). Human insecurity almost never has single causes, but instead emerges from the interaction of multiple factors. Climate change is an important factor in threats to human security through (1) undermining livelihoods, (2) compromising culture and identity, (3) increasing migration that people would rather have avoided, and (4) challenging the ability of states to provide the conditions necessary for human security. See Figure TS.11. [12.1 to 12.4, 12.6]

Climate change will compromise the cultural values that are important for community and individual well-being (*medium evidence, high agreement*). The effect of climate change on culture will vary across societies and over time, depending on cultural resilience and the mechanisms for maintaining and transferring knowledge. Changing weather and climatic conditions threaten cultural practices

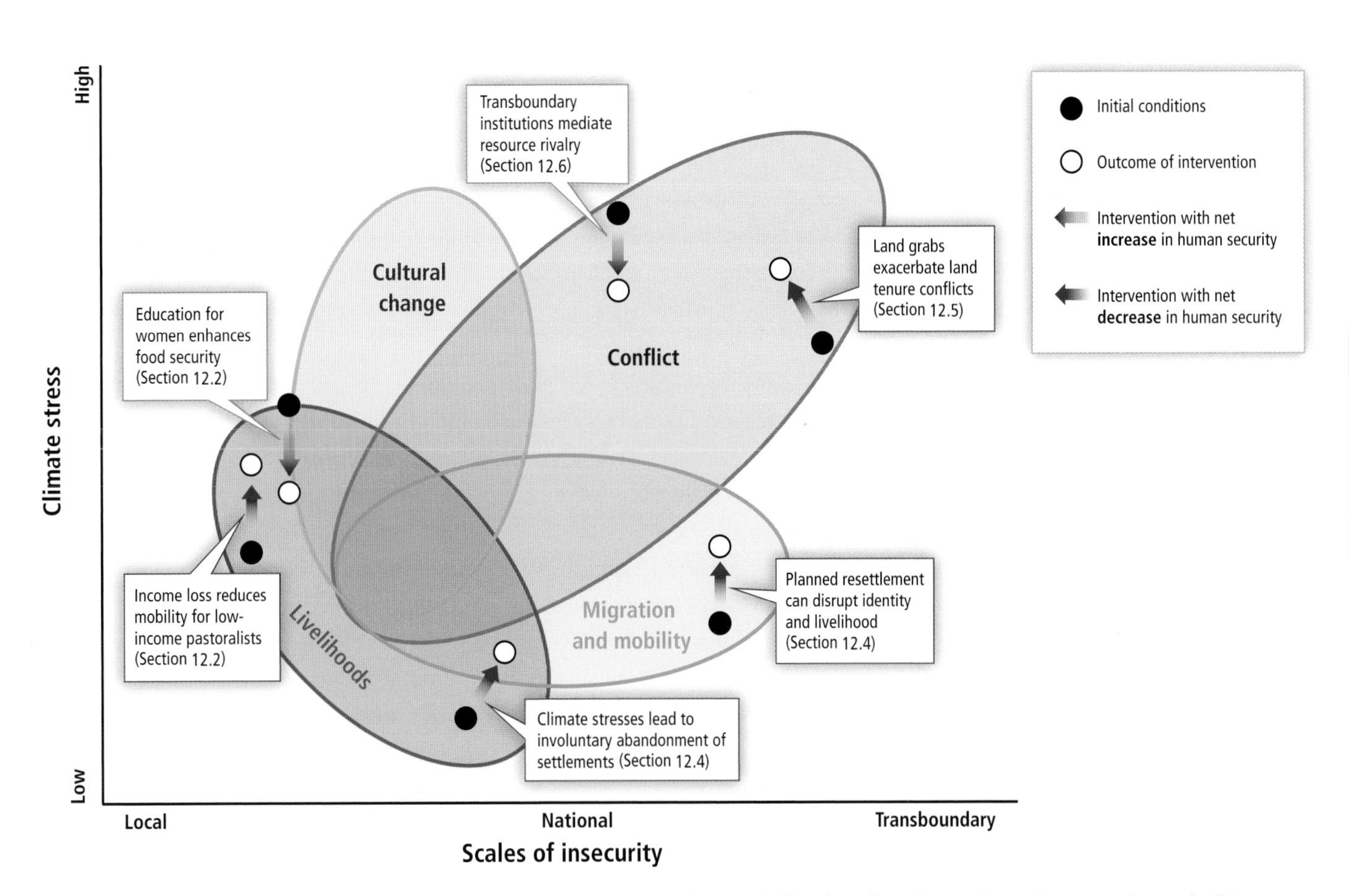

Figure TS.11 | Schematic of climate change risks for human security and the interactions between livelihoods, conflict, culture, and migration. Interventions and policies are indicated by the difference between initial conditions (solid black circles) and the outcome of intervention (white circles). Some interventions (blue arrows) show net increase in human security while others (red arrows) lead to net decrease in human security. [Figure 12-3]

embedded in livelihoods and expressed in narratives, worldviews, identity, community cohesion, and sense of place. Loss of land and displacement, for example, on small islands and coastal communities, have well documented negative cultural and well-being impacts. [12.3, 12.4]

Climate change over the 21st century is projected to increase displacement of people (*medium evidence, high agreement*). Displacement risk increases when populations that lack the resources for planned migration experience higher exposure to extreme weather events, in both rural and urban areas, particularly in developing countries with low income. Expanding opportunities for mobility can reduce vulnerability for such populations. Changes in migration patterns can be responses to both extreme weather events and longer-term climate variability and change, and migration can also be an effective adaptation strategy. There is *low confidence* in quantitative projections of changes in mobility, due to its complex, multi-causal nature. [9.3, 12.4, 19.4, 22.3, 25.9]

Climate change can indirectly increase risks of violent conflicts in the form of civil war and inter-group violence by amplifying well-documented drivers of these conflicts such as poverty and economic shocks (*medium confidence*). Multiple lines of evidence relate climate variability to these forms of conflict. [12.5, 13.2, 19.4]

The impacts of climate change on the critical infrastructure and territorial integrity of many states are expected to influence national security policies (*medium evidence, medium agreement*). For example, land inundation due to sea level rise poses risks to the territorial integrity of small island states and states with extensive coastlines. Some transboundary impacts of climate change, such as changes in sea ice, shared water resources, and pelagic fish stocks, have the potential to increase rivalry among states, but robust national and intergovernmental institutions can enhance cooperation and manage many of these rivalries. [12.5, 12.6, 23.9, 25.9]

Livelihoods and Poverty

Throughout the 21st century, climate-change impacts are projected to slow down economic growth, make poverty reduction more difficult, further erode food security, and prolong existing and create new poverty traps, the latter particularly in urban areas and emerging hotspots of hunger (*medium confidence*). Climate-change impacts are expected to exacerbate poverty in most developing countries and create new poverty pockets in countries with increasing inequality, in both developed and developing countries. In urban and rural areas, wage-labor-dependent poor households that are net buyers of food are expected to be particularly affected due to food price increases, including in regions with high food insecurity and high inequality (particularly in Africa), although the agricultural self-employed could benefit. Insurance programs, social protection measures, and disaster risk management may enhance long-term livelihood resilience

TS

Box TS.7 | Ocean Acidification

Anthropogenic ocean acidification and global warming share the same primary cause, which is the increase of atmospheric CO_2 (Box TS.7 Figure 1A). [WGI AR5 2.2] Eutrophication, upwelling, and deposition of atmospheric nitrogen and sulfur contribute to ocean acidification locally. [5.3, 6.1, 30.3] The fundamental chemistry of ocean acidification is well understood (*robust evidence, high agreement*). [30.3; WGI AR5 3.8, 6.4] It has been more difficult to understand and project changes within the more complex coastal systems. [5.3, 30.3]

Ocean acidification acts together with other global changes (e.g., warming, decreasing oxygen levels) and with local changes (e.g., pollution, eutrophication) (*high confidence*). Simultaneous drivers, such as warming and ocean acidification, can lead to interactive, complex, and amplified impacts for species and ecosystems. A pattern of positive and negative impacts of ocean acidification emerges for processes and organisms (*high confidence*; Box TS.7 Figure 1B), but key uncertainties remain from organismal to ecosystem levels. A wide range of sensitivities exists within and across organisms, with higher sensitivity in early life stages. [6.3] Lower pH decreases the rate of calcification of most, but not all, sea floor calcifiers, reducing their competitiveness with non-calcifiers (*robust evidence, medium agreement*). [5.4, 6.3] Ocean acidification stimulates dissolution of calcium carbonate (*very high confidence*). Growth and primary production are stimulated in seagrasses and some phytoplankton (*high confidence*), and harmful algal blooms could become more frequent (*limited evidence, medium agreement*). Serious behavioral disturbances have been reported in fishes

(A)

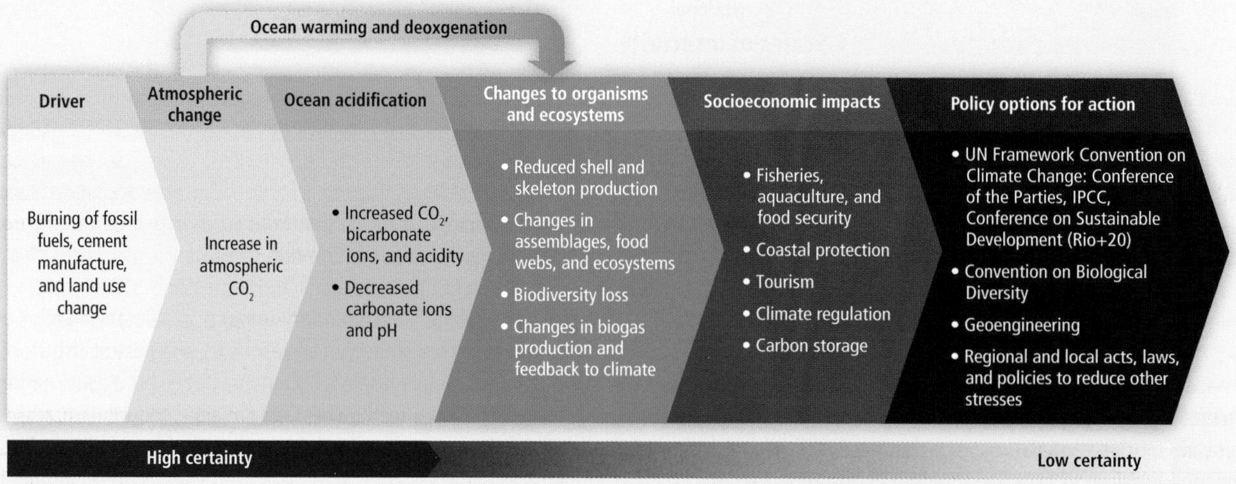

Ocean warming and deoxgenation

Driver	Atmospheric change	Ocean acidification	Changes to organisms and ecosystems	Socioeconomic impacts	Policy options for action
Burning of fossil fuels, cement manufacture, and land use change	Increase in atmospheric CO_2	• Increased CO_2, bicarbonate ions, and acidity • Decreased carbonate ions and pH	• Reduced shell and skeleton production • Changes in assemblages, food webs, and ecosystems • Biodiversity loss • Changes in biogas production and feedback to climate	• Fisheries, aquaculture, and food security • Coastal protection • Tourism • Climate regulation • Carbon storage	• UN Framework Convention on Climate Change: Conference of the Parties, IPCC, Conference on Sustainable Development (Rio+20) • Convention on Biological Diversity • Geoengineering • Regional and local acts, laws, and policies to reduce other stresses

High certainty — Low certainty

(B)

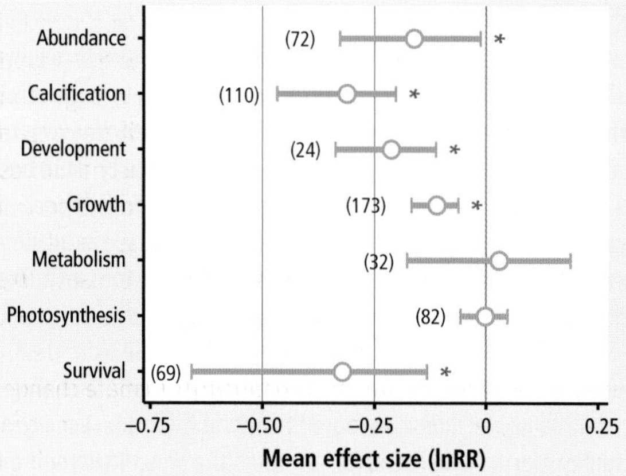

Abundance (72) *
Calcification (110) *
Development (24) *
Growth (173) *
Metabolism (32)
Photosynthesis (82)
Survival (69) *

−0.75 −0.50 −0.25 0 0.25

Mean effect size (lnRR)

Box TS.7 Figure 1 | (A) Overview of the chemical, biological, and socioeconomic impacts of ocean acidification and of policy options. (B) Effect of near-future acidification (seawater pH reduction of ≤0.5 units) on major response variables estimated using weighted random effects meta-analyses, with the exception of survival, which is not weighted. The log-transformed response ratio (lnRR) is the ratio of the mean effect in the acidification treatment to the mean effect in a control group. It indicates which process is most uniformly affected by ocean acidification, but large variability exists between species. Significance is determined when the 95% bootstrapped confidence interval does not cross zero. The number of experiments used in the analyses is shown in parentheses. The * denotes a statistically significant effect. [Figure OA-1, Box CC-OA]

Continued next page →

Box TS.7 (continued)

(*high confidence*). [6.3] Natural analogs at CO_2 vents indicate decreased species diversity, biomass, and trophic complexity. Shifts in organisms' performance and distribution will change both predator-prey and competitive interactions, which could impact food webs and higher trophic levels (*limited evidence, high agreement*). [6.3]

A few studies provide *limited evidence* for adaptation in phytoplankton and mollusks. However, mass extinctions in Earth history occurred during much slower rates of change in ocean acidification, combined with other drivers, suggesting that evolutionary rates may be too slow for sensitive and long-lived species to adapt to the projected rates of future change (*medium confidence*). [6.1]

The biological, ecological, and biogeochemical changes driven by ocean acidification will affect key ecosystem services. The oceans will become less efficient at absorbing CO_2 and hence moderating climate (*very high confidence*). [WGI AR5 Figure 6.26] The impacts of ocean acidification on coral reefs, together with those of thermal stress (driving mass coral bleaching and mortality) and sea level rise, will diminish their role in shoreline protection as well as their direct and indirect benefits to fishing and tourism industries (*limited evidence, high agreement*). [Box CC-CR] The global cost of production loss of mollusks could be over US$100 billion by 2100 (*low confidence*). The largest uncertainty is how the impacts on lower trophic levels will propagate through the food webs and to top predators. Models suggest that ocean acidification will generally reduce fish biomass and catch (*low confidence*) and complex additive, antagonistic, and/or synergistic interactions will occur with disruptive ramifications for ecosystems as well as for important ecosystem goods and services.

among poor and marginalized people, if policies address poverty and multidimensional inequalities. [8.1, 8.3, 8.4, 9.3, 10.9, 13.2 to 13.4, 22.3, 26.8]

B-3. Regional Risks and Potential for Adaptation

Risks will vary through time across regions and populations, dependent on myriad factors including the extent of adaptation and mitigation. A selection of key regional risks identified with *medium* to *high confidence* is presented in Table TS.5. Projected changes in climate and increasing atmospheric CO_2 will have positive effects for some sectors in some locations. For extended summary of regional risks and the more limited potential benefits, see introductory overviews for each region below and also WGII AR5 Part B: Regional Aspects, Chapters 21 to 30.

***Africa*. Climate change will amplify existing stress on water availability and on agricultural systems particularly in semi-arid environments (*high confidence*).** Increasing temperatures and changes in precipitation are *very likely* to reduce cereal crop productivity with strong adverse effects on food security (*high confidence*). Progress has been achieved on managing risks to food production from current climate variability and near-term climate change, but this will not be sufficient to address long-term impacts of climate change. Adaptive agricultural processes such as collaborative, participatory research that includes scientists and farmers, strengthened communication systems for anticipating and responding to climate risks, and increased flexibility in livelihood options provide potential pathways for strengthening adaptive capacities. Climate change is a multiplier of existing health vulnerabilities including insufficient access to safe water and improved sanitation, food insecurity, and limited access to health care and education. Strategies that integrate consideration of climate change risks with land and water management and disaster risk reduction bolster resilient development. [22.3 to 22.4, 22.6]

***Europe*. Climate change will increase the likelihood of systemic failures across European countries caused by extreme climate events affecting multiple sectors (*medium confidence*).** Sea level rise and increases in extreme rainfall are projected to further increase coastal and river flood risks and without adaptive measures will substantially increase flood damages (i.e., people affected and economic losses); adaptation can prevent most of the projected damages (*high confidence*). Heat-related deaths and injuries are *likely* to increase, particularly in southern Europe (*medium confidence*). Climate change is *likely* to increase cereal crop yields in northern Europe (*medium confidence*) but decrease yields in southern Europe (*high confidence*). Climate change will increase irrigation needs in Europe, and future irrigation will be constrained by reduced runoff, demand from other sectors, and economic costs, with integrated water management a strategy for addressing competing demands. Hydropower production is *likely* to decrease in all sub-regions except Scandinavia. Climate change is *very likely* to cause changes in habitats and species, with local extinctions (*high confidence*), continental-scale shifts in species distributions (*medium confidence*), and significantly reduced alpine plant habitat (*high confidence*). Climate change is *likely* to entail the loss or displacement of coastal wetlands. The introduction and expansion of invasive species, especially those with high migration rates, from outside Europe is *likely* to increase with climate change (medium confidence). [23.2 to 23.9]

Asia. **Climate change will cause declines in agricultural productivity in many sub-regions of Asia, for crops such as rice (*medium confidence*).** In Central Asia, cereal production in northern and eastern Kazakhstan could benefit from the longer growing season, warmer winters, and slight increase in winter precipitation, while droughts in western Turkmenistan and Uzbekistan could negatively affect cotton production, increase water demand for irrigation, and exacerbate desertification. The effectiveness of potential and practiced agricultural adaptation strategies is not well understood. Future projections of precipitation at sub-regional scales and thus of freshwater availability in most parts of Asia are uncertain (*low confidence* in projections), but increased water demand from population growth, increased water consumption per capita, and lack of good management will increase water scarcity challenges for most of the region (*medium confidence*). Adaptive responses include integrated water management strategies, such as development of water-saving technologies, increased water productivity, and water reuse. Extreme climate events will have an increasing impact on human health, security, livelihoods, and poverty, with the type and magnitude of impact varying across Asia (*high confidence*). In many parts of Asia, observed terrestrial impacts, such as permafrost degradation and shifts in plant species' distributions, growth rates, and timing of seasonal activities, will increase due to climate change projected during the 21st century. Coastal and marine systems in Asia, such as mangroves, seagrass beds, salt marshes, and coral reefs, are under increasing stress from climatic and non-climatic drivers. In the Asian Arctic, sea level rise interacting with projected changes in permafrost and the length of the ice-free season will increase rates of coastal erosion (*medium evidence, high agreement*). [24.4, 30.5]

Australasia. **Without adaptation, further changes in climate, atmospheric carbon dioxide, and ocean acidity are projected to have substantial impacts on water resources, coastal ecosystems, infrastructure, health, agriculture, and biodiversity (*high confidence*).** Freshwater resources are projected to decline in far

Table TS.5 | Key regional risks from climate change and the potential for reducing risks through adaptation and mitigation. Key risks have been identified based on assessment of the relevant scientific, technical, and socioeconomic literature detailed in supporting chapter sections. Identification of key risks was based on expert judgment using the following specific criteria: large magnitude, high probability, or irreversibility of impacts; timing of impacts; persistent vulnerability or exposure contributing to risks; or limited potential to reduce risks through adaptation or mitigation. Each key risk is characterized as very low to very high for three timeframes: the present, near term (here, assessed over 2030–2040), and longer term (here, assessed over 2080–2100). The risk levels integrate probability and consequence over the widest possible range of potential outcomes, based on available literature. These potential outcomes result from the interaction of climate-related hazards, vulnerability, and exposure. Each risk level reflects total risk from climatic and non-climatic factors. For the near-term era of committed climate change, projected levels of global mean temperature increase do not diverge substantially for different emission scenarios. For the longer-term era of climate options, risk levels are presented for two scenarios of global mean temperature increase (2°C and 4°C above preindustrial levels). These scenarios illustrate the potential for mitigation and adaptation to reduce the risks related to climate change. For the present, risk levels were estimated for current adaptation and a hypothetical highly adapted state, identifying where current adaptation deficits exist. For the two future timeframes, risk levels were estimated for a continuation of current adaptation and for a highly adapted state, representing the potential for and limits to adaptation. Climate-related drivers of impacts are indicated by icons. Key risks and risk levels vary across regions and over time, given differing socioeconomic development pathways, vulnerability and exposure to hazards, adaptive capacity, and risk perceptions. Risk levels are not necessarily comparable, especially across regions, because the assessment considers potential impacts and adaptation in different physical, biological, and human systems across diverse contexts. This assessment of risks acknowledges the importance of differences in values and objectives in interpretation of the assessed risk levels.

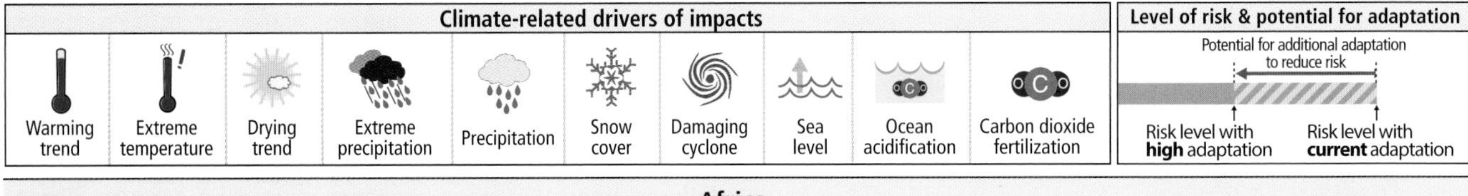

Africa

Key risk	Adaptation issues & prospects	Climatic drivers	Timeframe	Risk & potential for adaptation
Compounded stress on water resources facing significant strain from overexploitation and degradation at present and increased demand in the future, with drought stress exacerbated in drought-prone regions of Africa (*high confidence*) [22.3, 22.4]	• Reducing non-climate stressors on water resources • Strengthening institutional capacities for demand management, groundwater assessment, integrated water-wastewater planning, and integrated land and water governance • Sustainable urban development		Present Near term (2030–2040) Long term (2080–2100) 2°C / 4°C	Very low / Medium / Very high
Reduced crop productivity associated with heat and drought stress, with strong adverse effects on regional, national, and household livelihood and food security, also given increased pest and disease damage and flood impacts on food system infrastructure (*high confidence*) [22.3, 22.4]	• Technological adaptation responses (e.g., stress-tolerant crop varieties, irrigation, enhanced observation systems) • Enhancing smallholder access to credit and other critical production resources; Diversifying livelihoods • Strengthening institutions at local, national, and regional levels to support agriculture (including early warning systems) and gender-oriented policy • Agronomic adaptation responses (e.g., agroforestry, conservation agriculture)		Present Near term (2030–2040) Long term (2080–2100) 2°C / 4°C	Very low / Medium / Very high
Changes in the incidence and geographic range of vector- and water-borne diseases due to changes in the mean and variability of temperature and precipitation, particularly along the edges of their distribution (*medium confidence*) [22.3]	• Achieving development goals, particularly improved access to safe water and improved sanitation, and enhancement of public health functions such as surveillance • Vulnerability mapping and early warning systems • Coordination across sectors • Sustainable urban development		Present Near term (2030–2040) Long term (2080–2100) 2°C / 4°C	Very low / Medium / Very high

Continued next page →

Table TS.5 (continued)

Continued next page →

Europe

Key risk	Adaptation issues & prospects	Climatic drivers	Timeframe	Risk & potential for adaptation
Increased economic losses and people affected by flooding in river basins and coasts, driven by increasing urbanization, increasing sea levels, coastal erosion, and peak river discharges (*high confidence*) [23.2, 23.3, 23.7]	Adaptation can prevent most of the projected damages (*high confidence*). • Significant experience in hard flood-protection technologies and increasing experience with restoring wetlands • High costs for increasing flood protection • Potential barriers to implementation: demand for land in Europe and environmental and landscape concerns		Present Near term (2030–2040) Long term (2080–2100) 2°C / 4°C	Very low – Medium – Very high
Increased water restrictions. Significant reduction in water availability from river abstraction and from groundwater resources, combined with increased water demand (e.g., for irrigation, energy and industry, domestic use) and with reduced water drainage and runoff as a result of increased evaporative demand, particularly in southern Europe (*high confidence*) [23.4, 23.7]	• Proven adaptation potential from adoption of more water-efficient technologies and of water-saving strategies (e.g., for irrigation, crop species, land cover, industries, domestic use) • Implementation of best practices and governance instruments in river basin management plans and integrated water management		Present Near term (2030–2040) Long term (2080–2100) 2°C / 4°C	Very low – Medium – Very high
Increased economic losses and people affected by extreme heat events: impacts on health and well-being, labor productivity, crop production, air quality, and increasing risk of wildfires in southern Europe and in Russian boreal region (*medium confidence*) [23.3 to 23.7, Table 23-1]	• Implementation of warning systems • Adaptation of dwellings and workplaces and of transport and energy infrastructure • Reductions in emissions to improve air quality • Improved wildfire management • Development of insurance products against weather-related yield variations		Present Near term (2030–2040) Long term (2080–2100) 2°C / 4°C	Very low – Medium – Very high

Asia

Key risk	Adaptation issues & prospects	Climatic drivers	Timeframe	Risk & potential for adaptation
Increased riverine, coastal, and urban flooding leading to widespread damage to infrastructure, livelihoods, and settlements in Asia (*medium confidence*) [24.4]	• Exposure reduction via structural and non-structural measures, effective land-use planning, and selective relocation • Reduction in the vulnerability of lifeline infrastructure and services (e.g., water, energy, waste management, food, biomass, mobility, local ecosystems, telecommunications) • Construction of monitoring and early warning systems; Measures to identify exposed areas, assist vulnerable areas and households, and diversify livelihoods • Economic diversification		Present Near term (2030–2040) Long term (2080–2100) 2°C / 4°C	Very low – Medium – Very high
Increased risk of heat-related mortality (*high confidence*) [24.4]	• Heat health warning systems • Urban planning to reduce heat islands; Improvement of the built environment; Development of sustainable cities • New work practices to avoid heat stress among outdoor workers		Present Near term (2030–2040) Long term (2080–2100) 2°C / 4°C	Very low – Medium – Very high
Increased risk of drought-related water and food shortage causing malnutrition (*high confidence*) [24.4]	• Disaster preparedness including early-warning systems and local coping strategies • Adaptive/integrated water resource management • Water infrastructure and reservoir development • Diversification of water sources including water re-use • More efficient use of water (e.g., improved agricultural practices, irrigation management, and resilient agriculture)		Present Near term (2030–2040) Long term (2080–2100) 2°C / 4°C	Very low – Medium – Very high

southwest and far southeast mainland Australia (*high confidence*) and for some rivers in New Zealand (*medium confidence*). Rising sea levels and increasing heavy rainfall are projected to increase erosion and inundation, with consequent damages to many low-lying ecosystems, infrastructure, and housing (*high confidence*); increasing heat waves will increase risks to human health; rainfall changes and rising temperatures will shift agricultural production zones; and many native species will suffer from range contractions and some may face local or even global extinction. Uncertainty in projected rainfall changes remains large for many parts of Australia and New Zealand, which creates significant challenges for adaptation. Some sectors in some locations have the potential to benefit from projected changes in climate and increasing atmospheric CO_2, for example due to reduced energy demand for winter heating in New Zealand and southern parts of Australia, and due to forest growth in cooler regions except where soil nutrients or rainfall are limiting. Indigenous peoples in both Australia and New Zealand have higher than average exposure to climate change due to a heavy reliance on climate-sensitive primary industries and strong social connections to the natural environment, and face additional constraints to adaptation (*medium confidence*). [25.2, 25.3, 25.5 to 25.8, Boxes 25-1, 25-2, 25-5, and 25-8]

***North America*. Many climate-related hazards that carry risk, particularly related to severe heat, heavy precipitation, and declining snowpack, will increase in frequency and/or severity in North America in the next decades (*very high confidence*).** Climate

Table TS.5 (continued)

Continued next page →

Australasia

Key risk	Adaptation issues & prospects	Climatic drivers	Timeframe	Risk & potential for adaptation
Significant change in community composition and structure of coral reef systems in Australia (*high confidence*) [25.6, 30.5, Boxes CC-CR and CC-OA]	• Ability of corals to adapt naturally appears limited and insufficient to offset the detrimental effects of rising temperatures and acidification. • Other options are mostly limited to reducing other stresses (water quality, tourism, fishing) and early warning systems; direct interventions such as assisted colonization and shading have been proposed but remain untested at scale.		Present; Near term (2030–2040); Long term (2080–2100) 2°C / 4°C	Very low – Medium – Very high
Increased frequency and intensity of flood damage to infrastructure and settlements in Australia and New Zealand (*high confidence*) [Table 25-1, Boxes 25-8 and 25-9]	• Significant adaptation deficit in some regions to current flood risk. • Effective adaptation includes land-use controls and relocation as well as protection and accommodation of increased risk to ensure flexibility.		Present; Near term (2030–2040); Long term (2080–2100) 2°C / 4°C	Very low – Medium – Very high
Increasing risks to coastal infrastructure and low-lying ecosystems in Australia and New Zealand, with widespread damage towards the upper end of projected sea-level-rise ranges (*high confidence*) [25.6, 25.10, Box 25-1]	• Adaptation deficit in some locations to current coastal erosion and flood risk. Successive building and protection cycles constrain flexible responses. • Effective adaptation includes land-use controls and ultimately relocation as well as protection and accommodation.		Present; Near term (2030–2040); Long term (2080–2100) 2°C / 4°C	Very low – Medium – Very high

North America

Key risk	Adaptation issues & prospects	Climatic drivers	Timeframe	Risk & potential for adaptation
Wildfire-induced loss of ecosystem integrity, property loss, human morbidity, and mortality as a result of increased drying trend and temperature trend (*high confidence*) [26.4, 26.8, Box 26-2]	• Some ecosystems are more fire-adapted than others. Forest managers and municipal planners are increasingly incorporating fire protection measures (e.g., prescribed burning, introduction of resilient vegetation). Institutional capacity to support ecosystem adaptation is limited. • Adaptation of human settlements is constrained by rapid private property development in high-risk areas and by limited household-level adaptive capacity. • Agroforestry can be an effective strategy for reduction of slash and burn practices in Mexico.		Present; Near term (2030–2040); Long term (2080–2100) 2°C / 4°C	Very low – Medium – Very high
Heat-related human mortality (*high confidence*) [26.6, 26.8]	• Residential air conditioning (A/C) can effectively reduce risk. However, availability and usage of A/C is highly variable and is subject to complete loss during power failures. Vulnerable populations include athletes and outdoor workers for whom A/C is not available. • Community- and household-scale adaptations have the potential to reduce exposure to heat extremes via family support, early heat warning systems, cooling centers, greening, and high-albedo surfaces.		Present; Near term (2030–2040); Long term (2080–2100) 2°C / 4°C	Very low – Medium – Very high
Urban floods in riverine and coastal areas, inducing property and infrastructure damage; supply chain, ecosystem, and social system disruption; public health impacts; and water quality impairment, due to sea level rise, extreme precipitation, and cyclones (*high confidence*) [26.2 to 26.4, 26.8]	• Implementing management of urban drainage is expensive and disruptive to urban areas. • Low-regret strategies with co-benefits include less impervious surfaces leading to more groundwater recharge, green infrastructure, and rooftop gardens. • Sea level rise increases water elevations in coastal outfalls, which impedes drainage. In many cases, older rainfall design standards are being used that need to be updated to reflect current climate conditions. • Conservation of wetlands, including mangroves, and land-use planning strategies can reduce the intensity of flood events.		Present; Near term (2030–2040); Long term (2080–2100) 2°C / 4°C	Very low – Medium – Very high

change will amplify risks to water resources already affected by non-climatic stressors, with potential impacts associated with decreased snowpack, decreased water quality, urban flooding, and decreased water supplies for urban areas and irrigation (*high confidence*). More adaptation options are available to address water supply deficits than flooding and water quality concerns (*medium confidence*). Ecosystems are under increasing stress from rising temperatures, CO_2 concentrations, and sea levels, with particular vulnerability to climate extremes (*very high confidence*). In many cases, climate stresses exacerbate other anthropogenic influences on ecosystems, including land use changes, non-native species, and pollution. Projected increases in temperature, reductions in precipitation in some regions, and increased frequency of extreme events would result in net productivity declines in major North American crops by the end of the 21st century without adaptation, although some regions, particularly in the north, may benefit. Adaptation, often with mitigation co-benefits, could offset projected negative yield impacts for many crops at 2°C global mean temperature increase above preindustrial levels, with reduced effectiveness of adaptation at 4°C (*high confidence*). Although larger urban centers would have higher adaptive capacities, high population density, inadequate infrastructures, lack of institutional capacity, and degraded natural environments increase future climate risks from heat waves, droughts, storms, and sea level rise (*medium evidence, high agreement*). Future risks from climate extremes can be reduced, for example through targeted and sustainable air conditioning, more effective warning and response systems, enhanced pollution controls, urban

Table TS.5 (continued)

Continued next page →

Central and South America

Key risk	Adaptation issues & prospects	Climatic drivers	Timeframe	Risk & potential for adaptation
Water availability in semi-arid and glacier-melt-dependent regions and Central America; flooding and landslides in urban and rural areas due to extreme precipitation (*high confidence*) [27.3]	• Integrated water resource management • Urban and rural flood management (including infrastructure), early warning systems, better weather and runoff forecasts, and infectious disease control		Present Near term (2030–2040) Long term (2080–2100) 2°C / 4°C	Very low – Medium – Very high
Decreased food production and food quality (*medium confidence*) [27.3]	• Development of new crop varieties more adapted to climate change (temperature and drought) • Offsetting of human and animal health impacts of reduced food quality • Offsetting of economic impacts of land-use change • Strengthening traditional indigenous knowledge systems and practices		Present Near term (2030–2040) Long term (2080–2100) 2°C / 4°C	Very low – Medium – Very high
Spread of vector-borne diseases in altitude and latitude (*high confidence*) [27.3]	• Development of early warning systems for disease control and mitigation based on climatic and other relevant inputs. Many factors augment vulnerability. • Establishing programs to extend basic public health services		Present Near term (2030–2040) Long term (2080–2100) 2°C / 4°C	Very low – Medium – Very high Long term 2°C: not available Long term 4°C: not available

Polar Regions

Key risk	Adaptation issues & prospects	Climatic drivers	Timeframe	Risk & potential for adaptation
Risks for freshwater and terrestrial ecosystems (*high confidence*) and marine ecosystems (*medium confidence*), due to changes in ice, snow cover, permafrost, and freshwater/ocean conditions, affecting species' habitat quality, ranges, phenology, and productivity, as well as dependent economies [28.2 to 28.4]	• Improved understanding through scientific and indigenous knowledge, producing more effective solutions and/or technological innovations • Enhanced monitoring, regulation, and warning systems that achieve safe and sustainable use of ecosystem resources • Hunting or fishing for different species, if possible, and diversifying income sources		Present Near term (2030–2040) Long term (2080–2100) 2°C / 4°C	Very low – Medium – Very high
Risks for the health and well-being of Arctic residents, resulting from injuries and illness from the changing physical environment, food insecurity, lack of reliable and safe drinking water, and damage to infrastructure, including infrastructure in permafrost regions (*high confidence*) [28.2 to 28.4]	• Co-production of more robust solutions that combine science and technology with indigenous knowledge • Enhanced observation, monitoring, and warning systems • Improved communications, education, and training • Shifting resource bases, land use, and/or settlement areas		Present Near term (2030–2040) Long term (2080–2100) 2°C / 4°C	Very low – Medium – Very high
Unprecedented challenges for northern communities due to complex inter-linkages between climate-related hazards and societal factors, particularly if rate of change is faster than social systems can adapt (*high confidence*) [28.2 to 28.4]	• Co-production of more robust solutions that combine science and technology with indigenous knowledge • Enhanced observation, monitoring, and warning systems • Improved communications, education, and training • Adaptive co-management responses developed through the settlement of land claims		Present Near term (2030–2040) Long term (2080–2100) 2°C / 4°C	Very low – Medium – Very high

Small Islands

Key risk	Adaptation issues & prospects	Climatic drivers	Timeframe	Risk & potential for adaptation
Loss of livelihoods, coastal settlements, infrastructure, ecosystem services, and economic stability (*high confidence*) [29.6, 29.8, Figure 29-4]	• Significant potential exists for adaptation in islands, but additional external resources and technologies will enhance response. • Maintenance and enhancement of ecosystem functions and services and of water and food security • Efficacy of traditional community coping strategies is expected to be substantially reduced in the future.		Present Near term (2030–2040) Long term (2080–2100) 2°C / 4°C	Very low – Medium – Very high
The interaction of rising global mean sea level in the 21st century with high-water-level events will threaten low-lying coastal areas (*high confidence*) [29.4, Table 29-1; WGI AR5 13.5, Table 13.5]	• High ratio of coastal area to land mass will make adaptation a significant financial and resource challenge for islands. • Adaptation options include maintenance and restoration of coastal landforms and ecosystems, improved management of soils and freshwater resources, and appropriate building codes and settlement patterns.		Present Near term (2030–2040) Long term (2080–2100) 2°C / 4°C	Very low – Medium – Very high

Table TS.5 (continued)

The Ocean				
Key risk	**Adaptation issues & prospects**	**Climatic drivers**	**Timeframe**	**Risk & potential for adaptation**
Distributional shift in fish and invertebrate species, and decrease in fisheries catch potential at low latitudes, e.g., in equatorial upwelling and coastal boundary systems and sub-tropical gyres (*high confidence*) [6.3, 30.5, 30.6, Tables 6-6 and 30-3, Box CC-MB]	• Evolutionary adaptation potential of fish and invertebrate species to warming is limited as indicated by their changes in distribution to maintain temperatures. • Human adaptation options: Large-scale translocation of industrial fishing activities following the regional decreases (low latitude) vs. possibly transient increases (high latitude) in catch potential; Flexible management that can react to variability and change; Improvement of fish resilience to thermal stress by reducing other stressors such as pollution and eutrophication; Expansion of sustainable aquaculture and the development of alternative livelihoods in some regions.		Present Near term (2030–2040) Long term 2°C / 4°C (2080–2100)	Very low – Medium – Very high
Reduced biodiversity, fisheries abundance, and coastal protection by coral reefs due to heat-induced mass coral bleaching and mortality increases, exacerbated by ocean acidification, e.g., in coastal boundary systems and sub-tropical gyres (*high confidence*) [5.4, 6.4, 30.3, 30.5, 30.6, Tables 6-6 and 30-3, Box CC-CR]	• Evidence of rapid evolution by corals is very limited. Some corals may migrate to higher latitudes, but entire reef systems are not expected to be able to track the high rates of temperature shifts. • Human adaptation options are limited to reducing other stresses, mainly by enhancing water quality, and limiting pressures from tourism and fishing. These options will delay human impacts of climate change by a few decades, but their efficacy will be severely reduced as thermal stress increases.		Present Near term (2030–2040) Long term 2°C / 4°C (2080–2100)	Very low – Medium – Very high
Coastal inundation and habitat loss due to sea level rise, extreme events, changes in precipitation, and reduced ecological resilience, e.g., in coastal boundary systems and sub-tropical gyres (*medium* to *high confidence*) [5.5, 30.5, 30.6, Tables 6-6 and 30-3, Box CC-CR]	• Human adaptation options are limited to reducing other stresses, mainly by reducing pollution and limiting pressures from tourism, fishing, physical destruction, and unsustainable aquaculture. • Reducing deforestation and increasing reforestation of river catchments and coastal areas to retain sediments and nutrients • Increased mangrove, coral reef, and seagrass protection, and restoration to protect numerous ecosystem goods and services such as coastal protection, tourist value, and fish habitat		Present Near term (2030–2040) Long term 2°C / 4°C (2080–2100)	Very low – Medium – Very high

planning strategies, and resilient health infrastructure (*high confidence*). [26.3 to 26.6, 26.8]

***Central and South America*. Despite improvements, high and persistent levels of poverty in most countries result in high vulnerability to climate variability and change (*high confidence*).** Climate change impacts on agricultural productivity are expected to exhibit large spatial variability, for example with sustained or increased productivity through mid-century in southeast South America and decreases in productivity in the near term (by 2030) in Central America, threatening food security of the poorest populations (*medium confidence*). Reduced precipitation and increased evapotranspiration in semi-arid regions will increase risks from water-supply shortages, affecting cities, hydropower generation, and agriculture (*high confidence*). Ongoing adaptation strategies include reduced mismatch between water supply and demand, and water-management and coordination reforms (*medium confidence*). Conversion of natural ecosystems, a driver of anthropogenic climate change, is the main cause of biodiversity and ecosystem loss (*high confidence*). Climate change is expected to increase rates of species extinction (*medium confidence*). In coastal and marine systems, sea level rise and human stressors increase risks for fish stocks, corals, mangroves, recreation and tourism, and control of diseases (*high confidence*). Climate change will exacerbate future health risks given regional population growth rates and vulnerabilities due to pollution, food insecurity in poor regions, and existing health, water, sanitation, and waste collection systems (*medium confidence*). [27.2, 27.3]

***Polar Regions*. Climate change and often-interconnected non-climate-related drivers, including environmental changes, demography, culture, and economic development, interact in the Arctic to determine physical, biological, and socioeconomic risks, with rates of change that may be faster than social systems can adapt (*high confidence*).** Thawing permafrost and changing precipitation patterns have the potential to affect infrastructure and related services, with particular risks for residential buildings, for example in Arctic cities and small rural settlements. Climate change will especially impact Arctic communities that have narrowly based economies limiting adaptive choices. Increased Arctic navigability and expanded land- and freshwater-based transportation networks will increase economic opportunities. Impacts on the informal, subsistence-based economy will include changing sea ice conditions that increase the difficulty of hunting marine mammals. Polar bears have been and will be affected by loss of annual ice over continental shelves, decreased ice duration, and decreased ice thickness. Already, accelerated rates of change in permafrost thaw, loss of coastal sea ice, sea level rise, and increased intensity of weather extremes are forcing relocation of some indigenous communities in Alaska (*high confidence*). In the Arctic and Antarctic, some marine species will shift their ranges in response to changing ocean and sea ice conditions (*medium confidence*). Climate change will increase the vulnerability of terrestrial ecosystems to invasions by non-indigenous species (*high confidence*). [6.3, 6.5, 28.2 to 28.4]

***Small Islands*. Small islands have high vulnerability to climatic and non-climatic stressors (*high confidence*).** Diverse physical and human attributes and their sensitivity to climate-related drivers lead to variable climate change risk profiles and adaptation from one island region to another and among countries in the same region. Risks can originate from transboundary interactions, for example associated with existing and future invasive species and human health challenges. Sea level rise poses one of the most widely recognized climate change threats to low-lying coastal areas on islands and atolls. Projected sea level rise at the end of the 21st century, superimposed on extreme-sea-level events, presents severe coastal flooding and erosion risks for low-lying coastal areas and atoll islands. Wave over-wash will degrade groundwater resources. Coral reef ecosystem degradation associated with increasing sea surface temperature and ocean acidification will

Table TS.6 | Observed and projected future changes in some types of temperature and precipitation extremes over 26 sub-continental regions as defined in the IPCC *Special Report on Managing the Risks of Extreme Events and Disasters to Advance Climate Change Adaptation* (SREX). Confidence levels are indicated by symbol color. Likelihood terms are given only for *high* or *very high confidence* statements. Observed trends in temperature and precipitation extremes, including dryness and drought, are generally calculated from 1950, using 1961–1990 as the reference period, unless otherwise indicated. Future changes are derived from global and regional climate model projections for 2071–2100 compared with 1961–1990 or for 2080–2100 compared with 1980–2000. Table entries are summaries of information in SREX Tables 3-2 and 3-3 supplemented with or superseded by material from WGI AR5 2.6, 14.8, and Table 2.13 and WGII AR5 Table 25-1. The source(s) of information for each entry are indicated by superscripts: (a) SREX Table 3-2; (b) SREX Table 3-3; (c) WGI AR5 2.6 and Table 2.13; (d) WGI AR5 14.8; (e) WGII AR5 Table 25-1. [Tables 21-7 and SM21-2, Figure 21-4]

Region/ region code	Trends in daytime temperature extremes (frequency of hot and cool days)		Trends in heavy precipitation (rain, snow)		Trends in dryness and drought	
	Observed	**Projected**	**Observed**	**Projected**	**Observed**	**Projected**
West North America WNA, 3	*Very likely* large increases in hot days (large decreases in cool days)[a]	*Very likely* increase in hot days (decrease in cool days)[b]	Spatially varying trends. General increase, decrease in some areas[a]	Increase in 20-year return value of annual maximum daily precipitation and other metrics over northern part of the region (Canada)[b] Less confidence in southern part of the region, due to inconsistent signal in these other metrics[b]	No change or overall slight decrease in dryness[a]	Inconsistent signal[b]
Central North America CNA, 4	Spatially varying trends: small increases in hot days in the north, decreases in the south[a]	*Very likely* increase in hot days (decrease in cool days)[b]	*Very likely* increase since 1950[a]	Increase in 20-year return value of annual maximum daily precipitation[b] Inconsistent signal in other heavy precipitation days metrics[b]	*Likely* decrease[a, c]	Increase in consecutive dry days and soil moisture in southern part of central North America[b] Inconsistent signal in the rest of the region[b]
East North America ENA, 5	Spatially varying trends. Overall increases in hot days (decreases in cool days), opposite or insignificant signal in a few areas[a]	*Very likely* increase in hot days (decrease in cool days)[b]	*Very likely* increase since 1950[a]	Increase in 20-year return value of annual maximum daily precipitation. Additional metrics support an increase in heavy precipitation over northern part of the region.[b] No signal or inconsistent signal in these other metrics in the southern part of the region[b]	Slight decrease in dryness since 1950[a]	Inconsistent signal in consecutive dry days, some consistent decrease in soil moisture[b]
Alaska/ Northwest Canada ALA, 1	*Very likely* large increases in hot days (decreases in cool days)[a]	*Very likely* increase in hot days (decrease in cool days)[b]	Slight tendency for increase[a] No significant trend in southern Alaska[a]	*Likely* increase in heavy precipitation[b]	Inconsistent trends[a] Increase in dryness in part of the region[a]	Inconsistent signal[b]
East Canada, Greenland, Iceland CGI, 2	*Likely* increases in hot days (decreases in cool days) in some areas, decrease in hot days (increase in cool days) in others[a]	*Very likely* increase in hot days (decrease in cool days)[b]	Increase in a few areas[a]	*Likely* increase in heavy precipitation[b]	Insufficent evidence[a]	Inconsistent signal[b]
Northern Europe NEU, 11	Increase in hot days (decrease in cool days), but generally not significant at the local scale[a]	*Very likely* increase in hot days (decrease in cool days) [but smaller trends than in central and southern Europe][b]	Increase in winter in some areas, but often insignificant or inconsistent trends at sub-regional scale, particularly in summer[a]	*Likely* increase in 20-year return value of annual maximum daily precipitation. *Very likely* increases in heavy preciptation intensity and frequency in winter in the north[b]	Spatially varying trends. Overall only slight or no increase in dryness, slight decrease in dryness in part of the region[a]	No major changes in dryness[b]

Symbols

Increasing trend or signal

Decreasing trend or signal

Both increasing and decreasing trend or signal

Inconsistent trend or signal or insufficient evidence

No change or only slight change

Level of confidence in findings

Low confidence *Medium confidence* *High confidence*

Continued next page →

Table TS.6 (continued)

Region/ region code	Trends in daytime temperature extremes (frequency of hot and cool days)		Trends in heavy precipitation (rain, snow)		Trends in dryness and drought	
	Observed	Projected	Observed	Projected	Observed	Projected
Central Europe CEU, 12	*Likely* overall increase in hot days (decrease in cool days) in most regions. *Very likely* increase in hot days (*likely* decrease in cool days) in west-central Europe[a] Lower confidence in trends in east-central Europe (due to lack of literature, partial lack of access to observations, overall weaker signals, and change point in trends)[a]	*Very likely* increase in hot days (decrease in cool days)[b]	Increase in part of the region, in particular central western Europe and European Russia, especially in winter.[a] Insignificant or inconsistent trends elsewhere, in particular in summer[a]	*Likely* increase in 20-year return value of annual maximum daily precipitation. Additional metrics support an increase in heavy precipitation in large part of the region in winter.[b] Less confidence in summer, due to inconsistent evidence[b]	Spatially varying trends. Increase in dryness in part of the region but some regional variation in dryness trends and dependence of trends on studies considered (index, time period)[a]	Increase in dryness in central Europe and increase in short-term droughts[b]
Southern Europe and Mediterranean MED, 13	*Likely* increase in hot days (decrease in cool days) in most of the region. Some regional and temporal variations in the significance of the trends. *Likely* strongest and most significant trends in Iberian peninsula and southern France[a] Smaller or less significant trends in southeastern Europe and Italy due to change point in trends, strongest increase in hot days since 1976[a]	*Very likely* increase in hot days (decrease in cool days)[b]	Inconsistent trends across the region and across studies[a]	Inconsistent changes and/or regional variations[b]	Overall increase in dryness, *likely* increase in the Mediterranean[a, c]	Increase in dryness. Consistent increase in area of drought[b, d]
West Africa WAF, 15	Significant increase in temperature of hottest day and coolest day in some parts[a] Insufficient evidence in other parts[a]	*Likely* increase in hot days (decrease in cool days)[b]	Rainfall intensity increased[a]	Slight or no change in heavy precipitation indicators in most areas[b] Low model agreement in northern areas[b]	*Likely* increase but 1970s Sahel drought dominates the trend; greater inter-annual variation in recent years[a, c]	Inconsistent signal[b]
East Africa EAF, 16	Lack of evidence due to lack of literature and spatially non-uniform trends[a] Increases in hot days in southern tip (decreases in cool days)[a]	*Likely* increase in hot days (decrease in cool days)[b]	Insufficient evidence[a]	*Likely* increase in heavy precipitation[b]	Spatially varying trends in dryness[a]	Decreasing dryness in large areas[b]
Southern Africa SAF, 17	*Likely* increase in hot days (decrease in cool days)[a, c]	*Likely* increase in hot days (decrease in cool days)[b]	Increases in more regions than decreases but spatially varying trends[a, c]	Lack of agreement in signal for region as a whole[b] Some evidence of increase in heavy precipitation in southeast regions[b]	General increase in dryness[a]	Increase in dryness, except eastern part[b, d] Consistent increase in area of drought[b]
Sahara SAH, 14	Lack of literature[a]	*Likely* increase in hot days (decrease in cool days)[b]	Insufficient evidence[a]	*Low agreement*[b]	Limited data, spatial variation of the trends[a]	Inconsistent signal of change[b]
Central America and Mexico CAM, 6	Increases in the number of hot days, decreases in the number of cool days[a]	*Likely* increase in hot days (decrease in cool days)[b]	Spatially varying trends. Increase in many areas, decrease in a few others[a]	Inconsistent trends[b]	Varying and inconsistent trends[a]	Increase in dryness in Central America and Mexico, with less confidence in trend in extreme south of region[b]

Continued next page →

Table TS.6 (continued)

Region/ region code	Trends in daytime temperature extremes (frequency of hot and cool days)		Trends in heavy precipitation (rain, snow)		Trends in dryness and drought	
	Observed	Projected	Observed	Projected	Observed	Projected
Amazon AMZ, 7	Insufficient evidence to identify trends[a]	Hot days *likely* to increase (cool days *likely* to decrease)[b]	Increase in many areas, decrease in a few[a]	Tendency for increases in heavy precipitation events in some metrics[b]	Decrease in dryness for much of the region. Some opposite trends and inconsistencies[a]	Inconsistent signals[b]
Northeastern Brazil NEB, 8	Increases in the number of hot days[a]	Hot days *likely* to increase (cool days *likely* to decrease)[b]	Increase in many areas, decrease in a few[a]	Slight or no change[b]	Varying and inconsistent trends[a]	Increase in dryness[b]
Southeastern South America SSA, 10	Spatially varying trends (increases in hot days in some areas, decreases in others)[a]	Hot days *likely* to increase (cool days *likely* to decrease)[b]	Increase in northern areas[a] Insufficient evidence in southern areas[a]	Increases in northern areas[b] Insufficient evidence in southern areas[b]	Varying and inconsistent trends[a]	Inconsistent signals[b]
West Coast South America WSA, 9	Spatially varying trends (increases in hot days in some areas, decreases in others)[a]	Hot days *likely* to increase (cool days *likely* to decrease)[b]	Decrease in many areas, increase in a few areas[a]	Increases in tropics[b] *Low confidence* in extratropics[b]	Varying and inconsistent trends[a]	Decrease in consecutive dry days in the tropics, and increase in the extratropics[b] Increase in consecutive dry days and soil moisture in southwest South America[b]
North Asia NAS, 18	*Likely* increases in hot days (decreases in cool days)[a]	*Likely* increase in hot days (decrease in cool days)[b]	Increase in some regions, but spatial variation[a]	*Likely* increase in heavy precipitation for most regions[b]	Spatially varying trends[a]	Inconsistent signal of change[b]
Central Asia CAS, 20	*Likely* increases in hot days (decreases in cool days)[a]	*Likely* increase in hot days (decrease in cool days)[b]	Spatially varying trends[a]	Inconsistent signal in models[b]	Spatially varying trends[a]	Inconsistent signal of change[b]
East Asia EAS, 22	*Likely* increases in hot days (decreases in cool days)[a]	*Likely* increase in hot days (decrease in cool days)[b]	Spatially varying trends[a]	Increases in heavy precipitation across the region[b]	Tendency for increased dryness[a]	Inconsistent signal of change[b]
Southeast Asia SEA, 24	Increases in hot days (decreases in cool days) for northern areas[a] Insufficient evidence for Malay Archipelago[a]	*Likely* increase in hot days (decrease in cool days)[b]	Spatially varying trends, partial lack of evidence[a]	Increases in most metrics over most (especially non-continental) regions. One metric shows inconsistent signals of change.[b]	Spatially varying trends[a]	Inconsistent signal of change[b]
South Asia SAS, 23	Increase in hot days (decrease in cool days)[a]	*Likely* increase in hot days (decrease in cool days)[b]	Mixed signal in India[a]	More frequent and intense heavy precipitation days over parts of South Asia. Either no change or some consistent increases in other metrics[b]	Inconsistent signal for different studies and indices[a]	Inconsistent signal of change[b]
West Asia WAS, 19	*Very likely* increase in hot days (decrease in cool days *more likely than not*)[a]	*Likely* increase in hot days (decrease in cool days)[b]	Decrease in heavy precipitation events[a]	Inconsistent signal of change[b]	Lack of studies, mixed results[a]	Inconsistent signal of change[b]
Tibetan Plateau TIB, 21	*Likely* increase in hot days (decrease in cool days)[a]	*Likely* increase in hot days (decrease in cool days)[b]	Insufficient evidence[a]	Increase in heavy precipitation[b]	Insufficient evidence. Tendency to decreased dryness[a]	Inconsistent signal of change[b]
North Australia NAU, 25	*Likely* increase in hot days (decrease in cool days). Weaker trends in northwest[a]	*Very likely* increase in hot days (decrease in cool days)[b]	Spatially varying trends, which mostly reflect changes in mean rainfall[e]	Increase in most regions in the intensity of extreme (i.e., current 20-year return period) heavy rainfall events[e]	No significant change in drought occurrence over Australia (defined using rainfall anomalies)[e]	Inconsistent signal[b]

Continued next page →

Table TS.6 (continued)

Region/ region code	Trends in daytime temperature extremes (frequency of hot and cool days)		Trends in heavy precipitation (rain, snow)		Trends in dryness and drought	
	Observed	Projected	Observed	Projected	Observed	Projected
South Australia/ New Zealand SAU, 26	*Very likely* increase in hot days (decrease in cool days)[a]	*Very likely* increase in hot days (decrease in cool days)[b]	Spatially varying trends in southern Australia, which mostly reflect changes in mean rainfall[e] Spatially varying trends in New Zealand, which mostly reflect changes in mean rainfall[e]	Increase in most regions in the intensity of extreme (i.e., current 20-year return period) heavy rainfall events[e]	No significant change in drought occurrence over Australia (defined using rainfall anomalies)[e] No trend in drought occurrence over New Zealand (defined using a soil–water balance model) since 1972[e]	Increase in drought frequency in southern Australia, and in many regions of New Zealand[e]

negatively impact island communities and livelihoods, given the dependence of island communities on coral reef ecosystems for coastal protection, subsistence fisheries, and tourism. [29.3 to 29.5, 29.9, 30.5, Figure 29-1, Table 29-3, Box CC-CR]

***The Ocean*. Warming will increase risks to ocean ecosystems (*high confidence*).** Coral reefs within coastal boundary systems, semi-enclosed seas, and subtropical gyres are rapidly declining as a result of local non-climatic stressors (i.e., coastal pollution, overexploitation) and climate change. Projected increases in mass coral bleaching and mortality will alter or eliminate ecosystems, increasing risks to coastal livelihoods and food security (*medium* to *high confidence*). An analysis of the CMIP5 ensemble projects loss of coral reefs from most sites globally to be *very likely* by 2050 under mid to high rates of ocean warming. Reducing non-climatic stressors represents an opportunity to strengthen ecological resilience. The highly productive high-latitude spring bloom systems in the northeastern Atlantic are responding to warming (*medium evidence*, *high agreement*), with the greatest changes being observed since the late 1970s in the phenology, distribution, and abundance of plankton assemblages, and the reorganization of fish assemblages, with a range of consequences for fisheries (*high confidence*). Projected warming increases the likelihood of greater thermal stratification in some regions, which can lead to reduced O_2 ventilation and encourage the formation of hypoxic zones, especially in the Baltic and Black Seas (*medium confidence*). Changing surface winds and waves, sea level, and storm intensity will increase the vulnerability of ocean-based industries such as shipping, energy, and mineral extraction. New opportunities as well as international issues over access to resources and vulnerability may accompany warming waters particularly at high latitudes. [5.3, 5.4, 6.4, 28.2, 28.3, 30.3, 30.5, 30.6, Table 30-1, Figures 30-4 and 30-10, Boxes 6-1, CC-CR, and CC-MB]

Understanding of extreme events and their interactions with climate change is particularly important for managing risks in a regional context. Table TS.6 provides a summary of observed and projected trends in some types of temperature and precipitation extremes.

C: MANAGING FUTURE RISKS AND BUILDING RESILIENCE

Managing the risks of climate change involves adaptation and mitigation decisions with implications for future generations, economies, and environments. Figure TS.12 provides an overview of responses for addressing risk related to climate change.

Starting with principles for effective adaptation, this section evaluates the ways that interlinked human and natural systems can build resilience through adaptation, mitigation, and sustainable development. It describes understanding of climate-resilient pathways, of incremental versus transformational changes, and of limits to adaptation, and it considers co-benefits, synergies, and trade-offs among mitigation, adaptation, and development.

C-1. Principles for Effective Adaptation

The report assesses a wide variety of approaches for reducing and managing risks and building resilience. Strategies and approaches to climate change adaptation include efforts to decrease vulnerability or exposure and/or increase resilience or adaptive capacity. Mitigation is assessed in the WGIII AR5. Specific examples of responses to climate change are presented in Table TS.7.

Adaptation is place- and context-specific, with no single approach for reducing risks appropriate across all settings (*high confidence*). Effective risk reduction and adaptation strategies consider the dynamics of vulnerability and exposure and their linkages with socioeconomic processes, sustainable development, and climate change. [2.1, 8.3, 8.4, 13.1, 13.3, 13.4, 15.2, 15.3, 15.5, 16.2, 16.3, 16.5, 17.2, 17.4, 19.6, 21.3, 22.4, 26.8, 26.9, 29.6, 29.8]

Adaptation planning and implementation can be enhanced through complementary actions across levels, from individuals to governments (*high confidence*). National governments can coordinate adaptation efforts of local and subnational governments, for example by protecting vulnerable groups, by supporting economic diversification, and by providing information, policy and legal frameworks, and financial support (*robust evidence*, *high agreement*). Local government and the private sector are increasingly recognized as critical to progress in adaptation, given their roles in scaling up adaptation of communities, households, and civil society and in managing risk information and financing (*medium evidence*, *high agreement*). [2.1 to 2.4, 3.6, 5.5, 8.3, 8.4, 9.3, 9.4, 14.2, 15.2, 15.3, 15.5, 16.2 to 16.5, 17.2, 17.3, 22.4, 24.4, 25.4, 26.8, 26.9, 30.7, Tables 21-1, 21-5, and 21-6, Box 16-2]

A first step towards adaptation to future climate change is reducing vulnerability and exposure to present climate variability (*high confidence*). Strategies include actions with co-benefits for other objectives. Available strategies and actions can increase resilience across a range of possible future climates while helping to improve human health, livelihoods, social and economic well-being, and environmental quality. Examples of adaptation strategies that also strengthen livelihoods, enhance development, and reduce poverty include improved social protection, improved water and land governance, enhanced water storage and services, greater involvement in planning, and elevated attention to urban and peri-urban areas heavily affected by migration of poor people. See Table TS.7. [3.6, 8.3, 9.4, 14.3, 15.2, 15.3, 17.2, 20.4, 20.6, 22.4, 24.4, 24.5, 25.4, 25.10, 27.3 to 27.5, 29.6, Boxes 25-2 and 25-6]

Adaptation planning and implementation at all levels of governance are contingent on societal values, objectives, and risk perceptions (*high confidence*). Recognition of diverse interests, circumstances, social-cultural contexts, and expectations can

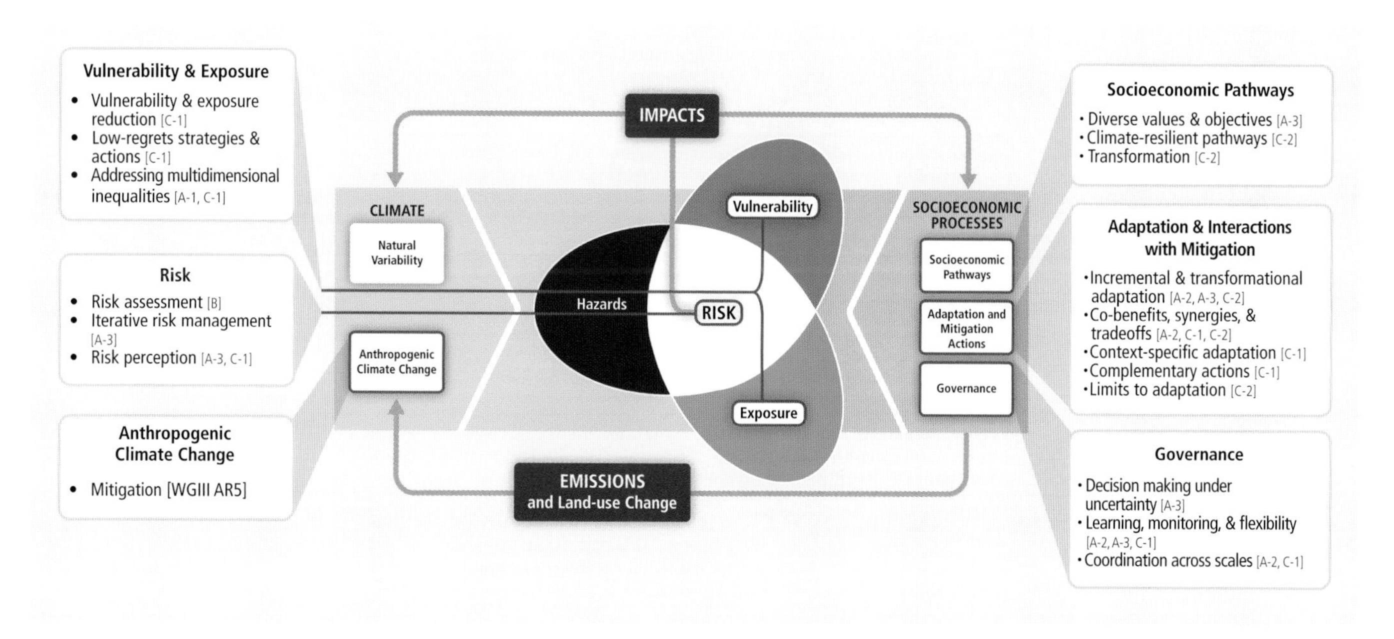

Figure TS.12 | The solution space. Core concepts of the WGII AR5, illustrating overlapping entry points and approaches, as well as key considerations, in managing risks related to climate change, as assessed in the report and presented throughout this summary. Bracketed references indicate sections of the summary with corresponding assessment findings.

Table TS.7 | Approaches for managing the risks of climate change. These approaches should be considered overlapping rather than discrete, and they are often pursued simultaneously. Mitigation is considered essential for managing the risks of climate change. It is not addressed in this table as mitigation is the focus of WGIII AR5. Examples are presented in no specific order and can be relevant to more than one category. [14.2, 14.3, Table 14-1]

Overlapping Approaches	Category	Examples	Chapter Reference(s)
Vulnerability & Exposure Reduction through development, planning, & practices including many low-regrets measures	Human development	Improved access to education, nutrition, health facilities, energy, safe housing & settlement structures, & social support structures; Reduced gender inequality & marginalization in other forms.	8.3, 9.3, 13.1 to 13.3, 14.2, 14.3, 22.4
	Poverty alleviation	Improved access to & control of local resources; Land tenure; Disaster risk reduction; Social safety nets & social protection; Insurance schemes.	8.3, 8.4, 9.3, 13.1 to 13.3
	Livelihood security	Income, asset, & livelihood diversification; Improved infrastructure; Access to technology & decision-making fora; Increased decision-making power; Changed cropping, livestock, & aquaculture practices; Reliance on social networks.	7.5, 9.4, 13.1 to 13.3, 22.3, 22.4, 23.4, 26.5, 27.3, 29.6, Table SM24-7
	Disaster risk management	Early warning systems; Hazard & vulnerability mapping; Diversifying water resources; Improved drainage; Flood & cyclone shelters; Building codes & practices; Storm & wastewater management; Transport & road infrastructure improvements.	8.2 to 8.4, 11.7, 14.3, 15.4, 22.4, 24.4, 26.6, 28.4, Table 3-3, Box 25-1
	Ecosystem management	Maintaining wetlands & urban green spaces; Coastal afforestation; Watershed & reservoir management; Reduction of other stressors on ecosystems & of habitat fragmentation; Maintenance of genetic diversity; Manipulation of disturbance regimes; Community-based natural resource management.	4.3, 4.4, 8.3, 22.4, Table 3-3, Boxes 4-3, 8-2, 15-1, 25-8, 25-9, & CC-EA
	Spatial or land-use planning	Provisioning of adequate housing, infrastructure, & services; Managing development in flood prone & other high risk areas; Urban planning & upgrading programs; Land zoning laws; Easements; Protected areas.	4.4, 8.1 to 8.4, 22.4, 23.7, 23.8, 27.3, Box 25-8
Adaptation including incremental & transformational adjustments	Structural/physical	***Engineered & built-environment options***: Sea walls & coastal protection structures; Flood levees; Water storage; Improved drainage; Flood & cyclone shelters; Building codes & practices; Storm & wastewater management; Transport & road infrastructure improvements; Floating houses; Power plant & electricity grid adjustments.	3.5, 3.6, 5.5, 8.2, 8.3, 10.2, 11.7, 23.3, 24.4, 25.7, 26.3, 26.8, Boxes 15-1, 25-1, 25-2, & 25-8
		Technological options: New crop & animal varieties; Indigenous, traditional, & local knowledge, technologies, & methods; Efficient irrigation; Water-saving technologies; Desalinization; Conservation agriculture; Food storage & preservation facilities; Hazard & vulnerability mapping & monitoring; Early warning systems; Building insulation; Mechanical & passive cooling; Technology development, transfer, & diffusion.	7.5, 8.3, 9.4, 10.3, 15.4, 22.4, 24.4, 26.3, 26.5, 27.3, 28.2, 28.4, 29.6, 29.7, Tables 3-3 & 15-1, Boxes 20-5 & 25-2
		Ecosystem-based options: Ecological restoration; Soil conservation; Afforestation & reforestation; Mangrove conservation & replanting; Green infrastructure (e.g., shade trees, green roofs); Controlling overfishing; Fisheries co-management; Assisted species migration & dispersal; Ecological corridors; Seed banks, gene banks, & other *ex situ* conservation; Community-based natural resource management.	4.4, 5.5, 6.4, 8.3, 9.4, 11.7, 15.4, 22.4, 23.6, 23.7, 24.4, 25.6, 27.3, 28.2, 29.7, 30.6, Boxes 15-1, 22-2, 25-9, 26-2, & CC-EA
		Services: Social safety nets & social protection; Food banks & distribution of food surplus; Municipal services including water & sanitation; Vaccination programs; Essential public health services; Enhanced emergency medical services.	3.5, 3.6, 8.3, 9.3, 11.7, 11.9, 22.4, 29.6, Box 13-2
	Institutional	***Economic options***: Financial incentives; Insurance; Catastrophe bonds; Payments for ecosystem services; Pricing water to encourage universal provision and careful use; Microfinance; Disaster contingency funds; Cash transfers; Public-private partnerships.	8.3, 8.4, 9.4, 10.7, 11.7, 13.3, 15.4, 17.5, 22.4, 26.7, 27.6, 29.6, Box 25-7
		Laws & regulations: Land zoning laws; Building standards & practices; Easements; Water regulations & agreements; Laws to support disaster risk reduction; Laws to encourage insurance purchasing; Defined property rights & land tenure security; Protected areas; Fishing quotas; Patent pools & technology transfer.	4.4, 8.3, 9.3, 10.5, 10.7, 15.2, 15.4, 17.5, 22.4, 23.4, 23.7, 24.4, 25.4, 26.3, 27.3, 30.6, Table 25-2, Box CC-CR
		National & government policies & programs: National & regional adaptation plans including mainstreaming; Sub-national & local adaptation plans; Economic diversification; Urban upgrading programs; Municipal water management programs; Disaster planning & preparedness; Integrated water resource management; Integrated coastal zone management; Ecosystem-based management; Community-based adaptation.	2.4, 3.6, 4.4, 5.5, 6.4, 7.5, 8.3, 11.7, 15.2 to 15.5, 22.4, 23.7, 25.4, 25.8, 26.8, 26.9, 27.3, 27.4, 29.6, Tables 9-2 & 17-1, Boxes 25-1, 25-2, & 25-9
	Social	***Educational options***: Awareness raising & integrating into education; Gender equity in education; Extension services; Sharing indigenous, traditional, & local knowledge; Participatory action research & social learning; Knowledge-sharing & learning platforms.	8.3, 8.4, 9.4, 11.7, 12.3, 15.2 to 15.4, 22.4, 25.4, 28.4, 29.6, Tables 15-1 & 25-2
		Informational options: Hazard & vulnerability mapping; Early warning & response systems; Systematic monitoring & remote sensing; Climate services; Use of indigenous climate observations; Participatory scenario development; Integrated assessments.	2.4, 5.5, 8.3, 8.4, 9.4, 11.7, 15.2 to 15.4, 22.4, 23.5, 24.4, 25.8, 26.6, 26.8, 27.3, 28.2, 28.5, 30.6, Table 25-2, Box 26-3
		Behavioral options: Household preparation & evacuation planning; Migration; Soil & water conservation; Storm drain clearance; Livelihood diversification; Changed cropping, livestock, & aquaculture practices; Reliance on social networks.	5.5, 7.5, 9.4, 12.4, 22.3, 22.4, 23.4, 23.7, 25.7, 26.5, 27.3, 29.6, Table SM24-7, Box 25-5
Transformation	Spheres of change	***Practical***: Social & technical innovations, behavioral shifts, or institutional & managerial changes that produce substantial shifts in outcomes.	8.3, 17.3, 20.5, Box 25-5
		Political: Political, social, cultural, & ecological decisions & actions consistent with reducing vulnerability & risk & supporting adaptation, mitigation, & sustainable development.	14.2, 14.3, 20.5, 25.4, 30.7, Table 14-1
		Personal: Individual & collective assumptions, beliefs, values, & worldviews influencing climate-change responses.	14.2, 14.3, 20.5, 25.4, Table 14-1

benefit decision-making processes. Awareness that climate change may exceed the adaptive capacity of some people and ecosystems may have ethical implications for mitigation decisions and investments. Economic analysis of adaptation is moving away from a unique emphasis on efficiency, market solutions, and benefit/cost analysis to include consideration of non-monetary and non-market measures, risks, inequities, behavioral biases, barriers and limits, and ancillary benefits and costs. [2.2 to 2.4, 9.4, 12.3, 13.2, 15.2, 16.2 to 16.4, 16.6, 16.7, 17.2, 17.3, 21.3, 22.4, 24.4, 24.6, 25.4, 25.8, 26.9, 28.2, 28.4, Table 15-1, Boxes 16-1, 16-4, and 25-7]

Indigenous, local, and traditional knowledge systems and practices, including indigenous peoples' holistic view of community and environment, are a major resource for adapting to climate change (*robust evidence, high agreement*). Natural resource dependent communities, including indigenous peoples, have a long history of adapting to highly variable and changing social and ecological conditions. But the salience of indigenous, local, and traditional knowledge will be challenged by climate change impacts. Such forms of knowledge have not been used consistently in existing adaptation efforts. Integrating such forms of knowledge with existing practices increases the effectiveness of adaptation. [9.4, 12.3, 15.2, 22.4, 24.4, 24.6, 25.8, 28.2, 28.4, Table 15-1]

Decision support is most effective when it is sensitive to context and the diversity of decision types, decision processes, and constituencies (*robust evidence, high agreement*). Organizations bridging science and decision making, including climate services, play an important role in the communication, transfer, and development of climate-related knowledge, including translation, engagement, and knowledge exchange (*medium evidence, high agreement*). [2.1 to 2.4, 8.4, 14.4, 16.2, 16.3, 16.5, 21.2, 21.3, 21.5, 22.4, Box 9-4]

Integration of adaptation into planning and decision making can promote synergies with development and disaster risk reduction (*high confidence*). Such mainstreaming embeds climate-sensitive thinking in existing and new institutions and organizations. Adaptation can generate larger benefits when connected with development activities and disaster risk reduction (*medium confidence*). [8.3, 9.3, 14.2, 14.6, 15.3, 15.4, 17.2, 20.2, 20.3, 22.4, 24.5, 29.6, Box CC-UR]

Existing and emerging economic instruments can foster adaptation by providing incentives for anticipating and reducing impacts (*medium confidence*). Instruments include public–private finance partnerships, loans, payments for environmental services, improved resource pricing, charges and subsidies, norms and regulations, and risk sharing and transfer mechanisms. Risk financing mechanisms in the public and private sector, such as insurance and risk pools, can contribute to increasing resilience, but without attention to major design challenges, they can also provide disincentives, cause market failure, and decrease equity. Governments often play key roles as regulators, providers, or insurers of last resort. [10.7, 10.9, 13.3, 17.4, 17.5, Box 25-7]

Constraints can interact to impede adaptation planning and implementation (*high confidence*). Common constraints on implementation arise from the following: limited financial and human resources; limited integration or coordination of governance; uncertainties about projected impacts; different perceptions of risks; competing values; absence of key adaptation leaders and advocates; and limited tools to monitor adaptation effectiveness. Another constraint includes insufficient research, monitoring, and observation and the finance to maintain them. Underestimating the complexity of adaptation as a social process can create unrealistic expectations about achieving intended adaptation outcomes. [3.6, 4.4, 5.5, 8.4, 9.4, 13.2, 13.3, 14.2, 14.5, 15.2, 15.3, 15.5, 16.2, 16.3, 16.5, 17.2, 17.3, 22.4, 23.7, 24.5, 25.4, 25.10, 26.8, 26.9, 30.6, Table 16-3, Boxes 16-1 and 16-3]

Poor planning, overemphasizing short-term outcomes, or failing to sufficiently anticipate consequences can result in maladaptation (*medium evidence, high agreement*). Maladaptation can increase the vulnerability or exposure of the target group in the future, or the vulnerability of other people, places, or sectors. Narrow focus on quantifiable costs and benefits can bias decisions against the poor, against ecosystems, and against those in the future whose values can be excluded or are understated. Some near-term responses to increasing risks related to climate change may also limit future choices. For example, enhanced protection of exposed assets can lock in dependence on further protection measures. [5.5, 8.4, 14.6, 15.5, 16.3, 17.2, 17.3, 20.2, 22.4, 24.4, 25.10, 26.8, Table 14-4, Box 25-1]

Limited evidence indicates a gap between global adaptation needs and funds available for adaptation (*medium confidence*). There is a need for a better assessment of global adaptation costs, funding, and investment. Studies estimating the global cost of adaptation are characterized by shortcomings in data, methods, and coverage (*high confidence*). [14.2, 17.4, Tables 17-2 and 17-3]

C-2. Climate-resilient Pathways and Transformation

Climate-resilient pathways are sustainable-development trajectories that combine adaptation and mitigation to reduce climate change and its impacts. They include iterative processes to ensure that effective risk management can be implemented and sustained. See Figure TS.13. [2.5, 20.3, 20.4]

Prospects for climate-resilient pathways for sustainable development are related fundamentally to what the world accomplishes with climate-change mitigation (*high confidence*). Since mitigation reduces the rate as well as the magnitude of warming, it also increases the time available for adaptation to a particular level of climate change, potentially by several decades. Delaying mitigation actions may reduce options for climate-resilient pathways in the future. [1.1, 19.7, 20.2, 20.3, 20.6, Figure 1-5]

Greater rates and magnitude of climate change increase the likelihood of exceeding adaptation limits (*high confidence*). See Box TS.8. Limits to adaptation occur when adaptive actions to avoid intolerable risks for an actor's objectives or for the needs of a system are not possible or are not currently available. Value-based judgments

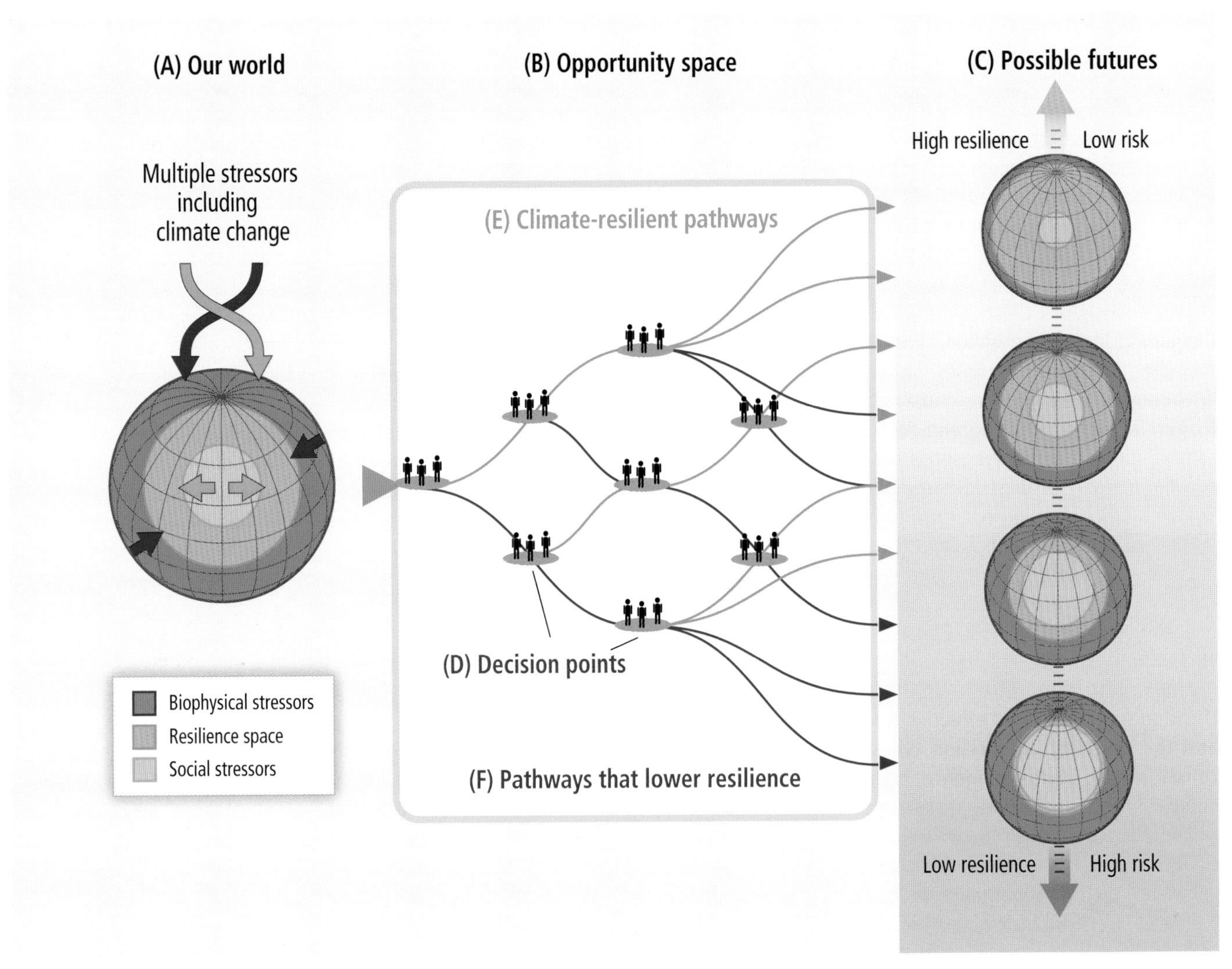

Figure TS.13 | Opportunity space and climate-resilient pathways. (A) Our world [Sections A-1 and B-1] is threatened by multiple stressors that impinge on resilience from many directions, represented here simply as biophysical and social stressors. Stressors include climate change, climate variability, land-use change, degradation of ecosystems, poverty and inequality, and cultural factors. (B) Opportunity space [Sections A-2, A-3, B-2, C-1, and C-2] refers to decision points and pathways that lead to a range of (C) possible futures [Sections C and B-3] with differing levels of resilience and risk. (D) Decision points result in actions or failures-to-act throughout the opportunity space, and together they constitute the process of managing or failing to manage risks related to climate change. (E) Climate-resilient pathways (in green) within the opportunity space lead to a more resilient world through adaptive learning, increasing scientific knowledge, effective adaptation and mitigation measures, and other choices that reduce risks. (F) Pathways that lower resilience (in red) can involve insufficient mitigation, maladaptation, failure to learn and use knowledge, and other actions that lower resilience; and they can be irreversible in terms of possible futures. [Figure 1-5]

of what constitutes an intolerable risk may differ. Limits to adaptation emerge from the interaction among climate change and biophysical and/or socioeconomic constraints. Opportunities to take advantage of positive synergies between adaptation and mitigation may decrease with time, particularly if limits to adaptation are exceeded. In some parts of the world, insufficient responses to emerging impacts are already eroding the basis for sustainable development. [1.1, 11.8, 13.4, 16.2 to 16.7, 17.2, 20.2, 20.3, 20.5, 20.6, 25.10, 26.5, Boxes 16-1, 16-3, and 16-4]

Transformations in economic, social, technological, and political decisions and actions can enable climate-resilient pathways (*high confidence*). Specific examples are presented in Table TS.7. See also Box TS.8. Strategies and actions can be pursued now that will move towards climate-resilient pathways for sustainable development, while at the same time helping to improve livelihoods, social and economic well-being, and responsible environmental management. Transformations in response to climate change may involve, for example, introduction of new technologies or practices, formation of new structures or systems of governance, or shifts in the types or locations of activities. The scale and magnitude of transformational adaptations depend on mitigation and on development processes. Transformational adaptation is an important consideration for decisions involving long life- or lead-times, and it can be a response to adaptation limits. At the national level, transformation is considered most effective when it reflects a country's own visions and approaches to achieving sustainable development in accordance with its national circumstances and priorities. Transformations to sustainability are considered to benefit from iterative learning, deliberative processes, and innovation. Societal debates about many aspects of transformation may place new and increased demands on governance structures. [1.1, 2.1, 2.5, 8.4, 14.1, 14.3, 16.2 to 16.7, 20.5, 22.4, 25.4, 25.10, Figure 1-5, Boxes 16-1 and 16-4]

Examples of Co-benefits, Synergies, and Trade-offs among Adaptation, Mitigation, and Sustainable Development

Significant co-benefits, synergies, and trade-offs exist between mitigation and adaptation and among different adaptation responses; interactions occur both within and across regions (*very high confidence*). Illustrative examples include the following.

- Increasing efforts to mitigate and adapt to climate change imply an increasing complexity of interactions, particularly at the intersections among water, energy, land use, and biodiversity, but tools to understand and manage these interactions remain limited (*very high confidence*). See Box TS.9. Widespread transformation of terrestrial ecosystems in order to mitigate climate change, such as carbon sequestration through planting fast-growing tree species into ecosystems where they did not previously occur, or the conversion of previously uncultivated or non-degraded land to bioenergy plantations, can lead to negative impacts on ecosystems and biodiversity (*high confidence*). [3.7, 4.2 to 4.4, 22.6, 24.6, 25.7, 25.9, 27.3, Boxes 25-10 and CC-WE]
- Climate policies such as increasing energy supply from renewable resources, encouraging bioenergy crop cultivation, or facilitating payments under REDD+ will affect some rural areas both positively (e.g., increasing employment opportunities) and negatively (e.g., land use changes, increasing scarcity of natural capital) (*medium confidence*). These secondary impacts, and trade-offs between mitigation and adaptation in rural areas, have implications for governance, including benefits of promoting participation of rural stakeholders. Mitigation policies with social co-benefits expected in their design, such as CDM and REDD+, have had limited or no effect in terms of poverty alleviation and sustainable development

TS

Box TS.8 | Adaptation Limits and Transformation

Adaptation can expand the capacity of natural and human systems to cope with a changing climate. Risk-based decision making can be used to assess potential limits to adaptation. Limits to adaptation occur when adaptive actions to avoid intolerable risks for an actor's objectives or for the needs of a system are not possible or are not currently available. Limits to adaptation are context-specific and closely linked to cultural norms and societal values. Value-based judgments of what constitutes an intolerable risk may differ among actors, but understandings of limits to adaptation can be informed by historical experiences, or by anticipation of impacts, vulnerability, and adaptation associated with different scenarios of climate change. The greater the magnitude or rate of climate change, the greater the likelihood that adaptation will encounter limits. [16.2 to 16.4, 20.5, 20.6, 22.4, 25.4, 25.10, Box 16-2]

Limits to adaptation may be influenced by the subjective values of societal actors, which can affect both the perceived need for adaptation and the perceived appropriateness of specific policies and measures. While limits imply that intolerable risks and the increased potential for losses and damages can no longer be avoided, the dynamics of social and ecological systems mean that there are both "soft" and "hard" limits to adaptation. For "soft" limits, there are opportunities in the future to alter limits and reduce risks, for example, through the emergence of new technologies or changes in laws, institutions, or values. In contrast, "hard" limits are those where there are no reasonable prospects for avoiding intolerable risks. Recent studies on tipping points, key vulnerabilities, and planetary boundaries provide some insights on the behavior of complex systems. [16.2 to 16.7, 25.10]

In cases where the limits to adaptation have been surpassed, losses and damage may increase and the objectives of some actors may no longer be achievable. There may be a need for transformational adaptation to change fundamental attributes of a system in response to actual or expected impacts of climate change. It may involve adaptations at a greater scale or intensity than previously experienced, adaptations that are new to a region or system, or adaptations that transform places or lead to a shift in the types or locations of activities. [16.2 to 16.4, 20.3, 20.5, 22.4, 25.10, Boxes 25-1 and 25-9]

The existence of limits to adaptation suggests transformational change may be a requirement for sustainable development in a changing climate—that is, not only for adapting to the impacts of climate change, but for altering the systems and structures, economic and social relations, and beliefs and behaviors that contribute to climate change and social vulnerability. However, just as there are ethical implications associated with some adaptation options, there are also legitimate concerns about the equity and ethical dimensions of transformation. Societal debates over risks from forced and reactive transformations as opposed to deliberate transitions to sustainability may place new and increased demands on governance structures at multiple levels to reconcile conflicting goals and visions for the future. [1.1, 16.2 to 16.7, 20.5, 25.10] (*medium confidence*). Mitigation efforts focused on land acquisition for biofuel production show preliminary negative impacts for the poor in many developing countries, and particularly for indigenous people and (women) smallholders. [9.3, 13.3, 22.6]

- Mangrove, seagrass, and salt marsh ecosystems offer important carbon storage and sequestration opportunities (*limited evidence, medium agreement*), in addition to ecosystem goods and services such as protection against coastal erosion and storm damage and maintenance of habitats for fisheries species. For ocean-related mitigation and adaptation in the context of anthropogenic ocean warming and acidification, international frameworks offer opportunities to solve problems collectively, for example, managing fisheries across national borders and responding to extreme events. [5.4, 25.6, 30.6, 30.7]

Table TS.8 | Illustrative examples of intra-regional interactions among adaptation, mitigation, and sustainable development.

Green infrastructure and green roofs	
Objectives	Storm water management, adaptation to increasing temperatures, reduced energy use, urban regeneration
Relevant sectors	Infrastructure, energy use, water management
Overview	Benefits of green infrastructure and roofs can include reduction of storm water runoff and the urban heat island effect, improved energy performance of buildings, reduced noise and air pollution, health improvements, better amenity value, increased property values, improved biodiversity, and inward investment. Trade-offs can result between higher urban density to improve energy efficiency and open space for green infrastructure. [8.3.3, 11.7.4, 23.7.4, 24.6, Tables 11-3 and 25-5]
Examples with interactions	**London:** The Green Grid for East London seeks to create interlinked and multi-purpose open spaces to support regeneration of the area. It aims to connect people and places, to absorb and store water, to cool the vicinity, and to provide a diverse mosaic of habitats for wildlife. [8.3.3] **New York:** In preparation for more intense storms, New York is using green infrastructure to capture rainwater before it can flood the combined sewer system, implementing green roofs, and elevating boilers and other equipment above ground. [8.3.3, 26.3.3, 26.8.4] **Singapore:** Singapore has used several anticipatory plans and projects to enhance green infrastructure, including its Streetscape Greenery Master Plan, constructed wetlands or drains, and community gardens. Under its Skyrise Greenery project, Singapore has provided subsidies and handbooks for rooftop and wall greening initiatives. [8.3.3] **Durban:** Ecosystem-based adaptation is part of Durban's climate change adaptation strategy. The approach seeks a more detailed understanding of the ecology of indigenous ecosystems and ways in which biodiversity and ecosystem services can reduce vulnerability of ecosystems and people. Examples include the Community Reforestation Programme, in which communities produce indigenous seedlings used in the planting and managing of restored forest areas. Development of ecosystem-based adaptation in Durban has demonstrated needs for local knowledge and data and the benefits of enhancing existing protected areas, land-use practices, and local initiatives contributing to jobs, business, and skill development. [8.3.3, Box 8-2]
Water management	
Primary objective	Water resource management given multiple stressors in a changing climate
Relevant sectors	Water use, energy production and use, biodiversity, carbon sequestration, biofuel production, food production
Overview	Water management in the context of climate change can encompass ecosystem-based approaches (e.g., watershed management or restoration, flood regulation services, and reduction of erosion or siltation), supply-side approaches (e.g., dams, reservoirs, groundwater pumping and recharge, and water capture), and demand-side approaches (e.g., increased use efficiency through water recycling, infrastructure upgrades, water-sensitive design, or more efficient allocation). Water may require significant amounts of energy for lifting, transport, distribution, and treatment. [3.7.2, 26.3, Tables 9-8 and 25-5, Boxes CC-EA and CC-WE]
Examples with interactions	**New York:** New York has a well-established program to protect and enhance its water supply through watershed protection. The Watershed Protection Program includes city ownership of land that remains undeveloped and coordination with landowners and communities to balance water-quality protection, local economic development, and improved wastewater treatment. The city government indicates it is the most cost-effective choice for New York given the costs and environmental impacts of a filtration plant. [8.3.3, Box 26-3] **Cape Town:** Facing challenges in ensuring future supplies, Cape Town responded by commissioning water management studies, which identified the need to incorporate climate change, as well as population and economic growth, in planning. During the 2005 drought, local authorities increased water tariffs to promote efficient water usage. Additional measures may include water restrictions, reuse of gray water, consumer education, or technological solutions such as low-flow systems or dual flush toilets. [8.3.3] **Capital cities in Australia:** Many Australian capital cities are reducing reliance on catchment runoff and groundwater—water resources most sensitive to climate change and drought—and are diversifying supplies through desalination plants, water reuse including sewage and storm water recycling, and integrated water cycle management that considers climate change impacts. Demand is being reduced through water conservation and water-sensitive urban design and, during severe shortfalls, through implementation of restrictions. The water augmentation program in Melbourne includes a desalination plant. Trade-offs beyond energy intensiveness have been noted, such as damage to sites significant to aboriginal communities and higher water costs that will disproportionately affect poorer households. [14.6.2, Tables 25-6 and 25-7, Box 25-2]
Payment for environmental services and green fiscal policies	
Primary objective	Management incorporating the costs of environmental externalities and the benefits of ecosystem services
Relevant sectors	Biodiversity, ecosystem services
Overview	Payment for environmental services (PES) is a market-based approach that aims to protect natural areas, and associated livelihoods and environmental services, by developing financial incentives for preservation. Mitigation-focused PES schemes are common, and there is emerging evidence of adaptation-focused PES schemes. Successful PES approaches can be difficult to design for services that are hard to define or quantify. [17.5.2, 27.6.2]
Examples with interactions	**Central and South America:** A variety of PES schemes have been implemented in Central and South America. For example, national-level programs have operated in Costa Rica and Guatemala since 1997 and in Ecuador since 2008. Examples to date have shown that PES can finance conservation, ecosystem restoration and reforestation, better land-use practices, mitigation, and more recently adaptation. Uniform payments for beneficiaries can be inefficient if, for example, recipients that promote greater environmental gains receive only the prevailing payment. [17.5.2, 27.3.2, 27.6.2, Table 27-8] **Brazil:** Municipal funding in Brazil tied to ecosystem-management quality is a form of revenue transfer important to funding local adaptation actions. State governments collect a value-added tax redistributed among municipalities, and some states allocate revenues in part based on municipality area set aside for protection. This mechanism has helped improve environmental management and increased creation of protected areas. It benefits relations between protected areas and surrounding inhabitants, as the areas can be perceived as opportunities for revenue generation rather than as obstacles to development. The approach builds on existing institutions and administrative procedures and thus has low transaction costs. [8.4.3, Box 8-4]

Continued next page →

Table TS.8 (continued)

Renewable energy	
Primary objective	Renewable energy production and reduction of emissions
Relevant sectors	Biodiversity, agriculture, food security
Overview	Renewable energy production can require significant land areas and water resources, creating the potential for both positive and negative interactions between mitigation policies and land management. [4.4.4, 13.3.1, 19.3.2, 19.4.1, Box CC-WE]
Examples with interactions	**Central and South America:** Renewable resources, especially hydroelectric power and biofuels, account for substantial fractions of energy production in countries such as Brazil. Where bioenergy crops compete for land with food crops, substantial trade-offs can exist. Land-use change to produce bioenergy can affect food crops, biodiversity, and ecosystem services. Lignocellulosic feedstocks, such as sugarcane second-generation technologies, do not compete with food. [19.3.2, 27.3.6, 27.6.1, Table 27-6] **Australia and New Zealand:** Mandatory renewable energy targets and incentives to increase carbon storage support increased biofuel production and increased biological carbon sequestration, with impacts on biodiversity depending on implementation. Benefits can include reduced erosion, additional habitat, and enhanced connectivity, with risks or lost opportunities associated with large-scale monocultures especially if replacing more diverse landscapes. Large-scale land cover changes can affect catchment yields and regional climate in complex ways. New crops such as oil mallees or other eucalypts may provide multiple benefits, especially in marginal areas, displacing fossil fuels or sequestering carbon, generating income for landholders (essential oils, charcoal, bio-char, biofuels), and providing ecosystem services. [Table 25-7, Box 25-10]

Disaster risk reduction and adaptation to climate extremes	
Primary objective	Increasing resilience to extreme weather events in a changing climate
Relevant sectors	Infrastructure, energy use, spatial planning
Overview	Synergies and tradeoffs among sustainable development, adaptation, and mitigation occur in preparing for and responding to climate extremes and disasters. [13.2 to 13.4, 20.3, 20.4]
Examples with interactions	**Philippines:** The Homeless People's Federation of the Philippines developed responses following disasters, including community-rooted data gathering (e.g., assessing destruction and victims' immediate needs); trust and contact building; savings support; community-organization registration; and identification of needed interventions (e.g., building-materials loans). Community surveys mapped inhabitants especially at risk in informal settlements, raising risk-awareness among the inhabitants and increasing community engagement in planning risk reduction and early warning systems. [8.3.2, 8.4.2] **London:** Within London, built form and other dwelling characteristics can have a stronger influence on indoor temperatures during heat waves than the urban heat island effect, and utilizing shade, thermal mass, ventilation control, and other passive-design features are effective adaptation options. Passive housing designs enhance natural ventilation and improve insulation, while also reducing household emissions. For example, in London the Beddington Zero Energy Development was designed to reduce or eliminate energy demand for heating, cooling, and ventilation for much of the year. [8.3.3, 11.7.4] **United States:** In the United States, post-disaster funds for loss reduction are added to funds provided for disaster recovery. They can be used, for instance, to buy out properties that have experienced repetitive flood losses and relocate residents to safer locations, to elevate structures, to assist communities with purchasing property and altering land-use patterns in flood-prone areas, and to undertake other activities designed to lessen the impacts of future disasters. [14.3.3]

- Geoengineering approaches involving manipulation of the ocean to ameliorate climate change (such as nutrient fertilization, binding of CO_2 by enhanced alkalinity, or direct CO_2 injection into the deep ocean) have very large environmental and associated socioeconomic consequences (*high confidence*). Alternative methods focusing on solar radiation management (SRM) leave ocean acidification unabated as they cannot mitigate rising atmospheric CO_2 emissions. [6.4]
- Some agricultural practices can reduce emissions and also increase resilience of crops to temperature and rainfall variability (*high confidence*). [23.8, Table 25-7]
- Many solutions for reducing energy and water consumption in urban areas with co-benefits for climate change adaptation (e.g., greening cities and recycling water) are already being implemented (*high confidence*). Transport systems promoting active transport and reduced motorized-vehicle use can improve air quality and increase physical activity (*medium confidence*). [11.9, 23.8, 24.4, 26.3, 26.8, Boxes 25-2 and 25-9]
- Improved energy efficiency and cleaner energy sources can lead to reduced emissions of health-damaging climate-altering air pollutants (*very high confidence*). [11.9, 23.8]
- In Africa, experience in implementing integrated adaptation–mitigation responses that leverage developmental benefits encompasses some participation of farmers and local communities in carbon offset systems and increased use of agroforestry and farmer-assisted tree regeneration (*high confidence*). [22.4, 22.6]
- In Asia, development of sustainable cities with fewer fossil-fuel-driven vehicles and with more trees and greenery would have a number of co-benefits, including improved public health (*high confidence*). [24.4 to 24.7]
- In Australasia, transboundary effects from climate change impacts and responses outside Australasia have the potential to outweigh some of the direct impacts within the region, particularly economic impacts on trade-intensive sectors such as agriculture (*medium confidence*) and tourism (*limited evidence, high agreement*), but they remain among the least-explored issues. [25.7, 25.9, Box 25-10]
- In North America, policies addressing local concerns (e.g., air pollution, housing for the poor, declines in agricultural production) can be adapted at low or no cost to fulfill adaptation, mitigation, and sustainability goals (*medium confidence*). [26.9]
- In Central and South America, biomass-based renewable energy can impact land use change and deforestation, and could be affected by climate change (*medium confidence*). The expansion of sugarcane, soy, and oil palm may have some effect on land use, leading to deforestation in parts of the Amazon and Central America, among other sub-regions, and to loss of employment in some countries. [27.3]
- For small islands, energy supply and use, tourism infrastructure and activities, and coastal wetlands offer opportunities for adaptation–mitigation synergies (*medium confidence*). [29.6 to 29.8]

Table TS.8 provides further specific examples of interactions among adaptation, mitigation, and sustainable development to complement the assessment findings above.

Box TS.9 | The Water–Energy–Food Nexus

Water, energy, and food/feed/fiber are linked through numerous interactive pathways affected by a changing climate (Box TS.9 Figure 1). [Box CC-WE] The depth and intensity of those linkages vary enormously among countries, regions, and production systems. Many energy sources require significant amounts of water and produce a large quantity of wastewater that requires energy for treatment. [3.7, 7.3, 10.2, 10.3, 22.3, 25.7, Box CC-WE] Food production, refrigeration, transport, and processing also require both energy and water. A major link between food and energy as related to climate change is the competition of bioenergy and food production for land and water, and the sensitivity of precipitation, temperature, and crop yields to climate change (*robust evidence*, *high agreement*). [7.3, Boxes 25-10 and CC-WE]

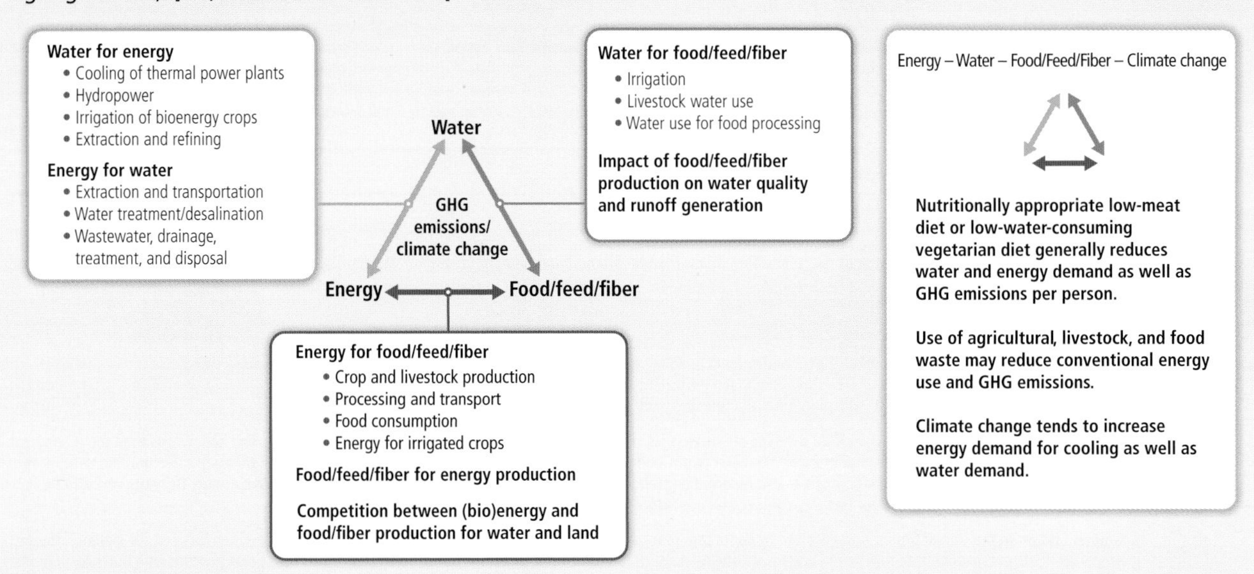

Box TS.9 Figure 1 | The water–energy–food nexus as related to climate change, with implications for both adaptation and mitigation strategies. [Figure WE-1, Box CC-WE]

Most energy production methods require significant amounts of water, either directly (e.g., crop-based energy sources and hydropower) or indirectly (e.g., cooling for thermal energy sources or other operations) (*robust evidence*, *high agreement*). [10.2, 10.3, 25.7, Box CC-WE] Water is required for mining, processing, and residue disposal of fossil fuels or their byproducts. [25.7] Water for energy currently ranges from a few percent in most developing countries to more than 50% of freshwater withdrawals in some developed countries, depending on the country. [Box CC-WE] Future water requirements will depend on electric demand growth, the portfolio of generation technologies, and water management options (*medium evidence*, *high agreement*). Future water availability for energy production will change due to climate change (*robust evidence*, *high agreement*). [3.4, 3.5, Box CC-WE]

Energy is also required to supply and treat water. Water may require significant amounts of energy for lifting (especially as aquifers continue to be depleted), transport, and distribution and for its treatment either to use it or to depollute it. Wastewater and even excess rainfall in cities requires energy to be treated or disposed. Some non-conventional water sources (wastewater or seawater) are often highly energy intensive. [Table 25-7, Box 25-2] Energy intensities per cubic meter of water vary by about a factor of 10 among different sources, for example, locally produced potable water from ground/surface water sources versus desalinated seawater. [Boxes 25-2 and CC-WE] Groundwater is generally more energy intensive than surface water. [Box CC-WE]

Linkages among water, energy, food/feed/fiber, and climate are strongly related to land use and management, such as afforestation, which can affect water as well as other ecosystem services, climate, and water cycles (*robust evidence*, *high agreement*). Land degradation often reduces efficiency of water and energy use (e.g., resulting in higher fertilizer demand and surface runoff), and many of these interactions can compromise food security. On the other hand, afforestation activities to sequester carbon have important co-benefits of reducing soil erosion and providing additional (even if only temporary) habitat, but may reduce renewable water resources. [3.7, 4.4, Boxes 25-10 and CC-WE]

Consideration of the interlinkages of energy, food/feed/fiber, water, land use, and climate change has implications for security of supplies of energy, food, and water; adaptation and mitigation pathways; air pollution reduction; and health and economic impacts. This nexus is increasingly recognized as critical to effective climate-resilient-pathway decision making (*medium evidence*, *high agreement*), although tools to support local- and regional-scale assessments and decision support remain very limited.

Working Group II
Frequently Asked Questions

Working Group II Frequently Asked Questions

These FAQs provide an entry point to the approach and scientific findings of the Working Group II contribution to the Fifth Assessment Report. For summary of the scientific findings, see the Summary for Policymakers (SPM) and Technical Summary (TS). These FAQs, presented in clear and accessible language, do not reflect formal assessment of the degree of certainty in conclusions, and they do not include calibrated uncertainty language presented in the SPM, TS, and underlying chapters. The sources of the relevant assessment in the report are noted by chapter numbers in square brackets.

FAQ 1: Are risks of climate change mostly due to changes in extremes, changes in average climate, or both?

[Chapters 3, 4, 5, 6, 7, 8, 9, 10, 11, 12, 13, 18, 19, 22, 23, 24, 25, 26, 27, 28, 29, 30; TS]

People and ecosystems across the world experience climate in many different ways, but weather and climate extremes strongly influence losses and disruptions. Average climate conditions are important. They provide a starting point for understanding what grows where and for informing decisions about tourist destinations, other business opportunities, and crops to plant. But the impacts of a change in average conditions often occur as a result of changes in the frequency, intensity, or duration of extreme weather and climate events. It is the extremes that place excessive and often unexpected demands on systems poorly equipped to deal with those extremes. For example, wet conditions lead to flooding when storm drains and other infrastructure for handling excess water are overwhelmed. Buildings fail when wind speeds exceed design standards. For many kinds of disruption, from crop failure caused by drought to sickness and death from heat waves, the main risks are in the extremes, with changes in average conditions representing a climate with altered timing, intensity, and types of extremes.

FAQ 2: How much can we say about what society will be like in the future, in order to plan for climate change impacts?

[Chapters 1, 2, 14, 15, 16, 17, 20, and 21; TS]

Overall characteristics of societies and economies, such as population size, economic activity, and land use, are highly dynamic. On the scale of just 1 or 2 decades, and sometimes in less time than that, technological revolutions, political movements, or singular events can shape the course of history in unpredictable ways. To understand potential impacts of climate change for societies and ecosystems, scientists use scenarios to explore implications of a range of possible futures. Scenarios are not predictions of what will happen, but they can be useful tools for researching a wide range of "what if" questions about what the world might be like in the future. They can be used to study future emissions of greenhouse gases and climate change. They can also be used to explore the ways climate-change impacts depend on changes in society, such as economic or population growth or progress in controlling diseases. Scenarios of possible decisions and policies can be used to explore the solution space for reducing greenhouse gas emissions and preparing for a changing climate. Scenario analysis creates a foundation for understanding risks of climate change for people, ecosystems, and economies across a range of possible futures. It provides important tools for smart decision making when both uncertainties and consequences are large.

FAQ 3: Why is climate change a particularly difficult challenge for managing risk?

[Chapters 1, 2, 16, 17, 19, 20, 21, and 25; TS]

Risk management is easier for nations, companies, and even individuals when the likelihood and consequences of possible events are readily understood. Risk management becomes much more challenging when the stakes are higher or when uncertainty is greater. As the WGII AR5 demonstrates, we know a great deal about the impacts of climate change that have already occurred, and we understand a great deal about expected impacts in the future. But many uncertainties remain, and will persist. In particular, future greenhouse gas emissions depend on societal choices, policies, and technology advancements not yet made, and climate-change impacts depend on both the amount of climate change that occurs and the effectiveness of development in reducing exposure and vulnerability. The real challenge of dealing effectively with climate change is recognizing the value of wise and timely decisions in a setting where complete knowledge is impossible. This is the essence of risk management.

FAQ 4: What are the timeframes for mitigation and adaptation benefits?

[Chapters 1, 2, 16, 19, 20, and 21; TS]

Adaptation can reduce damage from impacts that cannot be avoided. Mitigation strategies can decrease the amount of climate change that occurs, as summarized in the WGIII AR5. But the consequences of investments in mitigation emerge over time. The constraints of existing infrastructure, limited deployment of many clean technologies, and the legitimate aspirations for economic growth around the world all tend to slow the deviation from established trends in greenhouse gas emissions. Over the next few decades, the climate change we experience will be determined primarily by the combination of past actions and current trends. The near-term is thus an era where short-term risk reduction comes from adapting to the changes already underway. Investments in mitigation during both the near-term and the longer-term do, however, have substantial leverage on the magnitude of climate change in the latter decades of the century, making the second half of the 21st century and beyond an era of climate options. Adaptation will still be important during the era of climate options, but with opportunities and needs that will depend on many aspects of climate change and development policy, both in the near term and in the long term.

FAQ 5: Can science identify thresholds beyond which climate change is dangerous?

[Chapters 1, 2, 4, 5, 6, 16, 17, 18, 19, 20, and 25; TS]

Human activities are changing the climate. Climate-change impacts are already widespread and consequential. But while science can quantify climate change risks in a technical sense, based on the probability, magnitude, and nature of the potential consequences of climate change, determining what is dangerous is ultimately a judgment that depends on values and objectives. For example, individuals will value the present versus the future differently and will bring personal worldviews on the importance of assets like biodiversity, culture, and aesthetics. Values also influence judgments about the relative importance of global economic growth versus assuring the well-being of the most vulnerable among us. Judgments about dangerousness can depend on the extent to which one's livelihood, community, and family are directly exposed and vulnerable to climate change. An individual or community displaced by climate change might legitimately consider that specific impact dangerous, even though that single impact might not cross the global threshold of dangerousness. Scientific assessment of risk can provide an important starting point for such value judgments about the danger of climate change.

FAQ 6: Are we seeing impacts of recent climate change?
[Chapters 3, 4, 5, 6, 7, 11, 13, 18, 22, 23, 24, 25, 26, 27, 28, 29, and 30; SPM]
Yes, there is strong evidence of impacts of recent observed climate change on physical, biological, and human systems. Many regions have experienced warming trends and more frequent high-temperature extremes. Rising temperatures are associated with decreased snowpack, and many ecosystems are experiencing climate-induced shifts in the activity, range, or abundance of the species that inhabit them. Oceans are also displaying changes in physical and chemical properties that, in turn, are affecting coastal and marine ecosystems such as coral reefs, and other oceanic organisms such as mollusks, crustaceans, fishes, and zooplankton. Crop production and fishery stocks are sensitive to changes in temperature. Climate change impacts are leading to shifts in crop yields, decreasing yields overall and sometimes increasing them in temperate and higher latitudes, and catch potential of fisheries is increasing in some regions but decreasing in others. Some indigenous communities are changing seasonal migration and hunting patterns to adapt to changes in temperature.

FAQ 7: Are the future impacts of climate change only negative? Might there be positive impacts as well?
[Chapters 3, 4, 5, 6, 7, 8, 9, 10, 11, 12, 13, 19, 22, 23, 24, 25, 26, 27, and 30]
Overall, the report identifies many more negative impacts than positive impacts projected for the future, especially for high magnitudes and rates of climate change. Climate change will, however, have different impacts on people around the world and those effects will vary not only by region but over time, depending on the rate and magnitude of climate change. For example, many countries will face increased challenges for economic development, increased risks from some diseases, or degraded ecosystems, but some countries will probably have increased opportunities for economic development, reduced instances of some diseases, or expanded areas of productive land. Crop yield changes will vary with geography and by latitude. Patterns of potential catch for fisheries are changing globally as well, with both positive and negative consequences. Availability of resources such as usable water will also depend on changing rates of precipitation, with decreased availability in many places but possible increases in runoff and groundwater recharge in some regions like the high latitudes and wet tropics.

FAQ 8: What communities are most vulnerable to the impacts of climate change?
[Chapters 8, 9, 12, 13, 19, 22, 23, 26, 27, 29, and Box CC-GC]
Every society is vulnerable to the impacts of climate change, but the nature of that vulnerability varies across regions and communities, over time, and depends on unique socioeconomic and other conditions. Poorer communities tend to be more vulnerable to loss of health and life, while wealthier communities usually have more economic assets at risk. Regions affected by violence or governance failure can be particularly vulnerable to climate change impacts. Development challenges, such as gender inequality and low levels of education, and other differences among communities in age, race and ethnicity, socioeconomic status, and governance can influence vulnerability to climate change impacts in complex ways.

FAQ 9: Does climate change cause violent conflicts?
[Chapters 12, 19]
Some factors that increase risks from violent conflicts and civil wars are sensitive to climate change. For example, there is growing evidence that factors like low per capita incomes, economic contraction, and inconsistent state institutions are associated with the incidence of civil wars, and also seem to be sensitive to climate change. Climate-change policies, particularly those associated with changing rights to resources, can also increase risks from violent conflict. While statistical studies document a relationship between climate variability and conflict, there remains much disagreement about whether climate change directly causes violent conflicts.

FAQ 10: How are adaptation, mitigation, and sustainable development connected?
[Chapters 1, 2, 8, 9, 10, 11, 13, 17, 20, 22, 23, 24, 25, 26, 27, and 29]
Mitigation has the potential to reduce climate change impacts, and adaptation can reduce the damage of those impacts. Together, both approaches can contribute to the development of societies that are more resilient to the threat of climate change and therefore more sustainable. Studies indicate that interactions between adaptation and mitigation responses have both potential synergies and tradeoffs that vary according to context. Adaptation responses may increase greenhouse gas emissions (e.g., increased fossil-based air conditioning in response to higher temperatures), and mitigation may impede adaptation (e.g., increased use of land for bioenergy crop production negatively impacting ecosystems). There are growing examples of co-benefits of mitigation and development policies, like those which can potentially reduce local emissions of health-damaging and climate-altering air pollutants from energy systems. It is clear that adaptation, mitigation, and sustainable development will be connected in the future.

FAQ 11: Why is it difficult to be sure of the role of climate change in observed effects on people and ecosystems?
[Chapter 3, 4, 5, 6, 7, 11, 12, 13, 18, 22, 23, 24, 25, 26, 27, 28, 29, and 30]
Climate change is one of many factors impacting the Earth's complex human societies and natural ecosystems. In some cases the effect of climate change has a unique pattern in space or time, providing a fingerprint for identification. In others, potential effects of climate change are thoroughly mixed with effects of land use change, economic development, changes in technology, or other processes. Trends in human activities, health, and society often have many simultaneous causes, making it especially challenging to isolate the role of climate change.

Much climate-related damage results from extreme weather events and could be affected by changes in the frequency and intensity of these events due to climate change. The most damaging events are rare, and the level of damage depends on context. It can therefore be challenging to build statistical confidence in observed trends, especially over short time periods. Despite this, many climate change impacts on the physical environment and ecosystems have been identified, and increasing numbers of impacts have been found in human systems as well.

Cross-Chapter Boxes

Coral Reefs

Jean-Pierre Gattuso (France), Ove Hoegh-Guldberg (Australia), Hans-Otto Pörtner (Germany)

Coral reefs are shallow-water ecosystems that consist of reefs made of calcium carbonate which is mostly secreted by reef-building corals and encrusting macroalgae. They occupy less than 0.1% of the ocean floor yet play multiple important roles throughout the tropics, housing high levels of biological diversity as well as providing key ecosystem goods and services such as habitat for fisheries, coastal protection, and appealing environments for tourism (Wild et al., 2011). About 275 million people live within 30 km of a coral reef (Burke et al., 2011) and derive some benefits from the ecosystem services that coral reefs provide (Hoegh-Guldberg, 2011), including provisioning (food, livelihoods, construction material, medicine), regulating (shoreline protection, water quality), supporting (primary production, nutrient cycling), and cultural (religion, tourism) services. This is especially true for the many coastal and small island nations in the world's tropical regions (Section 29.3.3.1).

Coral reefs are one of the most vulnerable marine ecosystems (*high confidence*; Sections 5.4.2.4, 6.3.1, 6.3.2, 6.3.5, 25.6.2, and 30.5), and more than half of the world's reefs are under medium or high risk of degradation (Burke et al., 2011). Most human-induced disturbances to coral reefs were local until the early 1980s (e.g., unsustainable coastal development, pollution, nutrient enrichment, and overfishing) when disturbances from ocean warming (principally mass coral bleaching and mortality) began to become widespread (Glynn, 1984). Concern about the impact of ocean acidification on coral reefs developed over the same period, primarily over the implications of ocean acidification for the building and maintenance of the calcium carbonate reef framework (Box CC-OA).

A wide range of climatic and non-climatic drivers affect corals and coral reefs and negative impacts have already been observed (Sections 5.4.2.4, 6.3.1, 6.3.2, 25.6.2.1, 30.5.3, 30.5.6). Bleaching involves the breakdown and loss of endosymbiotic algae, which live in the coral tissues and play a key role in supplying the coral host with energy (see Section 6.3.1. for physiological details and Section 30.5 for a regional analysis). Mass coral bleaching and mortality, triggered by positive temperature anomalies (*high confidence*), is the most widespread and conspicuous impact of climate change (Figure CR-1A and B, Figure 5-3; Sections 5.4.2.4, 6.3.1, 6.3.5, 25.6.2.1, 30.5, and 30.8.2). For example, the level of thermal stress at most of the 47 reef sites where bleaching occurred during 1997–1998 was unmatched in the period 1903–1999 (Lough, 2000). Ocean acidification reduces biodiversity (Figure CR-1C and D) and the calcification rate of corals (*high confidence*; Sections 5.4.2.4, 6.3.2, 6.3.5) while at the same time increasing the rate of dissolution of the reef framework (*medium confidence*; Section 5.2.2.4) through stimulation of biological erosion and chemical dissolution. Taken together, these changes will tip the calcium carbonate balance of coral reefs toward net dissolution (*medium confidence*; Section 5.4.2.4).

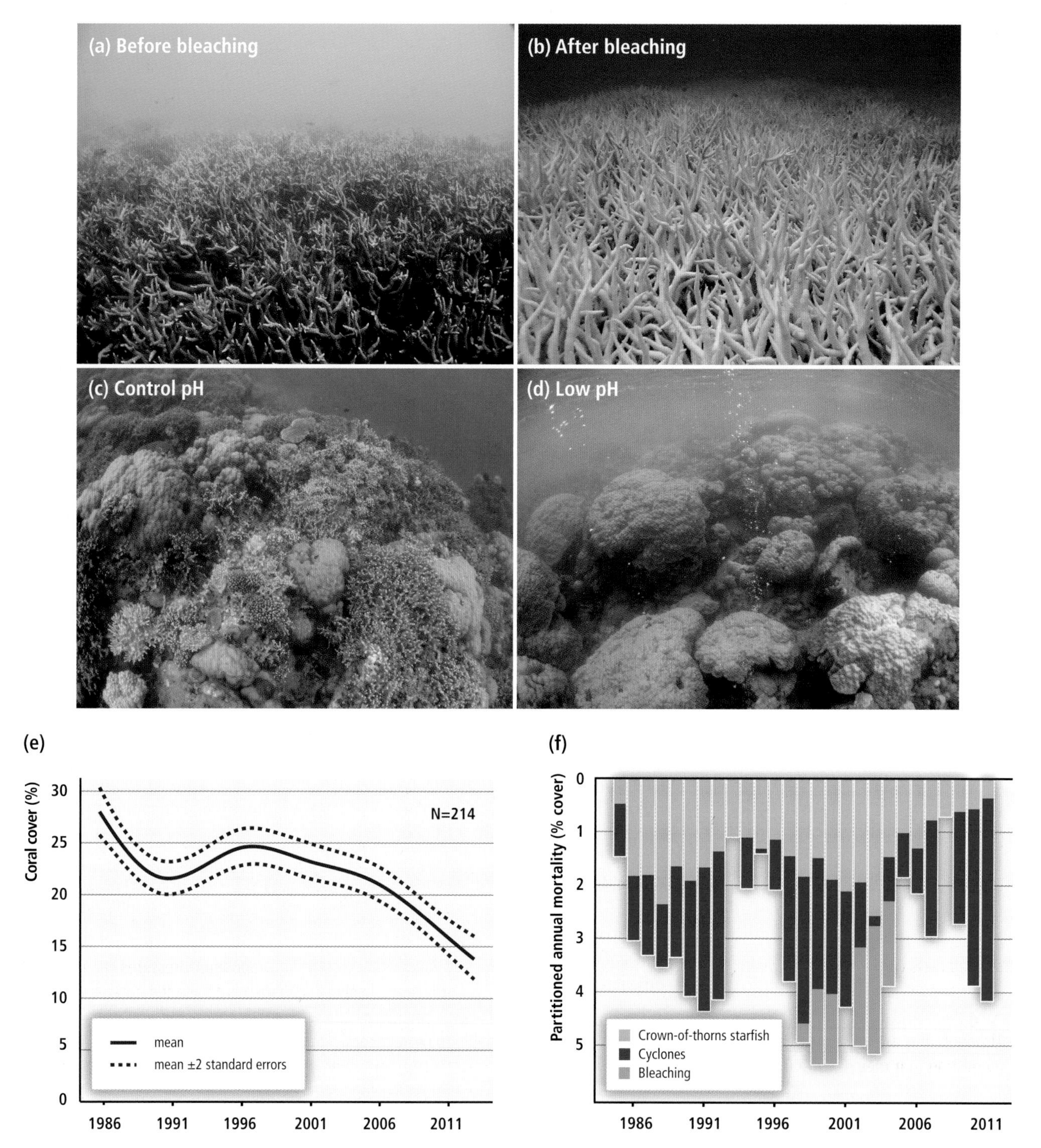

Figure CR-1 | (a, b) The same coral community before and after a bleaching event in February 2002 at 5 m depth, Halfway Island, Great Barrier Reef. Approximately 95% of the coral community was severely bleached in 2002 (Elvidge et al., 2004). Corals experience increasing mortality as the intensity of a heating event increases. A few coral species show the ability to shuffle symbiotic communities of dinoflagellates and appear to be more tolerant of warmer conditions (Berkelmans and van Oppen, 2006; Jones et al., 2008). (c, d) Three CO_2 seeps in Milne Bay Province, Papua New Guinea show that prolonged exposure to high CO_2 is related to fundamental changes in the ecology of coral reefs (Fabricius et al., 2011), including reduced coral diversity (–39%), severely reduced structural complexity (–67%), lower density of young corals (–66%), and fewer crustose coralline algae (–85%). At high CO_2 sites (d; median pH_T ~7.8, where pH_T is pH on the total scale), reefs are dominated by massive corals while corals with high morphological complexity are underrepresented compared with control sites (c; median pH_T ~8.0). Reef development ceases at pH_T values below 7.7. (e) Temporal trend in coral cover for the whole Great Barrier Reef over the period 1985–2012 (N=number of reefs, De'ath et al., 2012). (f) Composite bars indicate the estimated mean coral mortality for each year, and the sub-bars indicate the relative mortality due to crown-of-thorns starfish, cyclones, and bleaching for the whole Great Barrier Reef (De'ath et al., 2012). (Photo credit: R. Berkelmans (a and b) and K. Fabricius (c and d).)

Ocean warming and acidification have synergistic effects in several reef-builders (Section 5.2.4.2, 6.3.5). Taken together, these changes will erode habitats for reef-based fisheries, increase the exposure of coastlines to waves and storms, as well as degrading environmental features important to industries such as tourism (*high confidence*; Section 6.4.1.3, 25.6.2, 30.5).

A growing number of studies have reported regional scale changes in coral calcification and mortality that are consistent with the scale and impact of ocean warming and acidification when compared to local factors such as declining water quality and overfishing (Hoegh-Guldberg et al., 2007). The abundance of reef building corals is in rapid decline in many Pacific and Southeast Asian regions (*very high confidence*, 1 to 2% per year for 1968–2004; Bruno and Selig, 2007). Similarly, the abundance of reef-building corals has decreased by more than 80% on many Caribbean reefs (1977–2001; Gardner et al., 2003), with a dramatic phase shift from corals to seaweeds occurring on Jamaican reefs (Hughes, 1994). Tropical cyclones, coral predators, and thermal stress-related coral bleaching and mortality have led to a decline in coral cover on the Great Barrier Reef by about 51% between 1985 and 2012 (Figure CR-1E and F). Although less well documented, benthic invertebrates other than corals are also at risk (Przeslawski et al., 2008). Fish biodiversity is threatened by the permanent degradation of coral reefs, including in a marine reserve (Jones et al., 2004).

Future impacts of climate-related drivers (ocean warming, acidification, sea level rise as well as more intense tropical cyclones and rainfall events) will exacerbate the impacts of non-climate–related drivers (*high confidence*). Even under optimistic assumptions regarding corals being able to rapidly adapt to thermal stress, one-third (9 to 60%, 68% uncertainty range) of the world's coral reefs are projected to be subject to long-term degradation (next few decades) under the Representative Concentration Pathway (RCP)3-PD scenario (Frieler et al., 2013). Under the RCP4.5 scenario, this fraction increases to two-thirds (30 to 88%, 68% uncertainty range). If present-day corals have residual capacity to acclimate and/or adapt, half of the coral reefs may avoid high-frequency bleaching through 2100 (*limited evidence*, *limited agreement*; Logan et al., 2014). Evidence of corals adapting rapidly, however, to climate change is missing or equivocal (Hoegh-Guldberg, 2012).

Damage to coral reefs has implications for several key regional services:

- *Resources*: Coral reefs account for 10 to 12% of the fish caught in tropical countries, and 20 to 25% of the fish caught by developing nations (Garcia and de Leiva Moreno, 2003). More than half (55%) of the 49 island countries considered by Newton et al. (2007) are already exploiting their coral reef fisheries in an unsustainable way and the production of coral reef fish in the Pacific is projected to decrease 20% by 2050 under the Special Report on Emission Scenarios (SRES) A2 emissions scenario (Bell et al., 2013).
- *Coastal protection*: Coral reefs contribute to protecting the shoreline from the destructive action of storm surges and cyclones (Sheppard et al., 2005), sheltering the only habitable land for several island nations, habitats suitable for the establishment and maintenance of mangroves and wetlands, as well as areas for recreational activities. This role is threatened by future sea level rise, the decrease in coral cover, reduced rates of calcification, and higher rates of dissolution and bioerosion due to ocean warming and acidification (Sections 5.4.2.4, 6.4.1, 30.5).
- *Tourism*: More than 100 countries benefit from the recreational value provided by their coral reefs (Burke et al., 2011). For example, the Great Barrier Reef Marine Park attracts about 1.9 million visits each year and generates A$5.4 billion to the Australian economy and 54,000 jobs (90% in the tourism sector; Biggs, 2011).

Coral reefs make a modest contribution to the global gross domestic product (GDP) but their economic importance can be high at the country and regional scales (Pratchett et al., 2008). For example, tourism and fisheries represent 5% of the GDP of South Pacific islands (average for 2001–2011; Laurans et al., 2013). At the local scale, these two services provided in 2009–2011 at least 25% of the annual income of villages in Vanuatu and Fiji (Pascal, 2011; Laurans et al., 2013).

Isolated reefs can recover from major disturbance, and the benefits of their isolation from chronic anthropogenic pressures can outweigh the costs of limited connectivity (Gilmour et al., 2013). Marine protected areas (MPAs) and fisheries management have the potential to increase ecosystem resilience and increase the recovery of coral reefs after climate change impacts such as mass coral bleaching (McLeod et al., 2009). Although they are key conservation and management tools, they are unable to protect corals directly from thermal stress (Selig et al., 2012), suggesting that they need to be complemented with additional and alternative strategies (Rau et al., 2012; Billé et al., 2013). While MPA networks are a critical management tool, they should be established considering other forms of resource management (e.g., fishery catch limits and gear restrictions) and integrated ocean and coastal management to control land-based threats such as pollution and sedimentation. There is *medium confidence* that networks of highly protected areas nested within a broader management framework can contribute to preserving coral reefs under increasing human pressure at local and global scales (Salm et al., 2006). Locally, controlling the input of nutrients and sediment from land is an important complementary management strategy (Mcleod et al., 2009) because nutrient enrichment can increase the susceptibility of corals to bleaching (Wiedenmann et al., 2013) and coastal pollutants enriched with fertilizers can increase acidification (Kelly et al., 2011). In the long term, limiting the amount of ocean warming and acidification is central to ensuring the viability of coral reefs and dependent communities (*high confidence*; Section 5.2.4.4, 30.5).

References

Bell, J.D., A. Ganachaud, P.C. Gehrke, S.P. Griffiths, A.J. Hobday, O. Hoegh-Guldberg, J.E. Johnson, R. Le Borgne, P. Lehodey, J.M. Lough, R.J. Matear, T.D. Pickering, M.S. Pratchett, A. Sen Gupta, I. Senina and M. Waycott, 2013: Mixed responses of tropical Pacific fisheries and aquaculture to climate change. *Nature Climate Change*, **3(6)**, 591-599.

Berkelmans, R. and M.J.H. van Oppen, 2006: The role of zooxanthellae in the thermal tolerance of corals: a 'nugget of hope' for coral reefs in an era of climate change. *Proceedings of the Royal Society B: Biological Sciences*, **273(1599)**, 2305-2312.

Biggs, D., 2011: *Case study: the resilience of the nature-based tourism system on Australia's Great Barrier Reef*. Report prepared for the Australian Government Department of Sustainability Environment Water Population and Communities on behalf of the State of the Environment 2011 Committee, Canberra, 32 pp.

Billé, R., R. Kelly, A. Biastoch, E. Harrould-Kolieb, D. Herr, F. Joos, K.J. Kroeker, D. Laffoley, A. Oschlies and J.-P. Gattuso, 2013: Taking action against ocean acidification: a review of management and policy options. *Environmental Management*, **52**, 761-779.

Bruno, J.F. and E.R. Selig, 2007: Regional decline of coral cover in the Indo-Pacific: timing, extent, and subregional comparisons. PLoS ONE, 2(8), e711. doi: 10.1371/journal.pone.0000711.

Burke, L., K. Reytar, M. Spalding and A. Perry, 2011: *Reefs at risk revisited*. World Resources Institute, Washington D.C., 114 pp.

De'ath, G., K.E. Fabricius, H. Sweatman and M. Puotinen, 2012: The 27-year decline of coral cover on the Great Barrier Reef and its causes. *Proceedings of the National Academy of Sciences of the United States of America*, **109(44)**, 17995-17999.

Elvidge, C.D., J.B. Dietz, R. Berkelmans, S. Andréfouët, W. Skirving, A.E. Strong and B.T. Tuttle, 2004: Satellite observation of Keppel Islands (Great Barrier Reef) 2002 coral bleaching using IKONOS data. *Coral Reefs*, **23(1)**, 123-132.

Fabricius, K.E., C. Langdon, S. Uthicke, C. Humphrey, S. Noonan, G. De'ath, R. Okazaki, N. Muehllehner, M.S. Glas and J.M. Lough, 2011: Losers and winners in coral reefs acclimatized to elevated carbon dioxide concentrations. *Nature Climate Change*, **1(3)**, 165-169.

Frieler, K., M. Meinshausen, A. Golly, M. Mengel, K. Lebek, S.D. Donner and O. Hoegh-Guldberg, 2013: Limiting global warming to 2°C is unlikely to save most coral reefs. *Nature Climate Change*, **3(2)**, 165-170.

Garcia, S.M. and I. de Leiva Moreno, 2003: Global overview of marine fisheries. In: *Responsible Fisheries in the Marine Ecosystem* [Sinclair, M. and G. Valdimarsson (eds.)]. Wallingford: CABI, pp. 1-24.

Gardner, T.A., I.M. Côté, J.A. Gill, A. Grant and A.R. Watkinson, 2003: Long-term region-wide declines in Caribbean corals. *Science*, **301(5635)**, 958-960.

Gilmour, J.P., L.D. Smith, A.J. Heyward, A.H. Baird and M.S. Pratchett, 2013: Recovery of an isolated coral reef system following severe disturbance. *Science*, **340(6128)**, 69-71.

Glynn, P.W., 1984: Widespread coral mortality and the 1982-83 El Niño warming event. *Environmental Conservation*, **11(2)**, 133-146.

Hoegh-Guldberg, O., 2011: Coral reef ecosystems and anthropogenic climate change. *Regional Environmental Change*, **11**, 215-227.

Hoegh-Guldberg, O., 2012: The adaptation of coral reefs to climate change: Is the Red Queen being outpaced? *Scientia Marina*, **76(2)**, 403-408.

Hoegh-Guldberg, O., P.J. Mumby, A.J. Hooten, R.S. Steneck, P. Greenfield, E. Gomez, C.D. Harvell, P.F. Sale, A.J. Edwards, K. Caldeira, N. Knowlton, C.M. Eakin, R. Iglesias-Prieto, N. Muthiga, R.H. Bradbury, A. Dubi and M.E. Hatziolos, 2007: Coral reefs under rapid climate change and ocean acidification. *Science*, **318(5857)**, 1737-1742.

Hughes, T.P., 1994: Catastrophes, phase-shifts, and large-scale degradation of a Caribbean coral reef. *Science*, **265(5178)**, 1547-1551.

Jones, A.M., R. Berkelmans, M.J.H. van Oppen, J.C. Mieog and W. Sinclair, 2008: A community change in the algal endosymbionts of a scleractinian coral following a natural bleaching event: field evidence of acclimatization. Proceedings of the *Royal Society B: Biological Sciences*, **275(1641)**, 1359-1365.

Jones, G.P., M.I. McCormick, M. Srinivasan and J.V. Eagle, 2004: Coral decline threatens fish biodiversity in marine reserves. *Proceedings of the National Academy of Sciences of the United States of America*, **101(21)**, 8251-8253.

Kelly, R.P., M.M. Foley, W.S. Fisher, R.A. Feely, B.S. Halpern, G.G. Waldbusser and M.R. Caldwell, 2011: Mitigating local causes of ocean acidification with existing laws. *Science*, **332(6033)**, 1036-1037.

Laurans, Y., N. Pascal, T. Binet, L. Brander, E. Clua, G. David, D. Rojat and A. Seidl, 2013: Economic valuation of ecosystem services from coral reefs in the South Pacific: taking stock of recent experience. *Journal of Environmental Management*, **116**, 135-144.

Logan, C.A., J.P. Dunne, C.M. Eakin and S.D. Donner, 2014: Incorporating adaptive responses into future projections of coral bleaching. *Global Change Biology*, **20(1)**, 125-139.

Lough, J.M., 2000: 1997–98: Unprecedented thermal stress to coral reefs? *Geophysical Research Letters*, **27(23)**, 3901-3904.

McLeod, E., R. Salm, A. Green and J. Almany, 2009: Designing marine protected area networks to address the impacts of climate change. *Frontiers in Ecology and the Environment*, **7(7)**, 362-370.

Newton, K., I.M. Côté, G.M. Pilling, S. Jennings and N.K. Dulvy, 2007: Current and future sustainability of island coral reef fisheries. *Current Biology*, **17(7)**, 655-658.

Pascal, N., 2011: *Cost-benefit analysis of community-based marine protected areas: 5 case studies in Vanuatu, South Pacific*. CRISP Research Reports. CRIOBE (EPHE/CNRS). Insular Research Center and Environment Observatory, Mooréa, French Polynesia, 107 pp.

Pratchett, M.S., P.L. Munday, S.K. Wilson, N.A.J. Graham, J.E. Cinner, D.R. Bellwood, G.P. Jones, N.V.C. Polunin and T.R. McClanahan, 2008: Effects of climate-induced coral bleaching on coral-reef fishes - Ecological and economic consequences. *Oceanography and Marine Biology: An Annual Review*, **46**, 251-296.

Przeslawski, R., S. Ahyong, M. Byrne, G. Wörheide and P. Hutchings, 2008: Beyond corals and fish: the effects of climate change on noncoral benthic invertebrates of tropical reefs. *Global Change Biology*, **14(12)**, 2773-2795.

Rau, G.H., E.L. McLeod and O. Hoegh-Guldberg, 2012: The need for new ocean conservation strategies in a high-carbon dioxide world. *Nature Climate Change*, **2(10)**, 720-724.

Salm, R.V., T. Done and E. McLeod, 2006: Marine Protected Area planning in a changing climate. In: *Coastal and Estuarine Studies 61. Coral Reefs and Climate Change: Science and Management*. [Phinney, J.T., O. Hoegh-Guldberg, J. Kleypas, W. Skirving and A. Strong (eds.)]. American Geophysical Union, pp. 207-221.

Selig, E.R., K.S. Casey and J.F. Bruno, 2012: Temperature-driven coral decline: the role of marine protected areas. *Global Change Biology*, **18(5)**, 1561-1570.

Sheppard, C., D.J. Dixon, M. Gourlay, A. Sheppard and R. Payet, 2005: Coral mortality increases wave energy reaching shores protected by reef flats: Examples from the Seychelles. *Estuarine, Coastal and Shelf Science*, **64(2-3)**, 223-234.

Wiedenmann, J., C. D'Angelo, E.G. Smith, A.N. Hunt, F.-E. Legiret, A.D. Postle and E.P. Achterberg, 2013: Nutrient enrichment can increase the susceptibility of reef corals to bleaching. *Nature Climate Change*, **3(2)**, 160-164.

Wild, C., O. Hoegh-Guldberg, M.S. Naumann, M.F. Colombo-Pallotta, M. Ateweberhan, W.K. Fitt, R. Iglesias-Prieto, C. Palmer, J.C. Bythell, J.-C. Ortiz, Y. Loya and R. van Woesik, 2011: Climate change impedes scleractinian corals as primary reef ecosystem engineers. *Marine and Freshwater Research*, **62(2)**, 205-215.

This cross-chapter box should be cited as:

Gattuso, J.-P., O. Hoegh-Guldberg, and H.-O. Pörtner, 2014: Cross-chapter box on coral reefs. In: *Climate Change 2014: Impacts, Adaptation, and Vulnerability. Summaries, Frequently Asked Questions, and Cross-Chapter Boxes. Contribution of Working Group II to the Fifth Assessment Report of the Intergovernmental Panel on Climate Change* [Field, C.B., V.R. Barros, D.J. Dokken, K.J. Mach, M.D. Mastrandrea, T.E. Bilir, M. Chatterjee, K.L. Ebi, Y.O. Estrada, R.C. Genova, B. Girma, E.S. Kissel, A.N. Levy, S. MacCracken, P.R. Mastrandrea, and L.L. White (eds.)]. World Meteorological Organization, Geneva, Switzerland, pp. 99-102. (in Arabic, Chinese, English, French, Russian, and Spanish)

Ecosystem-Based Approaches to Adaptation—Emerging Opportunities

Rebecca Shaw (USA), Jonathan Overpeck (USA), Guy Midgley (South Africa)

Ecosystem-based adaptation (EBA), defined as the use of biodiversity and ecosystem services as part of an overall adaptation strategy to help people to adapt to the adverse effects of climate change (CBD, 2009), integrates the use of biodiversity and ecosystem services into climate change adaptation strategies (e.g., CBD, 2009; Munroe et al., 2011; see IPCC AR5 WGII Chapters 3, 4, 5, 8, 9, 13, 14, 15, 16, 19, 22, 25, and 27). EBA is implemented through the sustainable management of natural resources and conservation and restoration of ecosystems, to provide and sustain services that facilitate adaptation both to climate variability and change (Colls et al., 2009). It also sets out to take into account the multiple social, economic, and cultural co-benefits for local communities (CBD COP 10 Decision X/33).

EBA can be combined with, or even serve as a substitute for, the use of engineered infrastructure or other technological approaches. Engineered defenses such as dams, sea walls, and levees adversely affect biodiversity, potentially resulting in maladaptation due to damage to ecosystem regulating services (Campbell et al., 2009; Munroe et al., 2011). There is some evidence that the restoration and use of ecosystem services may reduce or delay the need for these engineering solutions (CBD, 2009). EBA offers lower risk of maladaptation than engineering solutions in that their application is more flexible and responsive to unanticipated environmental changes. Well-integrated EBA can be more cost effective and sustainable than non-integrated physical engineering approaches (Jones et al., 2012), and may contribute to achieving sustainable development goals (e.g., poverty reduction, sustainable environmental management, and even mitigation objectives), especially when they are integrated with sound ecosystem management approaches (CBD, 2009). In addition, EBA yields economic, social, and environmental co-benefits in the form of ecosystem goods and services (World Bank, 2009).

EBA is applicable in both developed and developing countries. In developing countries where economies depend more directly on the provision of ecosystem services (Vignola et al., 2009), EBA may be a highly useful approach to reduce risks to climate change impacts and ensure that development proceeds on a pathways that are resilient to climate change (Munang et al., 2013). EBA projects may be developed by enhancing existing initiatives, such as community-based adaptation and natural resource management approaches (e.g., Khan et al., 2012, Midgley et al., 2012; Roberts et al., 2012).

Examples of ecosystem based approaches to adaptation include:

- Sustainable water management, where river basins, aquifers, flood plains, and their associated vegetation are managed or restored to provide resilient water storage and

enhanced baseflows, flood regulation and protection services, reduction of erosion/siltation rates, and more ecosystem goods (e.g., Opperman et al., 2009; Midgley et al., 2012)
- Disaster risk reduction through the restoration of coastal habitats (e.g., mangroves, wetlands, and deltas) to provide effective measure against storm-surges, saline intrusion, and coastal erosion (Jonkman et al., 2013)
- Sustainable management of grasslands and rangelands to enhance pastoral livelihoods and increase resilience to drought and flooding
- Establishment of diverse and resilient agricultural systems, and adapting crop and livestock variety mixes to secure food provision. Traditional knowledge may contribute in this area through, for example, identifying indigenous crop and livestock genetic diversity, and water conservation techniques.
- Management of fire-prone ecosystems to achieve safer fire regimes while ensuring the maintenance of natural processes

Application of EBA, like other approaches, is not without risk, and risk/benefit assessments will allow better assessment of opportunities offered by the approach (CBD, 2009). The examples of EBA are too few and too recent to assess either the risks or the benefits comprehensively at this stage. EBA is still a developing concept but should be considered alongside adaptation options based more on engineering works or social change, and existing and new cases used to build understanding of when and where its use is appropriate.

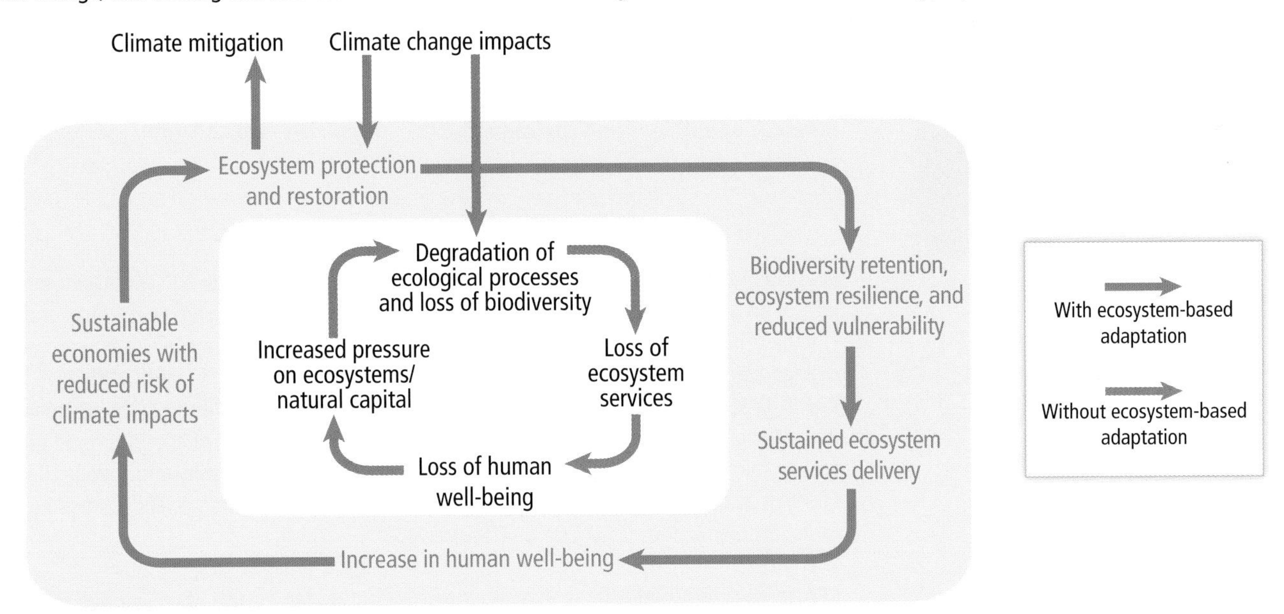

Figure EA-1 | Adapted from Munang et al. (2013). Ecosystem-based adaptation (EBA) uses the capacity of nature to buffer human systems from the adverse impacts of climate change. Without EBA, climate change may cause degradation of ecological processes (central white panel) leading to losses in human well-being. Implementing EBA (outer blue panel) may reduce or offset these adverse impacts resulting in a virtuous cycle that reduces climate-related risks to human communities, and may provide mitigation benefits.

References

Campbell, A., V. Kapos, J. Scharlemann, P. Bubb, A. Chenery, L. Coad, B. Dickson, N. Doswald, M. Khan, F. Kershaw, and M. Rashid, 2009: *Review of the Literature on the Links between Biodiversity and Climate Change: Impacts, Adaptation and Mitigation*. CBD Technical Series No. 42, Secretariat of the Convention on Biological Diversity (CBD), Montreal, QC, Canada, 124 pp.

CBD, 2009: *Connecting Biodiversity and Climate Change Mitigation and Adaptation: Report of the Second Ad Hoc Technical Expert Group on Biodiversity and Climate Change*. CBD Technical Series No. 41, Secretariat of the Convention on Biological Diversity (CBD), Montreal, QC, Canada, 126 pp.

Colls, A., N. Ash, and N. Ikkala, 2009: *Ecosystem-Based Adaptation: A Natural Response to Climate Change*. International Union for Conservation of Nature and Natural Resources (IUCN), Gland, Switzerland, 16 pp.

Jones, H.P., D.G. Hole, and E.S. Zavaleta, 2012: Harnessing nature to help people adapt to climate change. *Nature Climate Change*, **2(7)**, 504-509.

Jonkman, S.N., M.M. Hillen, R.J. Nicholls, W. Kanning, and M. van Ledden, 2013: Costs of adapting coastal defences to sea-level rise – new estimates and their implications. *Journal of Coastal Research*, **29(5)**, 1212-1226.

Khan, A.S., A. Ramachandran, N. Usha, S. Punitha, and V. Selvam, 2012: Predicted impact of the sea-level rise at Vellar-Coleroon estuarine region of Tamil Nadu coast in India: mainstreaming adaptation as a coastal zone management option. *Ocean & Coastal Management*, 69, 327-339.

Midgley, G.F., S. Marais, M. Barnett, and K. Wågsæther, 2012: *Biodiversity, Climate Change and Sustainable Development – Harnessing Synergies and Celebrating Successes*. Final Technical Report, The Adaptation Network Secretariat, hosted by Indigo Development & Change and The Environmental Monitoring Group, Nieuwoudtville, South Africa. 70 pp.

Munang, R, I. Thiaw, K. ALverson, M. Mumba, J. Liu, and M. Rivington, 2013: Climate change and ecosystem-based adaptation: a new pragmatic approach to buffering climate change impacts. *Current Opinion in Environmental Sustainability*, **5(1)**, 67-71.

Munroe, R., N. Doswald, D. Roe, H. Reid, A. Giuliani, I. Castelli, and I. Moller, 2011: *Does EbA Work? A Review of the Evidence on the Effectiveness of Ecosystem-Based Approaches to Adaptation*. Research collaboration between BirdLife International, United Nations Environment Programme-World Conservation Monitoring Centre (UNEP-WCMC), and the University of Cambridge, Cambridge, UK, and the International Institute for Environment and Development (IIED), London, UK, 4 pp.

Opperman, J.J., G.E. Galloway, J. Fargione, J.F. Mount, B.D. Richter, and S. Secchi, 2009: Sustainable floodplains through large-scale reconnection to rivers. *Science*, **326(5959)**, 1487-1488.

Roberts, D., R. Boon, N. Diederichs, E. Douwes, N. Govender, A. McInnes, C. McLean, S. O'Donoghue, and M. Spires, 2012: Exploring ecosystem-based adaptation in Durban, South Africa: "learning-by-doing" at the local government coal face. *Environment and Urbanization*, **24(1)**, 167-195.

Vignola, R., B. Locatelli, C. Martinez, and P. Imbach, 2009: Ecosystem-based adaptation to climate change: what role for policymakers, society and scientists? *Mitigation and Adaptation Strategies for Global Change*, **14(8)**, 691-696.

This cross-chapter box should be cited as:

Shaw, M.R., J.T. Overpeck, and G.F. Midgley, 2014: Cross-chapter box on ecosystem based approaches to adaptation—emerging opportunities. In: *Climate Change 2014: Impacts, Adaptation, and Vulnerability. Summaries, Frequently Asked Questions, and Cross-Chapter Boxes. Contribution of Working Group II to the Fifth Assessment Report of the Intergovernmental Panel on Climate Change* [Field, C.B., V.R. Barros, D.J. Dokken, K.J. Mach, M.D. Mastrandrea, T.E. Bilir, M. Chatterjee, K.L. Ebi, Y.O. Estrada, R.C. Genova, B. Girma, E.S. Kissel, A.N. Levy, S. MacCracken, P.R. Mastrandrea, and L.L. White (eds.)]. World Meteorological Organization, Geneva, Switzerland, pp. 103-105. (in Arabic, Chinese, English, French, Russian, and Spanish)

Gender and Climate Change

Katharine Vincent (South Africa), Petra Tschakert (U.S.A.), Jon Barnett (Australia), Marta G. Rivera-Ferre (Spain), Alistair Woodward (New Zealand)

Gender, along with sociodemographic factors of age, wealth, and class, is critical to the ways in which climate change is experienced. There are significant gender dimensions to impacts, adaptation, and vulnerability. This issue was raised in WGII AR4 and SREX reports (Adger et al., 2007; IPCC, 2012), but for the AR5 there are significant new findings, based on multiple lines of evidence on how climate change is differentiated by gender, and how climate change contributes to perpetuating existing gender inequalities. This new research has been undertaken in every region of the world (e.g. Brouwer et al., 2007; Buechler, 2009; Nelson and Stathers, 2009; Nightingale, 2009; Dankelman, 2010; MacGregor, 2010; Alston, 2011; Arora-Jonsson, 2011; Omolo, 2011; Resureccion, 2011).

Gender dimensions of vulnerability derive from differential access to the social and environmental resources required for adaptation. In many rural economies and resource-based livelihood systems, it is well established that women have poorer access than men to financial resources, land, education, health, and other basic rights. Further drivers of gender inequality stem from social exclusion from decision-making processes and labor markets, making women in particular less able to cope with and adapt to climate change impacts (Paavola, 2008; Djoudi and Brockhaus, 2011; Rijkers and Costa, 2012). These gender inequalities manifest themselves in gendered livelihood impacts and feminisation of responsibilities: whereas both men and women experience increases in productive roles, only women experience increased reproductive roles (Resureccion, 2011; Section 9.3.5.1.5, Box 13-1). A study in Australia, for example, showed how more regular occurrence of drought has put women under increasing pressure to earn off-farm income and contribute to more on-farm labor (Alston, 2011). Studies in Tanzania and Malawi demonstrate how women experience food and nutrition insecurity because food is preferentially distributed among other family members (Nelson and Stathers, 2009; Kakota et al., 2011).

AR4 assessed a body of literature that focused on women's relatively higher vulnerability to weather-related disasters in terms of number of deaths (Adger et al., 2007). Additional literature published since that time adds nuances by showing how socially constructed gender differences affect exposure to extreme events, leading to differential patterns of mortality for both men and women (*high confidence*; Section 11.3.3, Table 12-3). Statistical evidence of patterns of male and female mortality from recorded extreme events in 141 countries between 1981 and 2002 found that disasters kill women at an earlier age than men (Neumayer and Plümper, 2007; see also Box 13-1). Reasons for gendered differences in mortality include various socially and culturally determined gender roles. Studies in Bangladesh, for example, show that women do not learn to swim and so are vulnerable when exposed to flooding (Röhr, 2006) and that, in Nicaragua, the construction of gender roles means that middle-class women are expected to stay in the house, even during floods and in risk-prone areas (Bradshaw, 2010). Although the differential vulnerability of women to extreme events has long been understood, there is now increasing evidence to show how gender roles for men can affect their vulnerability. In particular, men are often expected to be brave and heroic, and engage in risky life-saving behaviors that increase their likelihood of mortality (Box 13-1). In Hai Lang district, Vietnam, for example, more men died than women as a result of their involvement in search and rescue and protection of fields during flooding (Campbell et al., 2009). Women and girls are more likely to become victims of domestic violence after a disaster, particularly when they are living in emergency accommodation, which has been documented in the USA and Australia (Jenkins and Phillips, 2008; Anastario et al., 2009; Alston, 2011; Whittenbury, 2013; see also Box 13-1).

Heat stress exhibits gendered differences, reflecting both physiological and social factors (Section 11.3.3). The majority of studies in European countries show women to be more at risk, but their usually higher physiological vulnerability can be offset in some circumstances by relatively lower social vulnerability (if they are well connected in supportive social networks, for example). During the Paris heat wave, unmarried men were at greater risk than unmarried women, and in Chicago elderly men were at greatest risk, thought to reflect their lack of connectedness in social support networks which led to higher social vulnerability (Kovats and Hajat, 2008). A multi-city study showed geographical variations in the relationship between sex and mortality due to heat stress: in Mexico City, women had a higher risk of mortality than men, although the reverse was true in Santiago and São Paulo (Bell et al., 2008).

Recognizing gender differences in vulnerability and adaptation can enable gender-sensitive responses that reduce the vulnerability of women and men (Alston, 2013). Evaluations of adaptation investments demonstrate that those approaches that are not sensitive to gender dimensions and other drivers of social inequalities risk reinforcing existing vulnerabilities (Vincent et al., 2010; Arora-Jonsson, 2011; Figueiredo and Perkins, 2012). Government-supported interventions to improve production through cash-cropping and non-farm enterprises in rural economies, for example, typically advantage men over women because cash generation is seen as a male activity in rural areas (Gladwin et al., 2001; see also Section 13.3.1). In contrast, rainwater and conservation-based adaptation initiatives may require additional labor, which women cannot necessarily afford to provide (Baiphethi et al., 2008). Encouraging gender-equitable access to education and strengthening of social capital are among the best means of improving adaptation of rural women farmers (Goulden et al., 2009; Vincent et al., 2010; Below et al., 2012) and could be used to complement existing initiatives mentioned above that benefit men. Rights-based approaches to development can inform adaptation efforts as they focus on addressing the ways in which institutional practices shape access to resources and control over decision-making processes, including through the social construction of gender and its intersection with other factors that shape inequalities and vulnerabilities (Tschakert and Machado, 2012; Bee et al., 2013; Tschakert, 2013; see also Section 22.4.3 and Table 22-5).

References

Adger, W.N., S. Agrawala, M.M.Q. Mirza, C. Conde, K. O'Brien, J. Pulhin, R. Pulwarty, B. Smit, and K. Takahashi, 2007: Chapter 17: Assessment of adaptation practices, options, constraints and capacity. In: *Climate Change 2007: Impacts, Adaptation and Vulnerability. Contribution of Working Group II to the Fourth Assessment Report of the Intergovernmental Panel on Climate Change* [Parry, M.L., O.F. Canziani, J.P. Palutikof, P.J. van der Linden, and C.E. Hanson (ed.)]. IPCC, Geneva, Switzerland, pp. 719-743.

Alston, M., 2011: Gender and climate change in Australia. *Journal of Sociology*, **47**(**1**), 53-70.

Alston, M., 2013: Women and adaptation. *Wiley Interdisciplinary Reviews: Climate Change*, (**4**)**5**, 351-358.

Anastario, M., N. Shebab, and L. Lawry, 2009: Increased gender-based violence among women internally displaced in Mississippi 2 years post-Hurricane Katrina. *Disaster Medicine and Public Health Preparedness*, **3**(**1**), 18-26.

Arora-Jonsson, S., 2011: Virtue and vulnerability: discourses on women, gender and climate change. *Global Environmental Change*, **21**, 744-751.

Baiphethi, M.N., M. Viljoen, and G. Kundhlande, 2008: Rural women and rainwater harvesting and conservation practices: anecdotal evidence from the Free State and Eastern Cape. *Agenda*, **22**(78), 163-171.

Bee, B., M. Biermann, and P. Tschakert, 2013: Gender, development, and rights-based approaches: lessons for climate change adaptation and adaptive social protection. In: *Research, Action and Policy: Addressing the Gendered Impacts of Climate Change* [Alston, M. and K. Whittenbury (eds.)]. Springer, Dordrecht, Netherlands, pp. 95-108.

Bell, M.L., M.S. O'Neill, N. Ranjit, V.H. Borja-Aburto, L.A. Cifuentes, and N.C. Gouveia, 2008: Vulnerability to heat-related mortality in Latin America: a case-crossover study in Sao Paulo, Brazil, Santiago, Chile and Mexico City, Mexico. *International Journal of Epidemiology*, **37**(**4**), 796-804.

Below, T.B., K.D. Mutabazi, D. Kirschke, C. Franke, S. Sieber, R. Siebert, and K. Tscherning, 2012: Can farmers' adaptation to climate change be explained by socio-economic household-level variables? *Global Environmental Change*, **22**(**1**), 223-235.

Bradshaw, S., 2010: Women, poverty, and disasters: exploring the links through Hurricane Mitch in Nicaragua. In: The International Handbook of Gender and Poverty: Concepts, Research, Policy [Chant, S. (ed.)]. Edward Elgar Publishing, Cheltenham, UK, pp. 627-632.

Brouwer, R., S. Akter, L. Brander, and E. Haque, 2007: Socioeconomic vulnerability and adaptation to environmental risk: a case study of climate change and flooding in Bangladesh. *Risk Analysis*, **27**(**2**), 313-326.

Campbell, B., S. Mitchell, and M. Blackett, 2009: *Responding to Climate Change in Vietnam. Opportunities for Improving Gender Equality*. A Policy Discussion Paper, Oxfam in Viet Nam and United Nations Development Programme-Viet Nam (UNDP-Viet Nam), Ha noi, Viet Nam, 62 pp.

Dankelman, I., 2010: Introduction: exploring gender, environment, and climate change. In: *Gender and Climate Change: An Introduction* [Dankelman, I. (ed.)]. Earthscan, London, UK and Washington, DC, USA, pp. 1-18.

Djoudi, H. and M. Brockhaus, 2011: Is adaptation to climate change gender neutral? Lessons from communities dependent on livestock and forests in northern Mali. *International Forestry Review*, **13**(**2**), 123-135.

Figueiredo, P. and P.E. Perkins, 2012: Women and water management in times of climate change: participatory and inclusive processes. *Journal of Cleaner Production*, **60**(**1**), 188-194.

Gladwin, C.H., A.M. Thomson, J.S. Peterson, and A.S. Anderson, 2001: Addressing food security in Africa via multiple livelihood strategies of women farmers. *Food Policy*, **26(2)**, 177-207.
Goulden, M., L.O. Naess, K. Vincent, and W.N. Adger, 2009: Diversification, networks and traditional resource management as adaptations to climate extremes in rural Africa: opportunities and barriers. In: *Adapting to Climate Change: Thresholds, Values and Governance* [Adger, W.N., I. Lorenzoni, and K. O'Brien (eds.)]. Cambridge University Press, Cambridge, UK, pp. 448-464.
IPCC, 2012: *Managing the Risks of Extreme Events and Disasters to Advance Climate Change Adaptation. A Special Report of Working Groups I and II of the Intergovernmental Panel on Climate Change* [Field, C.B., V. Barros, T.F. Stocker, D. Qin, D.J. Dokken, K.L. Ebi, M.D. Mastrandrea, K.J. Mach, G.-K. Plattner, S.K. Allen, M. Tignor, and P.M. Midgley (eds.)], Cambridge University Press, Cambridge, UK and New York, NY, USA, 582 pp.
Jenkins, P. and B. Phillips, 2008: Battered women, catastrophe, and the context of safety after Hurricane Katrina. *NWSA Journal*, **20(3)**, 49-68.
Kakota, T., D. Nyariki, D. Mkwambisi, and W. Kogi-Makau, 2011: Gender vulnerability to climate variability and household food insecurity. *Climate and Development*, **3(4)**, 298-309.
Kovats, R. and S. Hajat, 2008: Heat stress and public health: a critical review. *Public Health*, **29**, 41-55.
MacGregor, S., 2010: 'Gender and climate change': from impacts to discourses. *Journal of the Indian Ocean Region*, **6(2)**, 223-238.
Nelson, V. and T. Stathers, 2009: Resilience, power, culture, and climate: a case study from semi-arid Tanzania, and new research directions. *Gender & Development*, **17(1)**, 81-94.
Neumayer, E. and T. Plümper, 2007: The gendered nature of natural disasters: the impact of catastrophic events on the gender gap in life expectancy, 1981-2002. *Annals of the Association of American Geographers*, **97(3)**, 551-566.
Nightingale, A., 2009: Warming up the climate change debate: a challenge to policy based on adaptation. *Journal of Forest and Livelihood*, **8(1)**, 84-89.
Omolo, N., 2011: Gender and climate change-induced conflict in pastoral communities: case study of Turkana in northwestern Kenya. *African Journal on Conflict Resolution*, **10(2)**, 81-102.
Paavola, J., 2008: Livelihoods, vulnerability and adaptation to climate change in Morogoro, Tanzania. *Environmental Science & Policy*, **11(7)**, 642-654.
Resurreccion, B.P., 2011: *The Gender and Climate Debate: More of the Same or New Pathways of Thinking and Doing?* Asia Security Initiative Policy Series, Working Paper No. 10, RSIS Centre for Non-Traditional Security (NTS) Studies, Singapore, 19 pp.
Rijkers, B. and R. Costa, 2012: *Gender and Rural Non-Farm Entrepreneurship*. Policy Research Working Paper 6066, Macroeconomics and Growth Team, Development Research Group, The World Bank, Washington, DC, USA, 68 pp.
Röhr, U., 2006: Gender and climate change. *Tiempo*, **59**, 3-7.
Tschakert, P., 2013: From impacts to embodied experiences: tracing political ecology in climate change research. *Geografisk Tidsskrift/Danish Journal of Geography*, 112(2), 144-158.
Tschakert, P. and M. Machado, 2012: Gender justice and rights in climate change adaptation: opportunities and pitfalls. *Ethics and Social Welfare*, **6(3)**, 275-289, doi: 10.1080/17496535.2012.704929.
Vincent, K., T. Cull, and E. Archer, 2010: Gendered vulnerability to climate change in Limpopo province, South Africa. In: *Gender and Climate Change: An Introduction* [Dankelman, I. (ed.)]. Earthscan, London, UK and Washington, DC, USA, pp. 160-167.
Whittenbury, K., 2013: Climate change, women's health, wellbeing and experiences of gender-based violence in Australia. In: *Research, Action and Policy: Addressing the Gendered Impacts of Climate Change* [Alston, M. and K. Whittenbury (eds.)]. Springer Science, Dordrecht, Netherlands, pp. 207-222.

This cross-chapter box should be cited as:
K.E. Vincent, P. Tschakert, Barnett, J., M.G. Rivera-Ferre, and A. Woodward, 2014: Cross-chapter box on gender and climate change. In: *Climate Change 2014: Impacts, Adaptation, and Vulnerability. Summaries, Frequently Asked Questions, and Cross-Chapter Boxes. Contribution of Working Group II to the Fifth Assessment Report of the Intergovernmental Panel on Climate Change* [Field, C.B., V.R. Barros, D.J. Dokken, K.J. Mach, M.D. Mastrandrea, T.E. Bilir, M. Chatterjee, K.L. Ebi, Y.O. Estrada, R.C. Genova, B. Girma, E.S. Kissel, A.N. Levy, S. MacCracken, P.R. Mastrandrea, and L.L. White (eds.)]. World Meteorological Organization, Geneva, Switzerland, pp. 107-109. (in Arabic, Chinese, English, French, Russian, and Spanish)

Heat Stress and Heat Waves

Lennart Olsson (Sweden), Dave Chadee (Trinidad and Tobago), Ove Hoegh-Guldberg (Australia), Michael Oppenheimer (USA), John Porter (Denmark), Hans-O. Pörtner (Germany), David Satterthwaite (UK),Kirk R. Smith (USA), Maria Isabel Travasso (Argentina), Petra Tschakert (USA)

According to WGI, it is *very likely* that the number and intensity of hot days have increased markedly in the last three decades and *virtually certain* that this increase will continue into the late 21st century. In addition, it is *likely* (*medium confidence*) that the occurrence of heat waves (multiple days of hot weather in a row) has more than doubled in some locations, but *very likely* that there will be more frequent heat waves over most land areas after mid-century. Under a medium warming scenario, Coumou et al. (2013) predicted that the number of monthly heat records will be more than 12 times more common by the 2040s compared to a non-warming world. In a longer time perspective, if the global mean temperature increases to +7°C or more, the habitability of parts of the tropics and mid-latitudes will be at risk (Sherwood and Huber, 2010). Heat waves affect natural and human systems directly, often with severe losses of lives and assets as a result, and may act as triggers of tipping points (Hughes et al., 2013). Consequently, heat stress plays an important role in several key risks noted in Chapter 19 and CC-KR.

Economy and Society (Chapters 10, 11, 12, 13)

Environmental heat stress has already reduced the global labor capacity to 90% in peak months with a further predicted reduction to 80% in peak months by 2050. Under a high warming scenario (RCP8.5), labor capacity is expected to be less than 40% of present-day conditions in peak months by 2200 (Dunne et al., 2013). Adaptation costs for securing cooling capacities and emergency shelters during heat waves will be substantial.

Heat waves are associated with social predicaments such as increasing violence (Anderson, 2012) as well as overall health and psychological distress and low life satisfaction (Tawatsupa et al., 2012). Impacts are highly differential with disproportional burdens on poor people, elderly people, and those who are marginalized (Wilhelmi et al., 2012). Urban areas are expected to suffer more due to the combined effect of climate and the urban heat island effect (Fischer et al., 2012; see also Section 8.2.3.1). In low- and medium-income countries, adaptation to heat stress is severely restricted for most people in poverty and particularly those who are dependent on working outdoors in agriculture, fisheries, and construction. In small-scale agriculture, women and children are particularly at risk due to the gendered division of labor (Croppenstedt et al., 2013). The expected increase in wildfires as a result of heat waves (Pechony and Shindell, 2010) is a concern for human security, health, and ecosystems. Air pollution from wildfires already causes an estimated 339,000 premature deaths per year worldwide (Johnston et al., 2012).

Human Health (Chapter 11)

Morbidity and mortality due to heat stress is now common all over the world (Barriopedro et al., 2011; Nitschke et al., 2011; Rahmstorf and Coumou, 2011; Diboulo et al., 2012; Hansen et al., 2012). Elderly people and people with circulatory and respiratory diseases are also vulnerable even in developed countries; they can become victims even inside their own houses (Honda et al., 2011). People in physical work are at particular risk as such work produces substantial heat within the body, which cannot be released if the outside temperature and humidity is above certain limits (Kjellstrom et al., 2009). The risk of non-melanoma skin cancer from exposure to UV radiation during summer months increases with temperature (van der Leun, et al., 2008). High temperatures are also associated with an increase in air-borne allergens acting as triggers for respiratory illnesses such as asthma, allergic rhinitis, conjunctivitis, and dermatitis (Beggs, 2010).

Ecosystems (Chapters 4, 5, 6, 30)

Tree mortality is increasing globally (Williams et al., 2013) and can be linked to climate impacts, especially heat and drought (Reichstein et al., 2013), even though attribution to climate change is difficult owing to lack of time series and confounding factors. In the Mediterranean region, higher fire risk, longer fire season, and more frequent large, severe fires are expected as a result of increasing heat waves in combination with drought (Duguy et al., 2013; see also Box 4.2).

Marine ecosystem shifts attributed to climate change are often caused by temperature extremes rather than changes in the average (Pörtner and Knust, 2007). During heat exposure near biogeographical limits, even small (<0.5°C) shifts in temperature extremes can have large effects, often exacerbated by concomitant exposures to hypoxia and/or elevated CO_2 levels and associated acidification (*medium confidence*; Hoegh-Guldberg et al., 2007; see also Figure 6-5; Sections 6.3.1, 6.3.5, 30.4, 30.5; CC-MB).

Most coral reefs have experienced heat stress sufficient to cause frequent mass coral bleaching events in the last 30 years, sometimes followed by mass mortality (Baker et al., 2008). The interaction of acidification and warming exacerbates coral bleaching and mortality (*very high confidence*).Temperate seagrass and kelp ecosystems will decline with the increased frequency of heat waves and through the impact of invasive subtropical species (*high confidence*; Sections 5, 6, 30.4, 30.5, CC-CR, CC-MB).

Agriculture (Chapter 7)

Excessive heat interacts with key physiological processes in crops. Negative yield impacts for all crops past +3°C of local warming without adaptation, even with benefits of higher CO_2 and rainfall, are expected even in cool environments (Teixeira et al., 2013). For tropical systems where moisture availability or extreme heat limits the length of the growing season, there is a high potential for a decline in the length of the growing season and suitability for crops (*medium evidence*, *medium agreement*; Jones and Thornton, 2009). For example, half of the wheat-growing area of the Indo-Gangetic Plains could become significantly heat-stressed by the 2050s.

There is *high confidence* that high temperatures reduce animal feeding and growth rates (Thornton et al., 2009). Heat stress reduces reproductive rates of livestock (Hansen, 2009), weakens their overall performance (Henry et al., 2012), and may cause mass mortality of animals in feedlots during heat waves (Polley et al., 2013). In the USA, current economic losses due to heat stress of livestock are estimated at several billion US$ annually (St-Pierre et al., 2003).

References

Anderson, C.A., 2012: Climate change and violence. In: *The Encyclopedia of Peace Psychology* [Christie, D.J. (ed.)]. John Wiley & Sons/Blackwell, Chichester, UK, pp. 128-132.

Baker, A.C., P.W. Glynn, and B. Riegl, 2008: Climate change and coral reef bleaching: an ecological assessment of long-term impacts, recovery trends and future outlook. *Estuarine, Coastal and Shelf Science*, **80(4)**, 435-471.

Barriopedro, D., E.M. Fischer, J. Luterbacher, R.M. Trigo, and R. García-Herrera, 2011: The hot summer of 2010: redrawing the temperature record map of Europe. *Science*, **332(6026)**, 220-224.

Beggs, P.J., 2010: Adaptation to impacts of climate change on aeroallergens and allergic respiratory diseases. *International Journal of Environmental Research and Public Health*, **7(8)**, 3006-3021.

Coumou, D., A. Robinson, and S. Rahmstorf, 2013: Global increase in record-breaking monthly-mean temperatures. *Climatic Change*, **118(3-4)**, 771-782.

Croppenstedt, A., M. Goldstein, and N. Rosas, 2013: Gender and agriculture: inefficiencies, segregation, and low productivity traps. *The World Bank Research Observer*, **28(1)**, 79-109.

Diboulo, E., A. Sie, J. Rocklöv, L. Niamba, M. Ye, C. Bagagnan, and R. Sauerborn, 2012: Weather and mortality: a 10 year retrospective analysis of the Nouna Health and Demographic Surveillance System, Burkina Faso. *Global Health Action*, **5**, 19078, doi:10.3402/gha.v5i0.19078.

Duguy, B., S. Paula, J.G. Pausas, J.A. Alloza, T. Gimeno, and R.V. Vallejo, 2013: Effects of climate and extreme events on wildfire regime and their ecological impacts. In: *Regional Assessment of Climate Change in the Mediterranean, Volume 3: Case Studies* [Navarra, A. and L. Tubiana (eds.)]. Advances in Global Change Research Series: Vol. 52, Springer, Dordrecht, Netherlands, pp. 101-134.

Dunne, J.P., R.J. Stouffer, and J.G. John, 2013: Reductions in labour capacity from heat stress under climate warming. *Nature Climate Change*, **3**, 563-566.

Fischer, E., K. Oleson, and D. Lawrence, 2012: Contrasting urban and rural heat stress responses to climate change. *Geophysical Research Letters*, **39(3)**, L03705, doi:10.1029/2011GL050576.

Hansen, J., M. Sato, and R. Ruedy, 2012: Perception of climate change. *Proceedings of the National Academy of Sciences of the United States of America*, **109(37)**, E2415-E2423.

Hansen, P.J., 2009: Effects of heat stress on mammalian reproduction. *Philosophical Transactions of the Royal Society B*, **364(1534)**, 3341-3350.

Henry, B., R. Eckard, J.B. Gaughan, and R. Hegarty, 2012: Livestock production in a changing climate: adaptation and mitigation research in Australia. *Crop and Pasture Science*, **63(3)**, 191-202.

Hoegh-Guldberg, O., P. Mumby, A. Hooten, R. Steneck, P. Greenfield, E. Gomez, C. Harvell, P. Sale, A. Edwards, and K. Caldeira, 2007: Coral reefs under rapid climate change and ocean acidification. *Science*, **318(5857)**, 1737-1742.

Honda, Y., M. Ono, and K.L. Ebi, 2011: Adaptation to the heat-related health impact of climate change in Japan. In: *Climate Change Adaptation in Developed Nations: From Theory to Practice* [Ford, J.D. and L. Berrang-Ford (eds.)]. Springer, Dordrecht, Netherlands, pp. 189-203.

Hughes, T.P., S. Carpenter, J. Rockström, M. Scheffer, and B. Walker, 2013: Multiscale regime shifts and planetary boundaries. *Trends in Ecology & Evolution*, **28(7)**, 389-395.

Johnston, F.H., S.B. Henderson, Y. Chen, J.T. Randerson, M. Marlier, R.S. DeFries, P. Kinney, D.M. Bowman, and M. Brauer, 2012: Estimated global mortality attributable to smoke from landscape fires. *Environmental Health Perspectives*, **120(5)**, 695-701.

Jones, P.G. and P.K. Thornton, 2009: Croppers to livestock keepers: livelihood transitions to 2050 in Africa due to climate change. *Environmental Science & Policy*, **12(4)**, 427-437.

Kjellstrom, T., R. Kovats, S. Lloyd, T. Holt, and R. Tol, 2009: The direct impact of climate change on regional labor productivity. *Archives of Environmental & Occupational Health*, **64(4)**, 217-227.

Nitschke, M., G.R. Tucker, A.L. Hansen, S. Williams, Y. Zhang, and P. Bi, 2011: Impact of two recent extreme heat episodes on morbidity and mortality in Adelaide, South Australia: a case-series analysis. *Environmental Health*, **10**, 42, doi:10.1186/1476-069X-10-42.

Pechony, O. and D. Shindell, 2010: Driving forces of global wildfires over the past millennium and the forthcoming century. *Proceedings of the National Academy of Sciences of the United States of America*, **107(45)**, 19167-19170.

Polley, H.W., D.D. Briske, J.A. Morgan, K. Wolter, D.W. Bailey, and J.R. Brown, 2013: Climate change and North American rangelands: trends, projections, and implications. *Rangeland Ecology & Management*, **66(5)**, 493-511.

Pörtner, H.O. and R. Knust, 2007: Climate change affects marine fishes through the oxygen limitation of thermal tolerance. *Science*, **315(5808)**, 95-97.

Rahmstorf, S. and D. Coumou, 2011: Increase of extreme events in a warming world. *Proceedings of the National Academy of Sciences of the United States of America*, **108(44)**, 17905-17909.

Reichstein, M., M. Bahn, P. Ciais, D. Frank, M.D. Mahecha, S.I. Seneviratne, J. Zscheischler, C. Beer, N. Buchmann, and D.C. Frank, 2013: Climate extremes and the carbon cycle. *Nature*, **500(7462)**, 287-295.

Sherwood, S.C. and M. Huber, 2010: An adaptability limit to climate change due to heat stress. *Proceedings of the National Academy of Sciences of the United States of America*, **107(21)**, 9552-9555.

Smith, K.R., M. Jerrett, H.R. Anderson, R.T. Burnett, V. Stone, R. Derwent, R.W. Atkinson, A. Cohen, S.B. Shonkoff, and D. Krewski, 2010: Public health benefits of strategies to reduce greenhouse-gas emissions: health implications of short-lived greenhouse pollutants. *The Lancet*, **374(9707)**, 2091-2103.

St-Pierre, N., B. Cobanov, and G. Schnitkey, 2003: Economic losses from heat stress by US livestock industries. *Journal of Dairy Science*, **86**, E52-E77.

Tawatsupa, B., V. Yiengprugsawan, T. Kjellstrom, and A. Sleigh, 2012: Heat stress, health and well-being: findings from a large national cohort of Thai adults. *BMJ Open*, **2(6)**, e001396, doi:10.1136/bmjopen-2012-001396.

Teixeira, E.I., G. Fischer, H. van Velthuizen, C. Walter, and F. Ewert, 2013: Global hot-spots of heat stress on agricultural crops due to climate change. *Agricultural and Forest Meteorology*, **170**, 206-215.

Thornton, P., J. Van de Steeg, A. Notenbaert, and M. Herrero, 2009: The impacts of climate change on livestock and livestock systems in developing countries: a review of what we know and what we need to know. *Agricultural Systems*, **101(3)**, 113-127.

van der Leun, J.C., R.D. Piacentini, and F.R. de Gruijl, 2008: Climate change and human skin cancer. *Photochemical & Photobiological Sciences*, **7(6)**, 730-733.

Wilhelmi, O., A. de Sherbinin, and M. Hayden, 2012: Chapter 12. Exposure to heat stress in urban environments: current status and future prospects in a changing climate. In: *Ecologies and Politics of Health* [King, B. and K. Crews (eds.)]. Routledge Press, Abingdon, UK and New York, NY, USA, pp. 219-238.

Williams, A.P., C.D. Allen, A.K. Macalady, D. Griffin, C.A. Woodhouse, D.M. Meko, T.W. Swetnam, S.A. Rauscher, R. Seager, and H.D. Grissino-Mayer, 2013: Temperature as a potent driver of regional forest drought stress and tree mortality. *Nature Climate Change*, **3**, 292-297.

This cross-chapter box should be cited as:

Olsson, L., D.D. Chadee, O. Hoegh-Guldberg, M. Oppenheimer, J.R. Porter, H.-O. Pörtner, D. Satterthwaite, K.R. Smith, M.I. Travasso, and P. Tschakert, 2014: Cross-chapter box on heat stress and heat waves. In: *Climate Change 2014: Impacts, Adaptation, and Vulnerability. Summaries, Frequently Asked Questions, and Cross-Chapter Boxes. Contribution of Working Group II to the Fifth Assessment Report of the Intergovernmental Panel on Climate Change* [Field, C.B., V.R. Barros, D.J. Dokken, K.J. Mach, M.D. Mastrandrea, T.E. Bilir, M. Chatterjee, K.L. Ebi, Y.O. Estrada, R.C. Genova, B. Girma, E.S. Kissel, A.N. Levy, S. MacCracken, P.R. Mastrandrea, and L.L. White (eds.)]. World Meteorological Organization, Geneva, Switzerland, pp. 111-113. (in Arabic, Chinese, English, French, Russian, and Spanish)

A Selection of the Hazards, Key Vulnerabilities, Key Risks, and Emergent Risks Identified in the WGII Contribution to the Fifth Assessment Report

Birkmann, Joern (Germany), Rachel Licker (USA), Michael Oppenheimer (USA), Maximiliano Campos (Costa Rica), Rachel Warren (UK), George Luber (USA), Brian O'Neill (USA), and Kiyoshi Takahashi (Japan)

The accompanying table provides a selection of the hazards, key vulnerabilities, key risks, and emergent risks identified in various chapters in this report (Chapters 4, 6, 7, 8, 9, 11, 13, 19, 22, 23, 24, 25, 26, 27, 28, 29, 30). Key risks are determined by hazards interacting with vulnerability and exposure of human systems, and ecosystems or species. The table underscores the complexity of risks determined by various climate-related hazards, non-climatic stressors, and multifaceted vulnerabilities. The examples show that underlying phenomena, such as poverty or insecure land-tenure arrangements, unsustainable and rapid urbanization, other demographic changes, failure in governance and inadequate governmental attention to risk reduction, and tolerance limits of species and ecosystems that often provide important services to vulnerable communities, generate the context in which climatic change related harm and loss can occur. The table illustrates that current global megatrends (e.g., urbanization and other demographic changes) in combination and in specific development context (e.g., in low-lying coastal zones), can generate new systemic risks in their interaction with climate hazards that exceed existing adaptation and risk management capacities, particularly in highly vulnerable regions, such as dense urban areas of low-lying deltas. A representative set of lines of sight is provided from across WGI and WGII. See Section 19.6.2.1 for a full description of the methods used to select these entries.

Table KR-1 | Examples of hazards/stressors, key vulnerabilities, key risks, and emergent risks.

	Hazard	Key vulnerabilities	Key risks	Emergent risks
Terrestrial and Inland Water Systems (Chapter 4)	Rising air, soil, and water temperature (Sections 4.2.4, 4.3.2, 4.3.3)	Exceedance of eco-physiological climate tolerance limits of species (limited coping and adaptive capacities), increased viability of alien organisms	Risk of loss of native biodiversity, increase in non-native organism dominance	Cascades of native species loss due to interdependencies
		Health response to spread of temperature-sensitive vectors (insects)	Risk of novel and/or much more severe pest and pathogen outbreaks	Interactions among pests, drought, and fire can lead to new risks and large negative impacts on ecosystems.
	Change in seasonality of rain (Section 4.3.3)	Increasing susceptibility of plants and ecosystem services, due to mismatch between plant life strategy and growth opportunities	Changes in plant functional type mix leading to biome change with respective risks for ecosystems and ecosystem services	Fire-promoting grasses grow in winter-rainfall areas and provide fuel in dry summers.
Ocean Systems (Chapter 6)	Rising water temperature, increase of (thermal and haline) stratification, and marine acidification (Section 6.1.1)	Tolerance limits of endemic species surpassed (limited coping and adaptive capacities), increased abundance of invasive organisms, high susceptibility and sensitivity of warm water coral reefs and respective ecosystem services for coastal communities (Sections 6.3.1, 6.4.1)	Risk of loss of endemic species, mixing of ecosystem types, increased dominance of invasive organisms. Increasing risk of loss of coral cover and associated ecosystem with reduction of biodiversity and ecosystem services (Section 6.3.1)	Enhancement of risk as a result of interactions, e.g., acidification and warming on calcareous organisms (Section 6.3.5)
		New vulnerabilities can emerge as a result of shifted productivity zones and species distribution ranges, largely from low to high latitudes (Sections 6.3.4, 6.5.1), shifting fishery catch potential with species migration (Sections 6.3.1, 6.5.2, 6.5.3)	Risks due to unknown productivity and services of new ecosystem types (Sections 6.4.1, 6.5.3)	Enhancement of risk due to interactions of warming, hypoxia, acidification, new biotic interactions (Sections 6.3.5, 6.3.6)
	Expansion of oxygen minimum zones and coastal dead zones with stratification and eutrophication (Section 6.1.1)	Increasing susceptibility because hypoxia tolerance limits of larger animals surpassed, habitat contraction and loss for midwater fishes and benthic invertebrates (Section 6.3.3)	Risk of loss of larger animals and plants, shifts to hypoxia-adapted, largely microbial communities with reduced biodiversity (Section 6.3.3)	Enhancement of risk due to expanding hypoxia in warming and acidifying oceans (Section 6.3.5)
	Enhanced harmful algal blooms in coastal areas due to rising water temperature (Section 6.4.2.3)	Increasing susceptibility and limited adaptive capacities of important ecosystems and valuable services due to already existing multiple stresses (Sections 6.3.5, 6.4.1)	Increasing risk due to enhanced frequency of dinoflagellate blooms and respective potential losses and degradations of coastal ecosystems and ecosystem services (Section 6.4.2)	Disproportionate enhancement of risk due to interactions of various stresses (Section 6.3.5)
Food Security and Food Production Systems (Chapter 7)	Rising average temperatures and more frequent extreme temperatures (Sections 7.1, 7.2, 7.4, 7.5)	Susceptibility of all elements of the food system from production to consumption, particularly for key grain crops	Risk of crop failures, breakdown of food distribution and storage processes	Increase in the global population to about 9 billion combined with rising temperatures and other trace gases such as ozone affecting food production and quality. Upper temperature limit to the ability of some food systems to adapt
	Extreme precipitation and droughts (Section 7.4)	Crops, pasture, and husbandry are susceptible and sensitive to drought and extreme precipitation.	Risk of crop failure, risk of limited food access and quality	Flood and droughts affect crop yields and quality, and directly affect food access in most developing countries. (Section 7.4)
Urban Areas (Chapter 8)	Inland flooding (Sections 8.2.3, 8.2.4)	Large numbers of people exposed in urban areas to flood events. Particularly susceptible are people in low-income informal settlements with inadequate infrastructure (and often on flood plains or along river banks). These bring serious environmental health consequences from overwhelmed, aging, poorly maintained, and inadequate urban drainage infrastructure and widespread impermeable surfaces. Local governments are often unable or unwilling to give attention to needed flood-related disaster risk reduction. Much of the urban population unable to get or afford housing that protects against flooding, or insurance. Certain groups are more sensitive to ill health from flood impacts, which may include increased mosquito- and water-borne diseases.	Risks of deaths and injuries and disruptions to livelihoods/incomes, food supplies, and drinking water	In many urban areas, larger and more frequent flooding impacting much larger population. No insurance available or impacts reaching the limits of insurance. Shift in the burden of risk management from the state to those at risk, leading to greater inequality and property blight, abandonment of urban districts, and the creation of high-risk/high-poverty spatial traps
	Coastal flooding (including sea level rise and storm surge) (Sections 8.1.4, 8.2.3, 8.2.4)	High concentrations of people, businesses, and physical assets including critical infrastructure exposed in low-lying and unprotected coastal zones. Particularly susceptible is the urban population that is unable to get or afford housing that protects against flooding or insurance. The local government is unable or unwilling to give needed attention to disaster risk reduction.	Risks from deaths and injuries and disruptions to livelihoods/incomes, food supplies, and drinking water	Additional 2 billion or so urban dwellers expected over the next three decades Sea level rise means increasing risks over time, yet with high and often increasing concentrations of population and economic activities on the coasts. No insurance available or reaching the limits of insurance; shift in the burden of risk management from the state to those at risk leading to greater inequality and property blight, abandonment of urban districts, and the creation of high-risk/high-poverty spatial traps

Continued next page →

KR

Table KR-1 (continued)

	Hazard	Key vulnerabilities	Key risks	Emergent risks
Urban Areas (continued) (Chapter 8)	Heat and cold (including urban heat island effect) (Section 8.2.3)	Particularly susceptible is a large and often increasing urban population of infants, young children, older age groups, expectant mothers, people with chronic diseases or compromised immune system in settlements exposed to higher temperatures (especially in heat islands) and unexpected cold spells. Inability of local organizations for health, emergency, and social services to adapt to new risk levels and set up needed initiatives for vulnerable groups	Risk of mortality and morbidity increasing, including shifts in seasonal patterns and concentrations due to hot days with higher or more prolonged high temperatures or unexpected cold spells. Avoiding risks often most difficult for low-income groups	Duration and variability of heat waves increasing risks over time for most locations owing to interactions with multiple stressors such as air pollution
	Water shortages and drought in urban regions (Sections 8.2.3, 8.2.4)	Lack of piped water to homes of hundreds of millions of urban dwellers. Many urban areas subject to water shortages and irregular supplies, with constraints on increasing supplies. Lack of capacity and resilience in water management regimes including rural–urban linkages. Dependence on water resources in energy production systems	Risks from constraints on urban water provision services to people and industry with human and economic impacts. Risk of damage and loss to urban ecology and its services including urban and peri-urban agriculture.	Cities' viability may be threatened by loss or depletion of freshwater sources—including for cities dependent on distant glacier melt water or on depleting groundwater resources.
	Changes in urban meteorological regimes lead to enhanced air pollution. (Section 8.2.3)	Increases in exposure and in pollution levels with impacts most serious among physiologically susceptible populations. Limited coping and adaptive capacities, due to lacking implementation of pollution control legislation of urban governments	Increasing risk of mortality and morbidity, lowered quality of life. These risks can also undermine the competitiveness of global cities to attract key workers and investment.	Complex and compounding health crises
	Geo-hydrological hazards (salt water intrusion, mud/land slides, subsidence) (Sections 8.2.3, 8.2.4)	Local structures and networked infrastructure (piped water, sanitation, drainage, communications, transport, electricity, gas) particularly susceptible. Inability of many low-income households to move to housing on safer sites.	Risk of damage to networked infrastructure. Risk of loss of human life and property	Potential for large local and aggregate impacts Knock-on effects for urban activities and well-being
	Wind storms with higher intensity (Sections 8.1.4, 8.2.4)	Substandard buildings and physical infrastructure and the services and functions they support particularly susceptible. Old and difficult to retrofit buildings and infrastructure in cities Local government unable or unwilling to give attention to disaster risk reduction (limited coping and adaptive capacities)	Risk of damage to dwellings, businesses, and public infrastructure. Risk of loss of function and services. Challenges to recovery, especially where insurance is absent	Challenges to individuals, businesses, and public agencies where the costs of retrofitting are high and other sectors or interests capture investment budgets; potential for tensions between development and risk reduction investments
	Changing hazard profile including novel hazards and new multi-hazard complexes (Sections 8.1.4, 8.2.4)	Newly exposed populations and infrastructure, especially those with limited capacity for multi-hazard risk forecasting and where risk reduction capacity is limited, e.g., where risk management planning is overly hazard specific including where physical infrastructure is predesigned in anticipation of other risks (e.g., geophysical rather than hydrometeorological)	Risks from failures within coupled systems, e.g., reliance of drainage systems on electric pumps, reliance of emergency services on roads and telecommunications. Potential of psychological shock from unanticipated risks	Loss of faith in risk management institutions. Potential for extreme impacts that are magnified by a lack of preparation and capacity in response
	Compound slow-onset hazards including rising temperatures and variability in temperature and water (Sections 8.2.2, 8.2.4)	Large sections of the urban population in low- and middle-income nations with livelihoods or food supplies dependent on urban and peri-urban agriculture are especially susceptible.	Risk of damage to or degradation of soils, water catchment capacity, fuel wood production, urban and peri-urban agriculture, and other productive or protective ecosystem services. Risk of knock-on impacts for urban and peri-urban livelihoods and urban health	Collapsing of peri-urban economies and ecosystem services with wider implications for urban food security, service provision, and disaster risk reduction
	Climate change–induced or intensified hazard of more diseases and exposure to disease vectors (Sections 8.2.3, 8.2.4)	Large urban population that is exposed to food-borne and water-borne diseases and to malaria, dengue, and other vector-borne diseases that are influenced by climate change	Risk due to increases in exposure to these diseases	Lack of capacity of public health system to simultaneously address these health risks with other climate-related risks such as flooding
Rural Areas (Chapter 9)	Drought in pastoral areas (Sections 9.3.3.1, 9.3.5.2)	Increasing vulnerability due to encroachment on pastoral rangelands, inappropriate land policy, misperception and undermining of pastoral livelihoods, conflict over natural resources, all driven by remoteness and lack of voice	Risk of famine Risk of loss of revenues from livestock trade	Increasing risks for rural livelihoods through animal disease in pastoral areas combined with direct impacts of drought
	Effects of climate change on artisanal fisheries (Sections 9.3.3.1, 9.3.5.2)	Artisanal fisheries affected by pollution and mangrove loss, competition from aquaculture, and the neglect of the sector by governments and researchers as well as complex property rights	Risk of economic losses for artisanal fisherfolk, due to declining catches and incomes and damage to fishing gear and infrastructure	Reduced dietary protein for those consuming artisanally caught fish, combined with other climate-related risks

Continued next page →

KR

Table KR-1 (continued)

	Hazard	Key vulnerabilities	Key risks	Emergent risks
Rural Areas (continued) (Chapter 9)	Water shortages and drought in rural areas (Section 9.3.5.1.1)	Rural people lacking access to drinking and irrigation water. High dependence of rural people on natural resource-related activities. Lack of capacity and resilience in water management regimes (institutionally driven). Increased water demand from population pressure	Risk of reduced agricultural productivity of rural people, including those dependent on rainfed or irrigated agriculture, or high-yield varieties, forestry, and inland fisheries. Risk of food insecurity and decrease in incomes. Decreases in household nutritional status (Section 9.3.5.1)	Impacts on livelihoods driven by interaction with other factors (water management institutions, water demand, water used by non-food crops), including potential conflicts for access to water. Water-related diseases
Human Health (Chapter 11)	Increasing frequency and intensity of extreme heat	Older people living in cities are most susceptible to hot days and heat waves, as well as people with preexisting health conditions. (Section 11.3)	Risk of increased mortality and morbidity during hot days and heat waves. (Section 11.4.1) Risk of mortality, morbidity, and productivity loss, particularly among manual workers in hot climates	The number of elderly people is projected to triple from 2010 to 2050. This can result in overloading of health and emergency services.
	Increasing temperatures, increased variability in precipitation	Poorer populations are particularly susceptible to climate-induced reductions in local crop yields. Food insecurity may lead to undernutrition. Children are particularly vulnerable. (Section 11.3)	Risk of a larger burden of disease and increased food insecurity for particular population groups. Increasing risk that progress in reducing mortality and morbidity from undernutrition may slow or reverse. (Section 11.6.1)	Combined effects of climate impacts, population growth, plateauing productivity gains, land demand for livestock, biofuels, persistent inequality, and ongoing food insecurity for the poor
	Increasing temperatures, changing patterns of precipitation	Non-immune populations who are exposed to water- and vector-borne diseases that are sensitive to meteorological conditions (Section 11.3)	Increasing health risks due to changing spatial and temporal distribution of diseases strains public health systems, especially if this occurs in combination with economic downturn. (Section 11.5.1)	Rapid climate and other environmental change may promote emergence of new pathogens.
	Increased variability in precipitation	People exposed to diarrhea aggravated by higher temperatures, and unusually high or low precipitation (Section 11.3)	Risk that the progress to date in reducing childhood deaths from diarrheal disease is compromised (Section 11.5.2)	Increased rate of failure of water and sanitation infrastructure due to climate change leading to higher diarrhea risk
Livelihoods and Poverty (Chapter 13)	Increasing frequency and severity of droughts, coupled with decreasing rainfall and/or increased unpredictability of rainfall (Sections 13.2.1.2, 13.2.1.4, 13.2.2.2)	Poorly endowed farmers (high and persistent poverty), particularly in drylands, are susceptible to these hazards, since they have a very limited ability to compensate for losses in water-dependent farming systems and/or livestock.	Risk of irreversible harm due to short time for recovery between droughts, approaching tipping point in rainfed farming system and/or pastoralism	Deteriorating livelihoods stuck in poverty traps, heightened food insecurity, decreased land productivity, outmigration, and new urban poor in LICs and MICs
	Floods and flash floods in informal urban settlements and mountain environments, destroying physical assets (e.g., homes, roads, terraces, irrigation canals) (Sections 13.2.1.1, 13.2.1.3, 13.2.1.4)	High exposure and susceptibility of people, particularly children and elderly, as well as disabled in flood-prone areas. Inadequate infrastructure, culturally imposed gender roles, and limited ability to cope and adapt due to political and institutional marginalization and high poverty adds to the susceptibility of these people in informal urban settlements; limited political interest in development and building adaptive capacity	Risk of high morbidity and mortality due to floods and flash floods. Factors that further increase risk may include a shift from transient to chronic poverty due to eroded human and economic assets (e.g., labor market) and economic losses due to infrastructure damage.	Exacerbated inequality between better-endowed households able to invest in flood-control measures and/or insurance and increasingly vulnerable populations prone to eviction, erosion of livelihoods, and outmigration
	Increased variability of precipitation; shifts in mean climate and extreme events (Sections 13.2.1.1, 13.2.1.4)	Limited ability to cope owing to exhaustion of social networks, especially among the elderly and female-headed households; mobilization of labor and food no longer possible	Hazard combines with vulnerability to shift populations from transient to chronic poverty due to persistent and irreversible socioeconomic and political marginalization. In addition, the lack of governmental support, as well as limited effectiveness of response options, increase the risk.	Increasing yet invisible multidimensional vulnerability and deprivation at the convergence of climatic hazards and socioeconomic stressors
	Successive and extreme events (floods, droughts) coupled with increasing temperatures and rising water demand (Sections 13.2.1.1, 13.2.1.5)	Rural communities are particularly susceptible, due to the marginalization of rural water users to the benefit of urban users, given political and economic priorities (e.g., Australia, Andes, Himalayas, Caribbean).	Risk of loss of rural livelihoods, severe economic losses in agriculture, and damage to cultural values and identity; mental health impacts (including increased rates of suicide)	Loss of rural livelihoods that have existed for generations, heightened outmigration to urban areas; emergence of new poverty in MICs and HICs
	Sea level rise (Sections 13.1.4, 13.2.1.1, 13.2.2.1, 13.2.2.3)	High number of people exposed in low-lying areas coupled with high susceptibility due to multidimensional poverty, limited alternative livelihood options among poor households, and exclusion from institutional decision-making structures	Risk of severe harm and loss of livelihoods. Potential loss of common-pool resources; of sense of place, belonging, and identity, especially among indigenous populations	Loss of livelihoods and mental health risks due to radical change in landscape, disappearance of natural resources, and potential relocation; increased migration
	Increasing temperatures and heat waves (Sections 13.2.1.5, 13.2.2.3, 13.2.2.4)	Agricultural wage laborers, small-scale farmers in areas with multidimensional poverty and economic marginalization, children in urban slums, and the elderly are particularly susceptible.	Risk of increased morbidity and mortality due to heat stress, among male and female workers, children, and the elderly, limited protection due to socioeconomic discrimination and inadequate governmental responses	Declining labor pool for agriculture coupled with new challenges for rural health care systems in LICs and MICs; aging and low-income populations without safety nets in HICs at risk

Continued next page →

Table KR-1 (continued)

	Hazard	Key vulnerabilities	Key risks	Emergent risks
Livelihoods and Poverty (continued) (Chapter 13)	Increased variability of rainfall and/or extreme events (floods, droughts, heat waves) (Sections 13.2.1.1, 13.2.1.3, 13.2.1.4, 13.2.1.5)	People highly dependent on rainfed agriculture are particularly at risk. Persistent poverty among subsistence farmers and urban wage laborers who are net buyers of food with limited coping mechanisms	Risk of crop failure, spikes in food prices, reduction in consumption to protect household assets, risk of food insecurity, shifts from transient to chronic poverty due to limited ability to reduce risks	Food riots, child food poverty, global food crises, limits of insurance and other risk-spreading strategies
	Changing rainfall patterns (temporally and spatially)	Households or people with a high dependence on rainfed agriculture and little access to alternative modes of income	Risks of crop failure, food shortage, severe famine	Coincidence of hazard with periods of high global food prices leads to risk of failure of coping strategies and adaptation mechanisms such as crop insurance (risk spreading).
	Stressor from soaring demand (and prices) for biofuel feedstocks due to climate policies	Farmers and groups that have unclear and/or insecure land tenure arrangements are exposed to the dispossession of land due to land grabbing in developing countries.	Risk of harm and loss of livelihoods for some rural residents due to soaring demand for biofuel feedstocks and insecure land tenure and land grabbing	Creation of large groups of landless farmers unable to support themselves. Social unrest due to disparities between intensive energy production and neglected food production
	Increasing frequency of extreme events (droughts, floods), e.g., if 1:20 year drought/flood becomes 1:5 year drought/flood	Pastoralists and small farmers subject to damage to their productive assets (e.g., herds of livestock; dykes, fences, terraces)	Risk of the loss of livelihoods and harm due to shorter time for recovery between extremes. Pastoralists restocking after a drought may take several years; in terraced agriculture, need to rebuild terraces after flood, which may take several years	Collapse of coping strategies with risk of collapsing livelihoods. Adaptation mechanisms such as insurance fail due to increasing frequency of claims.
Emergent Risks and Key Vulnerabilities (Chapter 19)	Warming and drying (precipitation changes of uncertain magnitude) (WGI AR5 TS 5.3; SPM; Sections 11.3, 12.4)	Limits to coping capacity to deal with reduced water availability; increasing exposure and demand due to population increase; conflicting demands for alternative water uses; sociocultural constraints on some adaptation options (Sections 19.2.2, 19.3.2.2, 19.6.1.1, 19.6.3.4)	Risk of harm and loss due to livelihood degradation from systematic constraints on water resource use that lead to supply falling far below demand. In addition, limited coping and adaptation options increase the risk of harm and loss. (Sections 19.3.2.2, 19.6.3.4)	Competition for water from diverse sectors (e.g., energy, agriculture, industry) interacts with climate changes to produce locally severe shortages. (Sections 19.3.2.2, 19.6.3.4)
	Changes in regional and seasonal temperature and precipitation over land (WGI AR5 TS 5.3; SPM; Sections 11.3, 12.4)	Communities highly dependent on ecosystem services (Sections 19.2.2.1, 19.3.2.1) which are negatively affected by changes in regional and seasonal temperature	Risk of large-scale species richness loss over most of the global land surface. 57 ± 6% of widespread and common plants and 34 ± 7% of widespread and common animals are expected to lose ≥50% of their current climatic range by the 2080s leading to loss of services. (Section 19.3.2.1)	Widespread loss of ecosystem services, including: provisioning, such as food and water; regulating, such as the control of climate and disease; supporting, such as nutrient cycles and crop pollination; and cultural, such as spiritual and recreational benefit (Sections 19.3.2.1, 19.6.3.4)
Africa (Chapter 22)	Increasing temperature	Children, pregnant women, and those with compromised health status are particularly at risk for temperature-related changes in diarrheal and vector-borne diseases, and for temperature-related reductions in crop yields. Outdoor workers, older adults, and young children are most susceptible to hot weather and heat waves. (Sections 22.3.5.2, 22.3.5.4)	Risk of changes in the geographic distribution, seasonality, and incidence of infectious diseases, leading to increases in the health burden. Risk of increased burdens of stunting in children. Risk of increase in morbidity and mortality during hot days and heat waves	Interactions among factors lead to emerging and re-emerging epidemics.
		Populations dependent on aquatic systems and aquatic ecosystem services that are sensitive to increased water temperatures	Loss of aquatic ecosystems and risks for people who might depend on these resources; reduction in freshwater fisheries production (Sections 22.3.2.2, 22.3.4.4)	Risk of loss of livelihoods due to interactions of loss of ecosystem services and other climate-related stressors on poor communities
		Rural and urban populations whose food and livelihood security is diminished	Risk of harm and loss due to increased heat stress on crops and livestock resulting in reduced productivity; increased food storage losses due to spoilage (Sections 22.3.4.1, 22.3.4.2)	Range expansion of crop pests and diseases to high-elevation agroecosystems (Section 22.3.4.3)
	Extreme events, e.g., floods and flash floods (and drought)	Population groups living in informal settlements in highly exposed urban areas; women and children often the most vulnerable to disaster risk (Sections 22.3.6, 22.4.3)	Increasing risk of mortality, harm and losses due to water logging triggered by heavy rainfall events	Compounded risk of epidemics including diarrheal diseases (e.g., cholera)
		Susceptible groups include those who experience diminished access to food resulting from reduced capacity to transport, store, and market food, such as the urban poor.	Risk of food shortages and of damages to the food system due to storms and flooding	Food price spikes due to convergence of climatic and non-climatic forces that reduce food access for the poor whose income is disproportionately spent on food (Section 22.3.4.5)
		Children, pregnant women, and those with compromised health status are particularly vulnerable to reduced access to safe water and improved sanitation and increasing food insecurity. (Sections 22.3.5.2, 22.3.5.3)	Risk of crop and livestock losses from drought Risk of reduced water supply and quality for household use. (Sections 22.3.4.1, 22.3.4.2) Risk of increased incidence of food- and water-borne diseases (e.g., cholera) and undernutrition. Risk of drinking water contamination due to heavy precipitation events and flooding (Section 22.3.5.2)	Compound effects of high temperature and changes in rainfall on human and natural systems. Increased incidence of stunting in children (Section 22.3.5.3)

Continued next page →

Table KR-1 (continued)

	Hazard	Key vulnerabilities	Key risks	Emergent risks
Europe (Chapter 23)	Extreme weather events (Section 23.9)	Sectors with limited coping and adaptive capacity as well as high sensitivity to these extreme events, such as transport, energy, and health, are particularly susceptible.	Risk of new systemic threats due to stress on multiple and interconnected sectors. Risk of failure of service provision of one or more sectors	Disproportionate intensification of risk due to increasing interdependencies
	Climate change increases the spatial distribution and seasonality of pests and diseases. (Section 23.4.1, 23.4.3, 23.4.4)	High susceptibility of plants and animals that are exposed to pests and diseases	Risk of increases in crop losses and animal diseases or even fatalities of livestock	Increasing risks due to limited response options and various feedback processes in agriculture, e.g., use of pesticides or antibiotics to protect plants and livestock increases resistance of disease vectors
	Extreme weather events and reduced water availability due to climate change (Section 23.3.4)	Low adaptive capacity of power systems might lead to limited energy supply as well as higher supply costs during such extreme events and conditions.	Increasing risk of power shortages due to limited energy supply, e.g., of nuclear power plants due to limited cooling water during heat stress	Continued underinvestment in adaptive energy systems might increase the risk of mismatches between limited energy supply during these events and increased demands, e.g., during a heat wave.
Asia (Chapter 24)	Rising average temperatures and more frequent extreme temperatures, as well as changing rainfall patterns (temporally and spatially)	Food systems and food production systems for key grain crops, particularly rice and other cereal crop farming systems, are highly susceptible. (Section 24.4.4.3)	Risk of crop failures and lower crop yield also can increase the risk of major losses for farmers and rural livelihoods. (Section 24.4.4.3)	Increase in Asian population combined with rising temperatures affecting food production. Upper temperature limit to the ability of some food systems to adapt could be reached.
	Rising sea level	Paddy fields and farmers near the coasts are particularly susceptible. (Section 24.4.4.3)	Risk of loss of arable areas due to submergence (Section 24.4.4.3)	Migration of farming communities to higher elevation areas entails risks for migrants and receiving regions.
	Projected increase in frequency of various extreme events (heat wave, floods, and droughts) and sea level rise	Increasing exposure due to convergence of livelihood and properties into coastal megacities. People in areas that are not sufficiently protected against natural hazards are particularly susceptible.	Risk of loss of life and assets due to coastal floods accompanied by increasing vulnerabilities.	Projected increase in disruptions of basic services such as water supply, sanitation, energy provision, and transportation systems, which themselves could increase vulnerabilities
Australasia (Chapter 25)	Rising air and sea surface temperatures, drying trends, reduced snow cover, increased intensity of severe cyclones, ocean acidification (Section 25.2; Table 25-1; Figure 25-4; WGI AR5 Chapter 14 and Atlas)	Species that live in a limited climatic range and that suffer from habitat fragmentation as well as from external stressors (pollution, runoff, fishing, tourism, introduced predators, and pests) are especially susceptible. (Sections 25.6.1, 25.6.2)	Risk of significant change in community composition and structure of coral reefs and montane ecosystems and risk of loss of some native species in Australia (Sections 25.6.1, 25.6.2, 25.10.2)	Increasing risk from compound extreme events across time and space, and cumulative adaptation needs, with recovery and risk reduction measures hampered further by impacts and responses reaching across different levels of government (Sections 25.10.2, 25.10.3; Box 25-9)
	Increased extreme rainfall related to flood risk in many locations (Section 25.2; Table 25-1)	Adaptation deficit of existing infrastructure and settlements to current flood risk; expansion and densification of urban areas; effective adaptation includes transformative changes such as land-use controls and retreat. (Sections 25.3, 25.10.2; Box 25-8)	Increased frequency and intensity of flood damage to infrastructure and settlements in Australia and New Zealand (Box 25-8; Section 25.10.2)	
	Continuing sea level rise, with projections spanning a particularly large range and continuing beyond 2100, even under mitigation scenarios (Section 25.2; Box 25-1; WGI AR5 Chapter 13)	Long-lived and high asset value coastal infrastructure and low-lying ecosystems are highly susceptible. Expansion of coastal populations and assets into coastal zones increases the exposure. Conflicting priorities constrain adaptation options and limit effective response strategies. (25.3, Box 25-1)	Increasing risks to coastal infrastructure and low-lying ecosystems in Australia and New Zealand, with widespread damages toward the upper end of projected ranges (Box 25-1; Sections 25.6.1, 25.6.2, 25.10.2)	
North America (Chapter 26)	Increases in frequency and/or intensity of extreme events, such as heavy precipitation, river and coastal floods, heat waves, and droughts (Sections 26.2.2, 26.3.1, 26.8.1)	Physical infrastructure in a declining state in urban areas particularly susceptible. Also increases in income disparities and limited institutional capacities might result in larger proportions of people susceptible to these stressors due to limited economic resources. (Sections 26.7, 26.8.2)	Risk of harm and loss in urban areas, particularly in coastal and dry environments due to enhanced vulnerabilities of social groups, physical systems, and institutional settings combined with the increases of extreme weather events (Section 26.8.1)	Inability to reduce vulnerability in many areas results in an increase in risk more so than change in physical hazard. (Section 26.8.3)
	Higher temperatures, decreases in runoff, and lower soil moisture due to climate change (Sections 26.2, 26.3)	Vulnerability of small rural landholders, particularly in Mexican agriculture, and of the poor in rural settlements (Sections 26.5, 26.8.2.2)	Risk of increased losses and decreases in agricultural production. Risk of food and job insecurity for small landholders and social groups in regions exposed to these phenomena (Sections 26.5, 26.8.2.2)	Increasing risks of social instability and local economic disruption due to internal migration (Sections 26.2.1, 26.8.3)

Continued next page →

Table KR-1 (continued)

	Hazard	Key vulnerabilities	Key risks	Emergent risks
North America (continued) (Chapter 26)	Wildfires and drought conditions (Box 26-2)	Indigenous groups, low-income residents in peri-urban areas, and forest systems (Box 26-2; Section 26.8.2)	Risk of loss of ecosystem integrity, property loss, human morbidity, and mortality due to wildfires (Box 26-2; Section 26.8.3)	
	Extreme storm and heat events, air pollution, pollen, and infectious diseases (Section 26.6.1)	Susceptibility of individuals is determined by factors such as economic status, preexisting illness, age, and access to assets. (Section 26.6.1)	Increasing risk of extreme temperature-, storm-, pollen-, and infectious diseases–related human morbidity or mortality (Section 26.6.2)	
	River and coastal floods, and sea level rise (Sections 26.2.2, 26.4.2, 26.8.1)	Increasing exposure of populations, property, as well as ecosystems, partly resulting from overwhelmed drainage networks. Groups and economic sectors that highly depend on the functioning of different supply chains, public health institutions that can be disrupted, and groups that have limited coping capacities to deal with supply chain interruptions and disruptions to their livelihoods are particularly susceptible. (Sections 26.7, 26.8.1)	Risk of property damage, supply chain disruption, public health, water quality impairment, ecosystem disruption, infrastructure damage, and social system disruption from urban flooding due to river and coastal floods and floods of drainage networks (Sections 26.4.2, 26.8.1)	Multiple risks from interacting hazards on populations' livelihoods, infrastructure, and services (Sections 26.7, 26.8.3)
Central and South America (Chapter 27)	Reduced water availability in semi-arid regions and regions dependent on glacier meltwater; flooding in urban areas due to extreme precipitation (Sections 27.2.1, 27.3.3)	Groups that cannot keep agricultural livelihoods and are forced to migrate are especially vulnerable. Limited infrastructure and planning capacity can further increase the lack of coping and adaptive capacities to rapid changes expected (precipitation), especially in large cities.	Risk of loss of human lives, livelihood, and property	Increase in infectious diseases. Economic impacts due to reallocation of populations
	Ocean acidification and warming (Section 27.3.3; Box CC-OA)	Sensitivity of coral reef systems to ocean acidification and warming	Risk of loss of biodiversity (species) and risk of a reduced fishing capacity with respective impacts for coastal livelihoods	Economic losses and impact on food (fishery) production in certain regions
	Extremes of drought/ precipitation (Sections 27.2.1, 27.3.4)	Elevated CO_2 decreases nutrient contents in plants, especially nitrogen in relation to carbon in food products.	Risk of loss of (food) production and productivity in some regions where extreme events may occur. Need to adjust diet due to decrease in food quality (e.g., less protein due to lower nitrogen assimilation). Decrease in bioenergy production	Strong economic impacts related to the need to move crops to more suitable regions. Teleconnections (related to food quality) related to the intense exportation of food by the region. Impacts on energy system and carbon emissions with consequent increase in fossil fuel demand.
	Higher temperatures and humidity lead to a spread of vector-borne diseases in altitude and latitude. (Section 27.3.7)	People exposed and vulnerable to vector-borne diseases and an increase in mosquito biting rates that increase the probability of human infections	Risk of increase in morbidity and in disability-adjusted life years (DALYs); risk of loss of human lives; risk of decrease in school and labor productivity	High economic impacts owing to the necessity to increase the financing of health programs, as well as the costs of DALYs, increase in hospitals and medical infrastructure adequate to cope with increasing disease incidence rates, and the spread of diseases to newer regions
Polar Regions (Chapter 28)	Loss of multi-year ice and reductions in the spatial extent of summer sea ice (Sections 28.2.5, 28.3.2, 28.4.1)	Indigenous communities that depend on sea ice for traditional livelihoods are vulnerable to this hazard, particularly due to loss of breeding and foraging platforms for marine mammals.	Risk of loss of traditional livelihoods and food sources.	Top-down shifts in food webs
		Ecosystems are vulnerable owing to the shifts in the distribution and timing of ice algal and ocean phytoplankton blooms.	Risk of disruption of synchronized timing of zooplankton ontogeny and availability of prey. Increased variability in secondary production while zooplankton adapt to shifts in timing. Risks also to local marine food webs.	Bottom up shifts in food webs. Potential changes in pelagic and benthic coupling
	Ocean acidification (Sections 28.2.2, 28.3.2)	Tolerance limits of endemic species surpassed. Impacts on exoskeleton formation for some species and alteration of physiological and behavioral properties during larval development	Localized loss of endemic species, local impacts on marine food webs	Localized declines in commercial fisheries. Local declines in fish, shellfish, seabirds, and marine mammals
	Shifts in boundaries of marine eco-regions due to rising water temperature, shifts in mixed layer depth, changes in the distribution and intensity of ocean currents (Sections 28.2.2, 28.3.2)	Marine organisms that are susceptible to spatial shifts are particularly vulnerable.	Risk of changes in the structure and function of marine systems and potentially species invasions	Disputes over international fisheries and shared stocks

Continued next page →

Table KR-1 (continued)

	Hazard	Key vulnerabilities	Key risks	Emergent risks
Polar Regions (continued) (Chapter 28)	Declining sea ice, changes in snow and ice timing and state, decreasing predictability of weather (Sections 28.1, 28.4.1)	Many traditional subsistence food sources—especially for indigenous peoples—such as Arctic marine and land mammals, fish, and waterfowl. Various traditional livelihoods are susceptible to these hazards.	Risk of loss of habitats and changes in migration patterns of marine species	Enhancement of risk to food security and basic nutrition—especially for indigenous peoples—from loss of subsistence foods and increased risk to subsistence hunters', herders', and fishers' health and safety in changing ice conditions
	Increased river and coastal flooding and erosion and thawing of permafrost (Sections 28.2.4, 28.3.1, 28.3.4)	Rural and remote communities as well as urban communities in low-lying Arctic areas are exposed. Susceptibility and limited coping capacity of community water supplies due to potential damages to infrastructure.	Community and public health infrastructure damaged resulting in disease from contamination and sea water intrusion	Reduced water quality and quantity may result in increased rates of infection, other medical problems, and hospitalizations.
	Extreme and rapidly changing weather, intense weather and precipitation events, rapid snow and ice melt, changing river and sea ice conditions, permafrost thaw (Section 28.2.4)	People living from subsistence travel and hunting, herding, and fishing, for example indigenous peoples in remote and isolated communities, are particularly susceptible.	Accidents, physical/mental injuries, death, and cold-related exposure, injuries, and diseases	Enhanced risks to safe travel or subsistence hunting, herding, fishing activities affect livelihoods and well-being.
	Diminished sea ice; earlier sea ice melt-out; faster sea ice retreat; thinner, less predictable ice in general; greater variability in snow melt/freeze; ice, weather, winds, temperatures, precipitation (Sections 28.2.5, 28.2.6, 28.4.1)	Livelihoods of many indigenous peoples (e.g., Inuit and Saami) depend upon subsistence hunting and access to and favorable conditions for animals. These livelihoods are susceptible. Also marine ecosystems are susceptible (e.g., marine mammals).	Risk of loss of livelihoods and damage due to, e.g., more difficult access to marine mammals associated with diminishing sea ice (a risk to the Inuit), and loss of access by reindeer to their forage under snow due to ice layers formed by warming winter temperatures and "rain on snow" (a risk to the Saami).	Enhanced risk of loss of livelihoods and culture of increasing numbers of indigenous peoples, exacerbated by increasing loss of lands and sea ice for hunting, herding, fishing due to enhanced petroleum and mineral exploration, and increased maritime traffic
Small Islands (Chapter 29)	Increases in intensity of tropical cyclones (WGI AR5 Sections 14.6, 14.8.4)	Various countries and communities are vulnerable to these hazards because of their high dependence on natural and ecological systems for security of settlements and tourism (Section 29.3.3.1), human health (Section 29.3.3.2), and water resources (Section 29.3.2).	Risk of loss of ecosystems, settlements, and infrastructure, as well as negative impacts on human health and island economies (Figure 29-4)	Increased risk of interactions of damages to ecosystems, settlements, island economies, and risks to human life (Section 29.6; Figure 29-4)
	Ocean warming and acidification leading to coral bleaching (Sections 29.3.1.2, 30.5.4.2, 30.5.6.1.1, 30.5.6.2)	Tropical island communities are highly dependent on coral reef ecosystems for subsistence life styles, food security, coastal protection and beach, and reef-based tourist economic activity, and hence are highly susceptible to the hazard of coral bleaching. (Sections 29.3.1.2, 30.6.2.1.2)	Risk of decline and possible loss of coral reef ecosystems through thermal stress. Risk of serious harm and loss of subsistence lifestyles. Risk of loss of coastal protection and beaches, risk of loss of tourist revenue (Sections 29.3.1.1, 29.3.1.2)	Impacts on human health and loss of subsistence lifestyles. Potential increase in internal migration/urbanization (Section 29.3.3.3; Chapter 9)
	Sea level rise (Sections 29.3.1.1, 30.3.1.2; WGI AR5 Section 3.7.1)	Many small island communities and associated settlements and infrastructure are in low-lying coastal zones (high exposure) and are also vulnerable to increasing inundation, erosion and wave incursion. (Sections 5.3.2, 29.3.1.1; Figure 29-2)	Risk of loss and harm due to sea level rise in small island communities. Global mean sea level is likely to increase by 0.35 to 0.70 m for Representative Concentration Pathway (RCP) 4.5 during the 21st century, threatening low-lying coastal areas and atoll islands. (Section 29.4.3, Table 29-1; WGI AR5 Section 13.5.1, Table 13.5)	Incremental upwards shift in sea-level baselines results in increased frequency and extent of marine flooding during high tides and episodic storm surges. These events could render soils and fresh groundwater resources unfit for human use before permanent inundation of low-lying areas. (Sections 29.3.1.1, 29.3.2, 29.3.3.1, 29.5.1)

Continued next page →

Table KR-1 (continued)

	Hazard	Key vulnerabilities	Key risks	Emergent risks
The Ocean (Chapter 30)	Increasing ocean temperatures. Increased frequency of thermal extremes	Corals and other organisms whose tolerance limits are exceeded are particularly susceptible (especially CBS, STG, SES, and EUS ocean regions). (Sections 6.2.2.1, 6.2.2.2, 30.5.2, 30.5.4, 30.5.5; Boxes CC-CR, 30.5.6, CC-OA)	Risk of increased mass coral bleaching and mortality (loss of coral cover) with severe risks for coastal fisheries, tourism, and coastal protection (Sections 6.3.2. 6.3.5, 5.4.2.4, 7.2.1.2, 6.4.1.4, 29.3.1.2, 30.5.2, 30.5.3, 30.5.4, 30.5.5; Box CC-CR)	Loss of coastal reef systems, risk of decreased food security and reduced livelihoods, and reduced coastal protection (Sections 7.2.1.2, 30.6.2.1, 30.6.5)
		Marine species and ecosystems as well as fisheries and coastal livelihoods and tourism that cannot cope or adapt to changing temperatures and changes in the distribution are particularly vulnerable, especially for HLSBS, CBS, STG, and EBUE. (Sections 6.3.2, 6.3.4, 7.3.2.6, 30.5; Box CC-BIO)	Risk for fishery and coastal livelihoods. Fishery opportunity changes as stock abundance may rise or fall; increased risk of disease and invading species impacting ecosystems and fisheries (Sections 6.3.5, 6.4.1.1, 6.5.3, 7.3.2.6, 7.4.2, 29.5.3, 29.5.4)	Significant risk of fishery collapse may develop as the capacity of fisheries to resist the following is exceeded: a) fundamental change to fishery composition, and b) the increased migration of disease and other organisms. (Sections 6.5.3, 7.5.1.1.3)
		Coastal ecosystems and communities that might be exposed to phenomena of elevated rates of microbial respiration leading to reduced oxygen at depth and increased spread of dead zones are particularly vulnerable (particularly for EBUE, SES, EUS).	Risk of loss of habitats and fishery resources as well as losses of key fisheries species. Oxygen levels decrease, leading to impacts on ecosystems (e.g., loss of habitat) and organisms (e.g., physiological performance of fish) resulting in reduced capture of key fisheries species.	Increasing risk of loss of livelihoods
		Deep sea life is sensitive to hazards and to change given the very constant conditions under which it has evolved. (30.1.3.1.3, 30.5.2, 30.5.5)	Risk of fundamental changes in conditions associated with deep sea (e.g., oxygen, pH, carbonate, CO_2, temperature) drive fundamental changes that result in broad-scale changes throughout the ocean. (Sections 30.1.3.1.3, 30.5.2, 30.5.5; Boxes CC-UP, CC-NPP)	Changes in the deep ocean may be a prelude to ocean wide changes with planetary implications.
	Rising ocean acidification	Reef systems, corals, and coastal ecosystems that are exposed to a reduced rate of calcification and greater decalcification leading to potential loss of carbonate reef systems, corals, molluscs, and other calcifiers in key regions, such as the CBS, STG (Section 6.2.2.2)	Risk of the alteration of ecosystem services including risks to food provisioning with impacts on fisheries and aquaculture (Sections 6.2.5.3, 7.2.1.2, 7.3.2, 7.4.2,)	Income and livelihoods for communities are reduced as productivity of fisheries and aquaculture diminish. (Sections 7.5.1.1.3, 30.6)
		Marine organisms that are susceptible to changes in pH and carbonate chemistry imply a large number of changes to the physiology and ecology of marine organisms (particularly in CBS, STG, SES regions). (Sections 6.2.5, 6.3.4, 30.3.2.2)	Risk of fundamental shifts in ecosystems composition as well as organism function occur, leading to broad scale and fundamental change. Income and livelihoods from dependent communities are affected as ecosystem goods and services decline, with the prospect that recovery may take tens of thousands of years. (Section 6.1.1.2)	Risk to ecosystems and livelihoods is increased by the potential for interaction among ocean warming and acidification to create unknown impacts. (Section CC-OA)
		Coastal systems are increasingly exposed to upwelling in some areas, which results in periods of high CO_2, low O_2 and pH. (Box CC-UP; Sections 6.2.2.2, 6.2.5.3)	Risk of loss and harm to fishery and aquaculture operations and respective livelihoods (e.g., oyster cultivation), especially those exposed periodically to harmful conditions during elevated upwelling, which trigger adaptation responses. (Section 30.6.2.1.4)	Background pH and carbonate chemistry are also such that harmful conditions are always present (avoiding impacts via adaptation not possible any more). (Section 30.6.2.1.4)
	Increased stratification as a result of ocean warming; reduced ventilation	Ocean ecosystems are vulnerable due to the reduced regeneration of nutrients as mixing between the ocean and its surface is reduced (EUS, STG, and EBUE). (Sections 6.2, 6.3, 6.5, 30.5.2, 30.5.4, 30.5.5)	Risk of productivity losses of oceans and respective negative impacts on fisheries. The concentration of inorganic nutrients in the upper layers of the ocean is reduced, leading to lower rates of primary productivity. (Box CC-NPP)	Reduced primary productivity of the ocean impacts fisheries productivity leading to lower catch rates and effects on livelihoods (Section 6.4.1.1; Box CC-NPP)
		Ecosystems and organisms that are sensitive to decreasing oxygen levels (Sections 30.5.2, 30.5.3, 30.5.5, 30.5.6, 30.5.7)	Increased risk of dead (hypoxic) zones reducing key ecosystems and fisheries habitat (Sections 6.1.1.3, 30.3.2.3)	
	Changes to wind, wave height, and storm intensity	Shipping and industrial infrastructure is vulnerable to wave and storm intensity. (Section 30.6.2)	Risk of increasing losses and damages to shipping and industrial infrastructure	Risk of accidents increases for enterprises such as shipping, as well as deep sea oil gas and mineral extraction.

CBS = Coastal Boundary Systems; EBUE = Eastern Boundary Upwelling Ecosystems; EUS = Equatorial Upwelling Systems; HIC, LIC, MIC = high-, low-, and medium-income countries; HLSBS = High-Latitude Spring Bloom Systems; SES = Semi-Enclosed Seas; STG = Sub-Tropical Gyres.

This cross-chapter box should be cited as:

Birkmann, J., R. Licker, M. Oppenheimer, M. Campos, R. Warren, G. Luber, B.C. O'Neill, and K. Takahashi, 2014: Cross-chapter box on a selection of the hazards, key vulnerabilities, key risks, and emergent risks identified in the WGII contribution to the fifth assessment report. In: *Climate Change 2014: Impacts, Adaptation, and Vulnerability. Summaries, Frequently Asked Questions, and Cross-Chapter Boxes. Contribution of Working Group II to the Fifth Assessment Report of the Intergovernmental Panel on Climate Change* [Field, C.B., V.R. Barros, D.J. Dokken, K.J. Mach, M.D. Mastrandrea, T.E. Bilir, M. Chatterjee, K.L. Ebi, Y.O. Estrada, R.C. Genova, B. Girma, E.S. Kissel, A.N. Levy, S. MacCracken, P.R. Mastrandrea, and L.L. White (eds.)]. World Meteorological Organization, Geneva, Switzerland, pp. 115-123. (in Arabic, Chinese, English, French, Russian, and Spanish)

KR

Observed Global Responses of Marine Biogeography, Abundance, and Phenology to Climate Change

Elvira Poloczanska (Australia), Ove Hoegh-Guldberg (Australia), William Cheung (Canada), Hans-Otto Pörtner (Germany), Michael T. Burrows (UK)

IPCC WGII AR4 presented the detection of a global fingerprint on natural systems and its attribution to climate change (AR4, Chapter 1, SPM Figure 1), but studies from marine systems were mostly absent. Since AR4, there has been a rapid increase in studies that focus on climate change impacts on marine species, which represents an opportunity to move from more anecdotal evidence to examining and potentially attributing detected biological changes within the ocean to climate change (Section 6.3; Figure MB-1). Recent changes in populations of marine species and the associated shifts in diversity patterns are resulting, at least partly, from climate change–mediated biological responses across ocean regions (*robust evidence, high agreement, high confidence*; Sections 6.2, 30.5; Table 6-7).

Poloczanska et al. (2013) assess a potential pattern in responses of ocean life to recent climate change using a global database of 208 peer-reviewed papers. Observed responses (n = 1735) were recorded from 857 species or assemblages across regions and taxonomic groups, from phytoplankton to marine reptiles and mammals (Figure MB-1). Observations were defined as those where the authors of a particular paper assessed the change in a biological parameter (including distribution, phenology, abundance, demography, or community composition) and, if change occurred, the consistency of the change with that expected under climate change. Studies from the peer-reviewed literature were selected using three criteria: (1) authors inferred or directly tested for trends in biological and climatic variables; (2) authors included data after 1990; and (3) observations spanned at least 19 years, to reduce bias resulting from biological responses to short-term climate variability.

The results of this meta-analysis show that climate change has already had widespread impacts on species' distribution, abundance, phenology, and subsequently, species richness and community composition across a broad range of taxonomic groups (plankton to top predators). Of the observations that showed a response in either direction, changes in phenology, distribution and abundance were overwhelmingly (81%) in a direction that was consistent with theoretical responses to climate change (Section 6.2). Knowledge gaps exist, especially in equatorial sub-regions and the Southern Hemisphere (Figure MB-1).

The timing of many biological events (phenology) had an earlier onset. For example, over the last 50 years, spring events shifted earlier for many species with an average advancement of 4.4 ± 0.7 days per decade (mean ± SE) and summer events by 4.4 ± 1.1 days per decade (*robust evidence, high agreement, high confidence*) (Figure MB-2). Phenological observations included in the study range from shifts in peak abundance of phytoplankton and zooplankton, to reproduction and migration of invertebrates, fishes, and seabirds (Sections 6.3.2, 30.5).

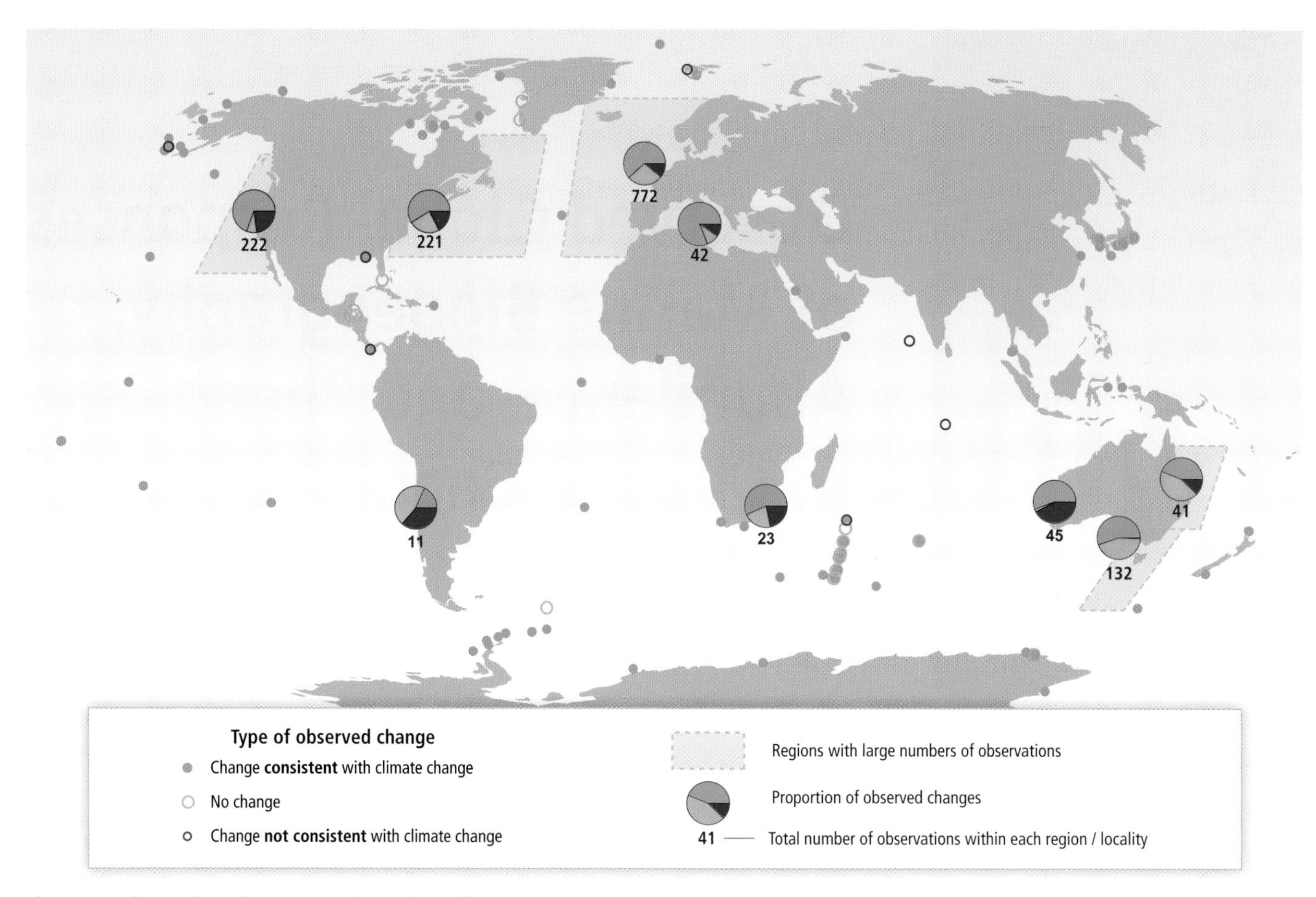

Figure MB-1 | 1735 observed responses to climate change from 208 single- and multi-species studies. Data shown include changes that are attributed (at least partly) to climate change (blue), changes that are inconsistent with climate change (red), and no change (orange). Each circle represents the center of a study area. Where points fall on land, it is because they are centroids of distributions that surround an island or peninsula. Studies encompass areas from single sites (e.g., seabird breeding colony) to large ocean regions (e.g., continuous plankton recorder surveys in north-east Atlantic). For regions (indicated by blue shading) and localities with large numbers of observations, pie charts summarize the relative proportions of the three types of observed changes (consistent with climate change, inconsistent with climate change, and no change) in those regions or localities. The numbers indicate the total observations within each region or locality. Note: 57% of the studies included were published since AR4. (From Poloczanska et al., 2013).

The distributions of benthic, pelagic, and demersal species and communities have shifted by up to a thousand kilometers, although the range shifts have not been uniform across taxonomic groups or ocean regions (Sections 6.3.2, 30.5) (*robust evidence, high agreement, high confidence*). Overall, leading range edges expanded in a poleward direction at 72.0 ± 13.5 km per decade and trailing edges contracted in a poleward direction at 15.8 ± 8.7 km per decade (Figure MB-2), revealing much higher current rates of migration than the potential maximum rates reported for terrestrial species (Figure 4-6) despite slower warming of the ocean than land surface (WGI Section 3.2).

Poleward distribution shifts have resulted in increased species richness in mid- to high-latitude regions (Hiddink and ter Hofstede, 2008) and changing community structure (Simpson et al., 2011; see also Section 28.2.2). Increases in warm-water components of communities concurrent with regional warming have been observed in mid- to high-latitude ocean regions including the Bering Sea, Barents Sea, Nordic Sea, North Sea, and Tasman Sea (Box 6.1; Section 30.5). Observed changes in species composition of catches from 1970–2006 that are partly attributed to long-term ocean warming suggest increasing dominance of warmer water species in subtropical and higher latitude regions, and reduction in abundance of subtropical species in equatorial waters (Cheung et al., 2013), with implications for fisheries (Sections 6.5, 7.4.2, 30.6.2.1).

The magnitude and direction of distribution shifts can be related to temperature velocities (i.e., the speed and direction at which isotherms propagate across the ocean's surface (Section 30.3.1.1; Burrows et al., 2011). Pinsky et al. (2013) showed that shifts in both latitude and depth of benthic fish and crustaceans could be explained by climate velocity with remarkable accuracy, using a database of 128 million individuals across 360 marine taxa from surveys of North American coastal waters conducted over 1968–2011. Poloczanska et al. (2013) found that faster distribution shifts generally occur in regions of highest surface temperature velocity, such as the North Sea and sub-Arctic Pacific Ocean. Observed marine species shifts, since approximately the 1950s, have generally been able to track observed velocities (Figure MB-3), with phyto- and zooplankton distribution shifts vastly exceeding climate velocities observed over most of the ocean surface, but with considerable variability within and among taxonomic groups (Poloczanska et al., 2013).

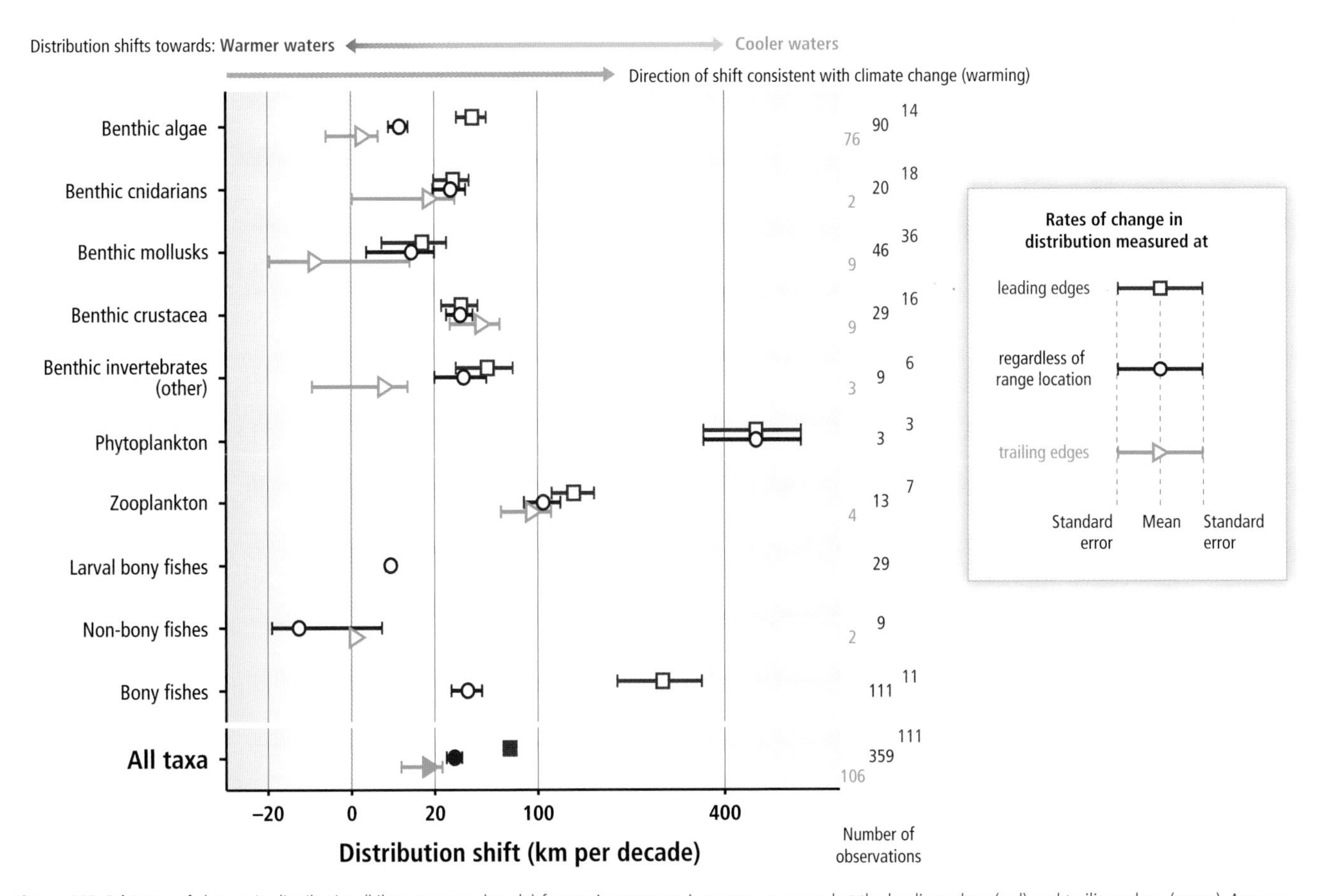

Figure MB-2 | Rates of change in distribution (kilometers per decade) for marine taxonomic groups, measured at the leading edges (red) and trailing edges (green). Average distribution shifts were calculated using all data, regardless of range location, and are in dark blue. Distribution shifts have been square-root transformed; standard errors may be asymmetric as a result. Positive distribution changes are consistent with warming (into previously cooler waters, generally poleward). Means ± standard error are shown, along with number of observations. Non-bony fishes include sharks, rays, lampreys, and hagfish. (From Poloczanska et al., 2013).

Biogeographic shifts are also influenced by other factors such as currents, nutrient and stratification changes, light levels, sea ice, species' interactions, habitat availability and fishing, some of which can be independently influenced by climate change (Section 6.3). Rate and pattern of biogeographic shifts in sedentary organisms and benthic macroalgae are complicated by the influence of local dynamics and topographic features (islands, channels, coastal lagoons, e.g., of the Mediterranean (Bianchi, 2007), coastal upwelling e.g., (Lima et al., 2007)). Geographical barriers constrain range shifts and may cause a loss of endemic species (Ben Rais Lasram et al., 2010), with associated niches filled by alien species, either naturally migrating or artificially introduced (Philippart et al., 2011).

Whether marine species can continue to keep pace as rates of warming, hence climate velocities, increase (Figure MB-3b) is a key uncertainty. Climate velocities on land are expected to outpace the ability of many terrestrial species to track climate velocities this century (Section 4.3.2.5; Figure 4-6). For marine species, the observed rates of shift are generally much faster than those for land species, particularly for primary producers and lower trophic levels (Poloczanska et al., 2013). Phyto- and zooplankton communities (excluding larval fish) have extended distributions at remarkable rates (Figure MB-3b), such as in the Northeast Atlantic (Section 30.5.1) with implications for marine food webs.

Geographical range shifts and depth distribution vary between coexisting marine species (Genner et al., 2004; Perry et al., 2005; Simpson et al., 2011) as a consequence of the width of species-specific thermal windows and associated vulnerabilities (Figure 6-5). Warming therefore causes differential changes in growth, reproductive success, larval output, early juvenile survival, and recruitment, implying shifts in the relative performance of animal species and, thus, their competitiveness (Pörtner and Farrell, 2008; Figure 6-7A). Such effects may underlie abundance losses or local extinctions, "regime shifts" between coexisting species, or critical mismatches between predator and prey organisms, resulting in changes in local and regional species richness, abundance, community composition, productivity, energy flows, and invasion resistance. Even among Antarctic stenotherms, differences in biological responses related to mode of life, phylogeny and associated metabolic capacities exist (Section 6.3.1.4). As a consequence, marine ecosystem functions may be substantially reorganized at the regional scale, potentially triggering a range of cascading effects (Hoegh-Guldberg and Bruno, 2010). A focus on understanding the mechanisms underpinning the nature and magnitude of responses of marine organisms to climate change can help forecast impacts and the associated costs to society as well as facilitate adaptive management strategies formitigating these impacts (Sections 6.3, 6.4).

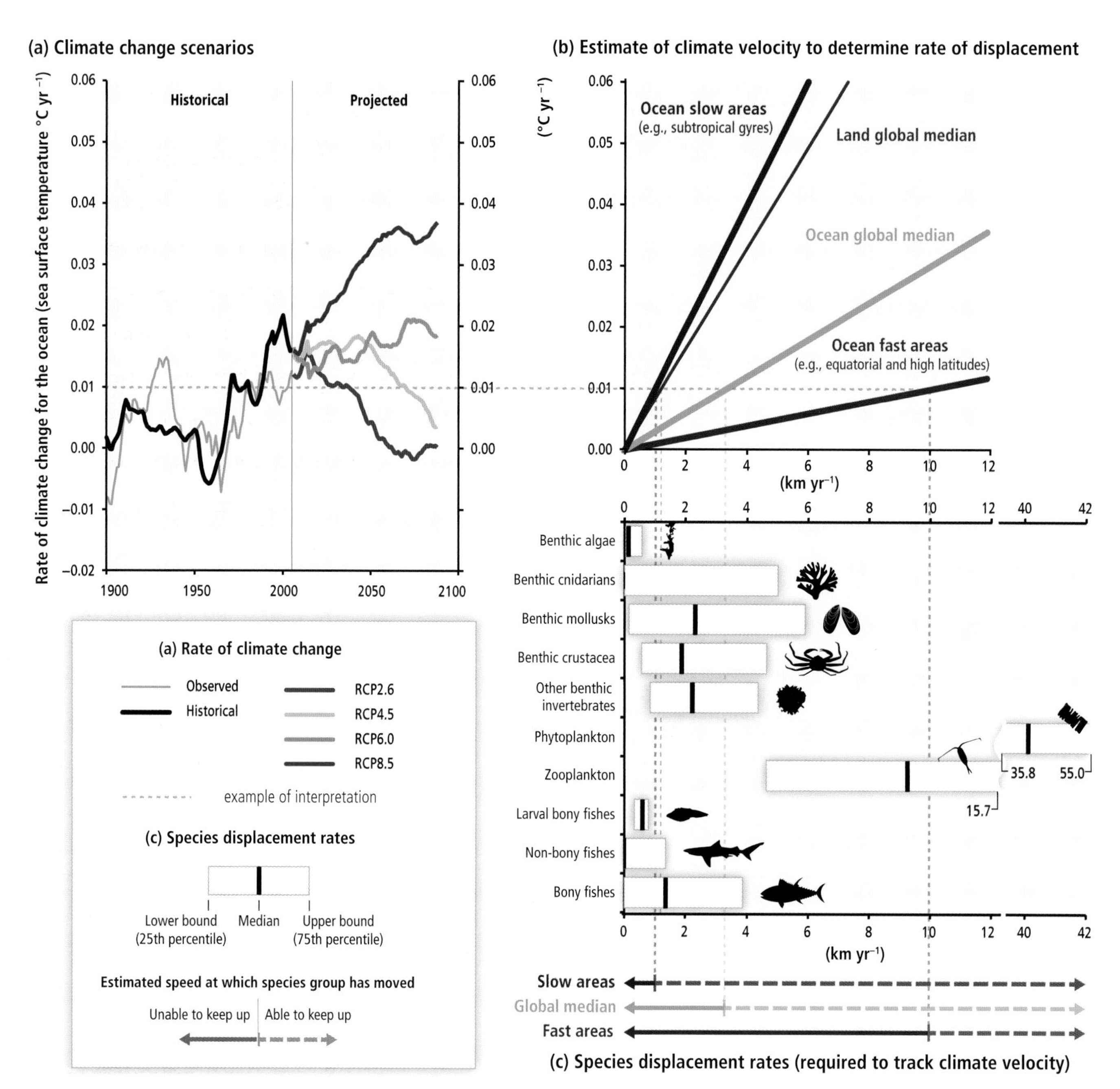

Figure MB-3 | (a) Rate of climate change for the ocean (sea surface temperature (SST) °C yr⁻¹). (b) Corresponding climate velocities for the ocean and median velocity from land (adapted from Burrows et al., 2011). (c) Observed rates of displacement of marine taxonomic groups based on observations over 1900–2010. The dotted bands give an example of interpretation. Rates of climate change of 0.01 °C yr⁻¹ correspond to approximately 3.3 km yr⁻¹ median climate velocity in the ocean. When compared to observed rates of displacement (c), many marine taxonomic groups have been able to track these velocities. For phytoplankton and zooplankton the rates of displacement greatly exceed median climate velocity for the ocean and, for phytoplankton exceed velocities in fast areas of the ocean approximately 10.0 km yr⁻¹. All values are calculated for ocean surface with the exclusion of polar seas (Figure 30-1a). (a) Observed rates of climate change for ocean SST (green line) are derived from the Hadley Centre Interpolated SST 1.1 (HadISST1.1) data set, and all other rates are calculated based on the average of the Coupled Model Intercomparison Project Phase 5 (CMIP5) climate model ensembles (Table SM30-3) for the historical period and for the future based on the four Representative Concentration Pathway (RCP) scenarios. Data were smoothed using a 20-year sliding window. (b) Median climate velocity over the global ocean surface (light blue line; excluding polar seas) calculated from HadSST1.1 data set over 1960–2009 using the methods of Burrows et al. (2011). Median velocities representative of ocean regions of slow velocities such as the Pacific subtropical gyre (dark blue line) and of high velocities such as the Coral Triangle or the North Sea (purple line) shown. Median rates over global land surface (red line) over 1960–2009 calculated using Climate Research Unit data set CRU TS3.1. Figure 30-3 shows climate velocities over the ocean surface calculated over 1960–2009. (c) Rates of displacement for marine taxonomic groups estimated by Poloczanska et al. (2013) using published studies. Note the displacement rates for phytoplankton exceed the axis, so values are given.

References

Ben Rais Lasram, F., F. Guilhaumon, C. Albouy, S. Somot, W. Thuiller, and D. Mouillot, 2010: The Mediterranean Sea as a 'cul-de-sac' for endemic fishes facing climate change. *Global Change Biology*, **16**, 3233-3245.

Bianchi, C.N., 2007: Biodiversity issues for the forthcoming Mediterranean Sea. *Hydrobiologia*, **580**, 7-21.

Burrows, M.T., D. S. Schoeman, L.B. Buckley, P.J. Moore, E.S. Poloczanska, K. Brander, K, C.J. Brown, J.F. Bruno, C.M. Duarte, B.S. Halpern, J. Holding, C.V. Kappel, W. Kiessling, M.I. O'Connor, J.M. Pandolfi, C. Parmesan, F. Schwing, W.J. Sydeman, and A.J. Richardson, 2011: The pace of shifting climate in marine and terrestrial ecosystems. *Science*, **334**, 652-655.

Cheung, W.W.L., R. Watson, and D. Pauly, 2013: Signature of ocean warming in global fisheries catch. *Nature*, **497(7449)**, 365-368.

Genner, M.J., D.W. Sims, V.J. Wearmouth, E.J. Southall, A.J. Southward, P.A. Henderson, and S.J. Hawkins, 2004: Regional climatic warming drives long-term community changes of British marine fish. *Proceedings of the Royal Society B*, **271(1539)**, 655-661.

Hiddink, J.G. and R. ter Hofstede, 2008: Climate induced increases in species richness of marine fishes. *Global Change Biology*, **14**, 453-460.

Hoegh-Guldberg, O. and J.F. Bruno, 2010: The impact of climate change on the world's marine ecosystems. *Science*, **328**, 1523-1528.

Lima, F.P., P.A. Ribeiro, N. Queiroz, S.J. Hawkins, and A.M. Santos, 2007: Do distributional shifts of northern and southern species of algae match the warming pattern? *Global Change Biology*, **13**, 2592-2604.

Perry, A.L., P.J. Low, J.R. Ellis, and J.D. Reynolds, 2005: Climate change and distribution shifts in marine fishes. *Science*, **308(5730)**, 1912-1915.

Philippart, C.J.M., R. Anadon, R. Danovaro, J.W. Dippner, K.F. Drinkwater, S.J. Hawkins, T. Oguz, G. O'Sullivan, and P.C. Reid, 2011: Impacts of climate change on European marine ecosystems: observations, expectations and indicators. *Journal of Experimental Marine Biology and Ecology*, **400**, 52-69.

Pinksy, M.L., B. Worm, M.J. Fogarty, J.L. Sarmiento, and S.A. Levin, 2013: Marine taxa track local climate velocities. *Science*, **341**, 1239-1242.

Pörtner, H.O. and A.P. Farrell, 2008: Physiology and climate change. *Science*, **322(5902)**, 690-692.

Poloczanska, E.S., C.J. Brown, W.J. Sydeman, W. Kiessling, D.S. Schoeman, P.J. Moore, K. Brander, J.F. Bruno, L.B. Buckley, M.T. Burrows, C.M. Duarte, B.S. Halpern, J. Holding, C.V. Kappel, M.I. O'Connor, J.M. Pandolfi, C. Parmesan, F. Schwing, S.A.Thompson, and A.J. Richardson, 2013: Global imprint of climate change on marine life. *Nature Climate Change*, **3**, 919-925.

Simpson, S.D., S. Jennings, M.P. Johnson, J.L. Blanchard, P.J. Schon, D.W. Sims, and M.J. Genner, 2011: Continental shelf-wide response of a fish assemblage to rapid warming of the sea. *Current Biology*, **21**, 1565-1570.

This cross-chapter box should be cited as:

Poloczanska, E.S., O. Hoegh-Guldberg, W. Cheung, H.-O. Pörtner, and M. Burrows, 2014: Cross-chapter box on observed global responses of marine biogeography, abundance, and phenology to climate change. In: *Climate Change 2014: Impacts, Adaptation, and Vulnerability. Summaries, Frequently Asked Questions, and Cross-Chapter Boxes. Contribution of Working Group II to the Fifth Assessment Report of the Intergovernmental Panel on Climate Change* [Field, C.B., V.R. Barros, D.J. Dokken, K.J. Mach, M.D. Mastrandrea, T.E. Bilir, M. Chatterjee, K.L. Ebi, Y.O. Estrada, R.C. Genova, B. Girma, E.S. Kissel, A.N. Levy, S. MacCracken, P.R. Mastrandrea, and L.L. White (eds.)]. World Meteorological Organization, Geneva, Switzerland, pp. 125-129. (in Arabic, Chinese, English, French, Russian, and Spanish)

Ocean Acidification

Jean-Pierre Gattuso (France), Peter G. Brewer (USA), Ove Hoegh-Guldberg (Australia), Joan A. Kleypas (USA), Hans-Otto Pörtner (Germany), Daniela N. Schmidt (UK)

Anthropogenic ocean acidification and global warming share the same primary cause, which is the increase of atmospheric CO_2 (Figure OA-1A; WGI, Section 2.2.1). Eutrophication, loss of sea ice, upwelling and deposition of atmospheric nitrogen and sulfur all exacerbate ocean acidification locally (Sections 5.3.3.6, 6.1.1, 30.3.2.2).

Chemistry and Projections

The fundamental chemistry of ocean acidification is well understood (*robust evidence, high agreement*). Increasing atmospheric concentrations of CO_2 result in an increased flux of CO_2 into a mildly alkaline ocean, resulting in a reduction in pH, carbonate ion concentration, and the capacity of seawater to buffer changes in its chemistry (*very high confidence*). The changing chemistry of the surface layers of the open ocean can be projected at the global scale with high accuracy using projections of atmospheric CO_2 levels (Figure CC-OA-1B). Observations of changing upper ocean CO_2 chemistry over time support this linkage (WGI Table 3.2 and Figure 3.18; Figures 30-8, 30-9). Projected changes in open ocean, surface water chemistry for the year 2100 based on representative concentration pathways (WGI, Figure 6.28) compared to pre-industrial values range from a pH change of –0.14 units with Representative Concentration Pathway (RCP)2.6 (421 ppm CO_2, +1°C, 22% reduction of carbonate ion concentration) to a pH change of –0.43 units with RCP8.5 (936 ppm CO_2, +3.7°C, 56% reduction of carbonate ion concentration). Projections of regional changes, especially in the highly complex coastal systems (Sections 5.3.3.5, 30.3.2.2), in polar regions (WGI Section 6.4.4), and at depth are more difficult but generally follow similar trends.

Biological, Ecological, and Biogeochemical Impacts

Investigations of the effect of ocean acidification on marine organisms and ecosystems have a relatively short history, recently analyzed in several meta-analyses (Sections 6.3.2.1, 6.3.5.1). A wide range of sensitivities to projected rates of ocean acidification exists within and across diverse groups of organisms, with a trend for greater sensitivity in early life stages (*high confidence*; Sections 5.4.2.2, 5.4.2.4, 6.3.2). A pattern of positive and negative impacts emerges (*high confidence*; Figure OA-1C) but key uncertainties remain in our understanding of the impacts on organisms, life histories, and ecosystems. Responses can be influenced, often exacerbated by other drivers, such as warming, hypoxia, nutrient concentration, and light availability (*high confidence*; Sections 5.4.2.4, 6.3.5).

Growth and primary production are stimulated in seagrass and some phytoplankton (*high confidence*; Sections 5.4.2.3, 6.3.2.2, 6.3.2.3, 30.5.6). Harmful algal blooms could become more frequent (*limited evidence, medium agreement*). Ocean acidification may stimulate nitrogen fixation (*limited evidence, low agreement*; 6.3.2.2). It decreases the rate of calcification of most, but not all, sea floor calcifiers (*medium agreement, robust evidence*) such as reef-building corals (Box CC-CR), coralline algae, bivalves, and gastropods, reducing the competitiveness with non-calcifiers (Sections 5.4.2.2, 5.4.2.4, 6.3.2.5). Ocean warming and acidification promote higher rates of calcium carbonate dissolution resulting in the net dissolution of carbonate sediments and frameworks and loss of associated habitat (*medium confidence*; 5.4.2.4, 6.3.2.5, 6.3.5.4). Some corals and temperate fishes experience disturbances to behavior, navigation, and their ability to tell conspecifics from predators (Section 6.3.2.4). However, there is no evidence for these effects to persist on evolutionary time scales in the few groups analyzed (Section 6.3.2).

Some phytoplankton and molluscs displayed adaptation to ocean acidification in long-term experiments (*limited evidence, medium agreement*; Section 6.3.2.1), indicating that the long-term responses could be less than responses obtained in short-term experiments. However, mass extinctions in Earth history occurred during much slower rates of ocean acidification, combined with other drivers changing, suggesting that evolutionary rates are not fast enough for sensitive animals and plants to adapt to the projected rate of future change (*medium confidence*; Section 6.1.2).

Projections of ocean acidification effects at the ecosystem level are made difficult by the diversity of species-level responses. Differential sensitivities and associated shifts in performance and distribution will change predator–prey relationships and competitive interactions (Sections

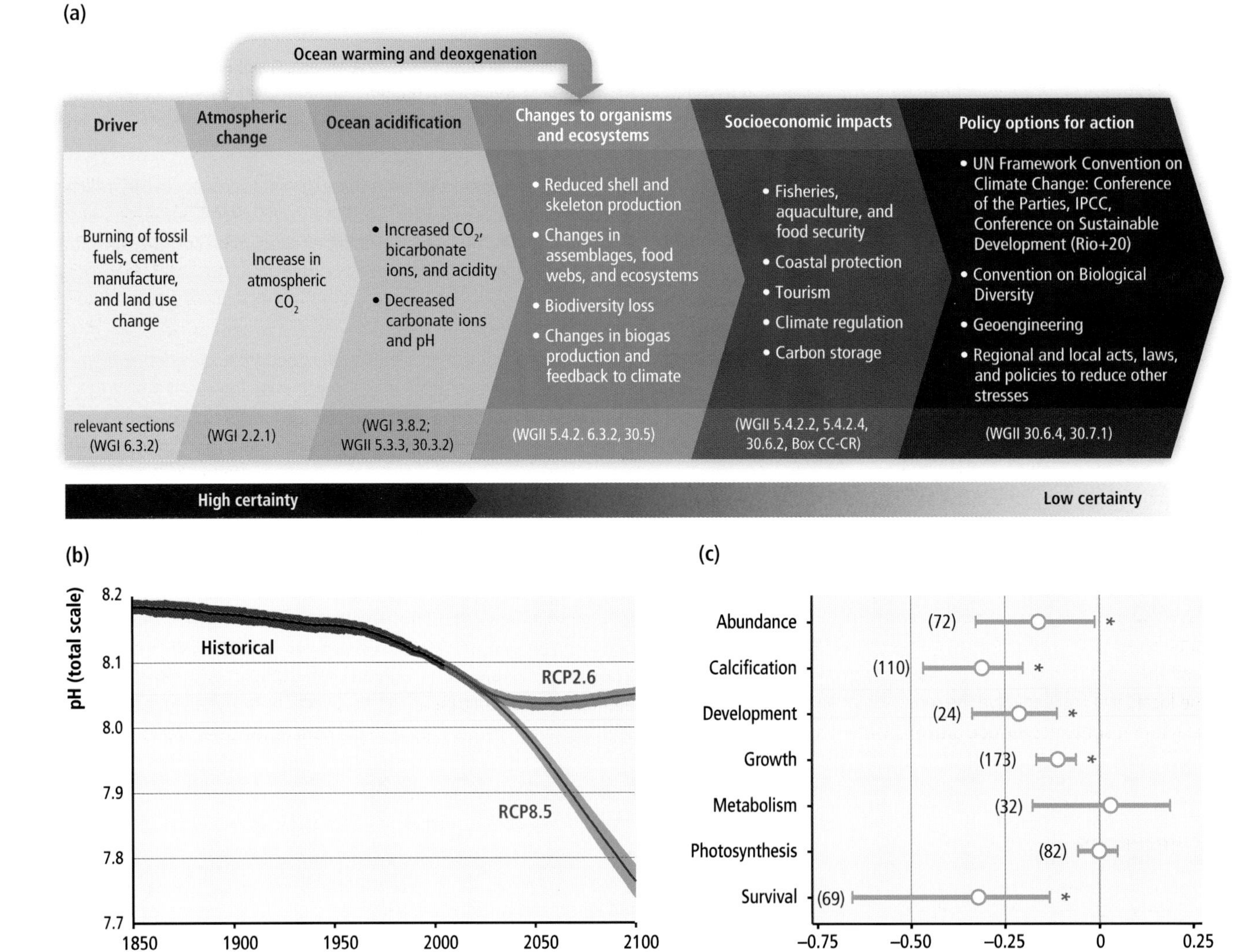

Figure OA-1 | (a) Overview of the chemical, biological, and socio-economic impacts of ocean acidification and of policy options (adapted from Turley and Gattuso, 2012). (b) Multi-model simulated time series of global mean ocean surface pH (on the total scale) from Coupled Model Intercomparison Project Phase 5 (CMIP5) climate model simulations from 1850 to 2100. Projections are shown for emission scenarios Representative Concentration Pathway (RCP)2.6 (blue) and RCP8.5 (red) for the multi-model mean (solid lines) and range across the distribution of individual model simulations (shading). Black (gray shading) is the modeled historical evolution using historical reconstructed forcings. The models that are included are those from CMIP5 that simulate the global carbon cycle while being driven by prescribed atmospheric CO_2 concentrations (WGI AR5 Figures SPM.7 and TS.20). (c) Effect of near-future acidification (seawater pH reduction of ≤0.5 units) on major response variables estimated using weighted random effects meta-analyses, with the exception of survival, which is not weighted (Kroeker et al., 2013). The log-transformed response ratio (lnRR) is the ratio of the mean effect in the acidification treatment to the mean effect in a control group. It indicates which process is most uniformly affected by ocean acidification, but large variability exists between species. Significance is determined when the 95% bootstrapped confidence interval does not cross zero. The number of experiments used in the analyses is shown in parentheses. The * denotes a statistically significant effect.

6.3.2.5, 6.3.5, 6.3.6), which could impact food webs and higher trophic levels (*limited evidence, high agreement*). Natural analogues at CO_2 vents indicate decreased species diversity, biomass, and trophic complexity of communities (Box CC-CR; Sections 5.4.2.3, 6.3.2.5, 30.3.2.2, 30.5). Shifts in community structure have also been documented in regions with rapidly declining pH (Section 5.4.2.2).

Owing to an incomplete understanding of species-specific responses and trophic interactions, the effect of ocean acidification on global biogeochemical cycles is not well understood (*limited evidence, low agreement*) and represents an important knowledge gap. The additive, synergistic, or antagonistic interactions of factors such as temperature, concentrations of oxygen and nutrients, and light are not sufficiently investigated yet.

Risks, Socioeconomic Impacts, and Costs

The risks of ocean acidification to marine organisms, ecosystems, and ultimately to human societies, include both the probability that ocean acidification will affect fundamental physiological and ecological processes of organisms (Section 6.3.2.1), and the magnitude of the resulting impacts on ecosystems and the ecosystem services they provide to society (Box 19-2). For example, ocean acidification under RCP4.5 to RCP8.5 will impact formation and maintenance of coral reefs (*high confidence*; Box CC-CR, Section 5.4.2.4) and the goods and services that they provide such as fisheries, tourism, and coastal protection (*limited evidence, high agreement*; Box CC-CR; Sections 6.4.1.1,19.5.2, 27.3.3, 30.5, 30.6). Ocean acidification poses many other potential risks, but these cannot yet be quantitatively assessed because of the small number of studies available, particularly on the magnitude of the ecological and socioeconomic impacts (Section 19.5.2).

Global estimates of observed or projected economic costs of ocean acidification do not exist. The largest uncertainty is how the impacts on lower trophic levels will propagate through the food webs and to top predators. However, there are a number of instructive examples that illustrate the magnitude of potential impacts of ocean acidification. A decrease of the production of commercially exploited shelled molluscs (Section 6.4.1.1) would result in a reduction of USA production of 3 to 13% according to the Special Report on Emission Scenarios (SRES) A1FI emission scenario (*low confidence*). The global cost of production loss of molluscs could be more than US$100 billion by 2100 (*limited evidence, medium agreement*). Models suggest that ocean acidification will generally reduce fish biomass and catch (*low confidence*) and that complex additive, antagonistic, and/or synergistic interactions will occur with other environmental (warming) and human (fisheries management) factors (Section 6.4.1.1). The annual economic damage of ocean-acidification–induced coral reef loss by 2100 has been estimated, in 2012, to be US$870 and 528 billion, respectively for the A1 and B2 SRES emission scenarios (*low confidence*; Section 6.4.1). Although this number is small compared to global gross domestic product (GDP), it can represent a very large GDP loss for the economies of many coastal regions or small islands that rely on the ecological goods and services of coral reefs (Sections 25.7.5, 29.3.1.2).

Mitigation and Adaptation

Successful management of the impacts of ocean acidification includes two approaches: mitigation of the source of the problem (i.e., reduce anthropogenic emissions of CO_2) and/or adaptation by reducing the consequences of past and future ocean acidification (Section 6.4.2.1). Mitigation of ocean acidification through reduction of atmospheric CO_2 is the most effective and the least risky method to limit ocean acidification and its impacts (Section 6.4.2.1). Climate geoengineering techniques based on solar radiation management will not abate ocean acidification and could increase it under some circumstances (Section 6.4.2.2). Geoengineering techniques to remove CO_2 from the atmosphere could directly address the problem but are very costly and may be limited by the lack of CO_2 storage capacity (Section 6.4.2.2). In addition, some ocean-based approaches, such as iron fertilization, would only relocate ocean acidification from the upper ocean to the ocean interior, with potential ramifications on deep water oxygen levels (Sections 6.4.2.2, 30.3.2.3, 30.5.7). A low-regret approach, with relatively limited effectiveness, is to limit the number and the magnitude of drivers other than CO_2, such as nutrient pollution (Section 6.4.2.1). Mitigation of ocean acidification at the local level could involve the reduction of anthropogenic inputs of nutrients and organic matter in the coastal ocean (Section 5.3.4.2). Some adaptation strategies include drawing water for aquaculture from local watersheds only when pH is in the right range, selecting for less sensitive species or strains, or relocating industries elsewhere (Section 6.4.2.1).

References

Kroeker, K., R.C. Kordas, A. Ryan, I. Hendriks, L. Ramajo, G. Singh, C. Duarte, and J.-P. Gattuso, 2013: Impacts of ocean acidification on marine organisms: quantifying sensitivities and interaction with warming. *Global Change Biology*, **19**, 1884-1896.

Turley, C. and J.-P. Gattuso, 2012: Future biological and ecosystem impacts of ocean acidification and their socioeconomic-policy implications. *Current Opinion in Environmental Sustainability*, **4**, 278-286.

This cross-chapter box should be cited as:

Gattuso, J.-P., P.G. Brewer, O. Hoegh-Guldberg, J.A. Kleypas, H.-O. Pörtner, and D.N. Schmidt, 2014: Cross-chapter box on ocean acidification. In: *Climate Change 2014: Impacts, Adaptation, and Vulnerability. Summaries, Frequently Asked Questions, and Cross-Chapter Boxes. Contribution of Working Group II to the Fifth Assessment Report of the Intergovernmental Panel on Climate Change* [Field, C.B., V.R. Barros, D.J. Dokken, K.J. Mach, M.D. Mastrandrea, T.E. Bilir, M. Chatterjee, K.L. Ebi, Y.O. Estrada, R.C. Genova, B. Girma, E.S. Kissel, A.N. Levy, S. MacCracken, P.R. Mastrandrea, and L.L. White (eds.)]. World Meteorological Organization, Geneva, Switzerland, pp. 131-133. (in Arabic, Chinese, English, French, Russian, and Spanish)

Net Primary Production in the Ocean

Philip W. Boyd (New Zealand), Svein Sundby (Norway), Hans-Otto Pörtner (Germany)

Net Primary Production (NPP) is the rate of photosynthetic carbon fixation minus the fraction of fixed carbon used for cellular respiration and maintenance by autotrophic planktonic microbes and benthic plants (Sections 6.2.1, 6.3.1). Environmental drivers of NPP include light, nutrients, micronutrients, CO_2, and temperature (Figure PP-1a). These drivers, in turn, are influenced by oceanic and atmospheric processes, including cloud cover; sea ice extent; mixing by winds, waves, and currents; convection; density stratification; and various forms of upwelling induced by eddies, frontal activity, and boundary currents. Temperature has multiple roles as it influences rates of phytoplankton physiology and heterotrophic bacterial recycling of nutrients, in addition to stratification of the water column and sea ice extent (Figure PP-1a). Climate change is projected to strongly impact NPP through a multitude of ways that depend on the regional and local physical settings (WGI AR5, Chapter 3), and on ecosystem structure and functioning (*medium confidence*; Sections 6.3.4, 6.5.1). The influence of environmental drivers on NPP causes as much as a 10-fold variation in regional productivity with nutrient-poor subtropical waters and light-limited Arctic waters at the lower range and productive upwelling regions and highly eutrophic coastal regions at the upper range (Figure PP-1b).

The oceans currently provide ~50 × 10^{15} g C yr^{-1}, or about half of global NPP (Field et al., 1998). Global estimates of NPP are obtained mainly from satellite remote sensing (Section 6.1.2), which provides unprecedented spatial and temporal coverage, and may be validated regionally against oceanic measurements. Observations reveal significant changes in rates of NPP when environmental controls are altered by episodic natural perturbations, such as volcanic eruptions enhancing iron supply, as observed in high-nitrate low-chlorophyll waters of the Northeast Pacific (Hamme et al., 2010). Climate variability can drive pronounced changes in NPP (Chavez et al., 2011), such as from El Niño to La Niña transitions in Equatorial Pacific, when vertical nutrient and trace element supply are enhanced (Chavez et al., 1999).

Multi-year time series records of NPP have been used to assess spatial trends in NPP in recent decades. Behrenfeld et al. (2006), using satellite data, reported a prolonged and sustained global NPP decrease of 190 × 10^{12} g C yr^{-1}, for the period 1999–2005—an annual reduction of 0.57% of global NPP. In contrast, a time series of directly measured NPP between 1988 and 2007 by Saba et al. (2010) (i.e., *in situ* incubations using the radiotracer ^{14}C-bicarbonate) revealed an increase (2% yr^{-1}) in NPP for two low-latitude open ocean sites. This discrepancy between *in situ* and remotely sensed NPP trends points to uncertainties in either the methodology used and/or the extent to which discrete sites are representative of oceanic provinces (Saba et al., 2010, 2011). Modeling studies have subsequently revealed that the <15-year archive of satellite-

(a)

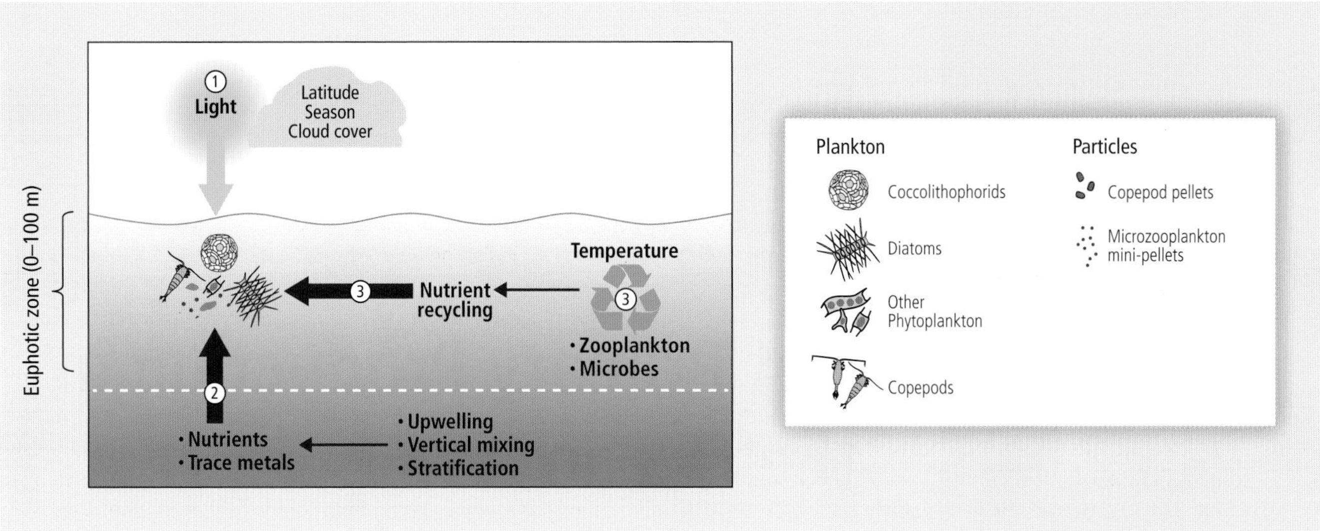

(b)

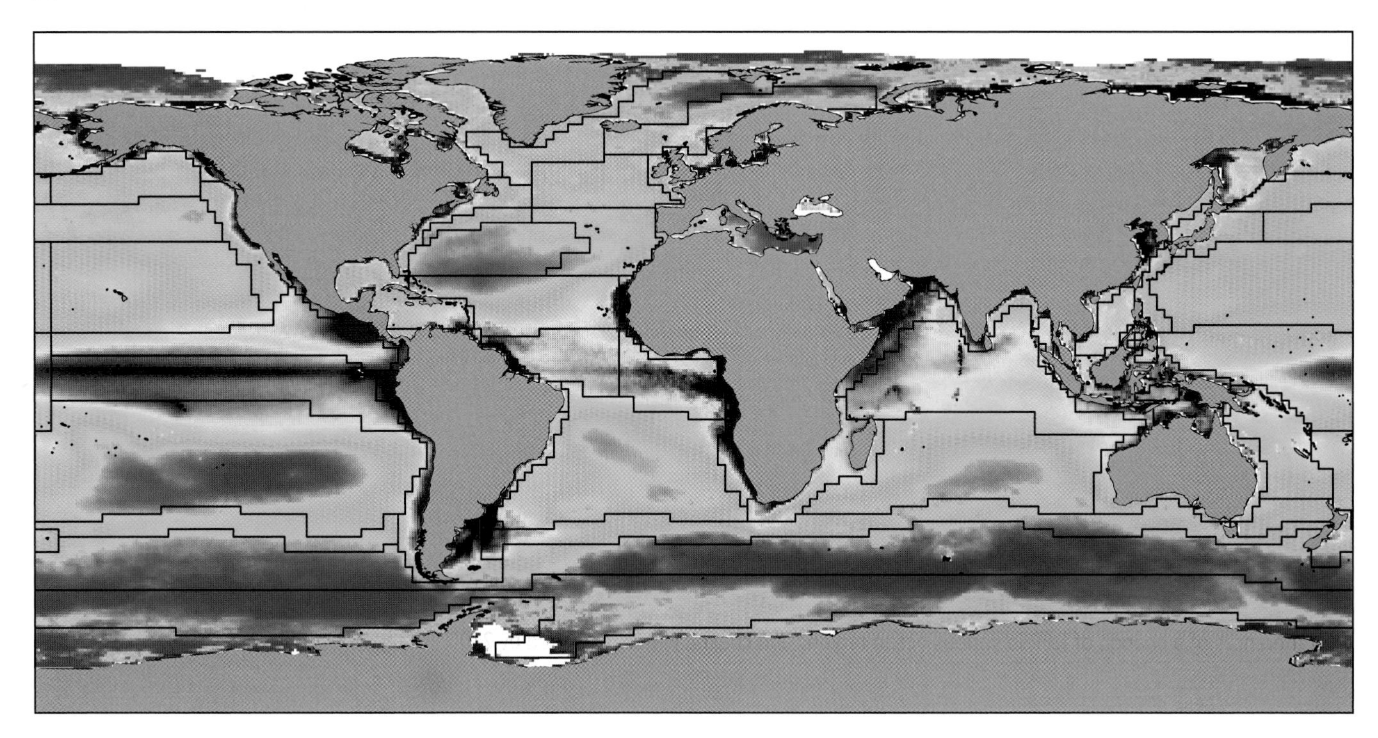

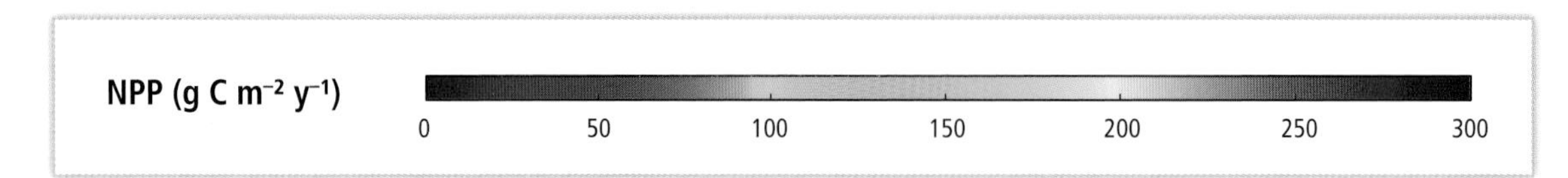

Figure PP-1 | (a) Environmental factors controlling Net Primary Production (NPP). NPP is controlled mainly by three basic processes: (1) light conditions in the surface ocean, that is, the photic zone where photosynthesis occurs; (2) upward flux of nutrients and micronutrients from underlying waters into the photic zone, and (3) regeneration of nutrients and micronutrients via the breakdown and recycling of organic material before it sinks out of the photic zone. All three processes are influenced by physical, chemical, and biological processes and vary across regional ecosystems. In addition, water temperature strongly influences the upper rate of photosynthesis for cells that are resource-replete. Predictions of alteration of primary productivity under climate change depend on correct parameterizations and simulations of each of these variables and processes for each region. (b) Annual composite map of global areal NPP rates (derived from Moderate Resolution Imaging Spectroradiometer (MODIS) Aqua satellite climatology from 2003–2012; NPP was calculated with the Carbon-based Productivity Model (CbPM; Westberry et al., 2008)). Overlaid is a grid of (thin black lines) that represent 51 distinct global ocean biogeographical provinces (after Longhurst, 1998 and based on Boyd and Doney, 2002). The characteristics and boundaries of each province are primarily set by the underlying regional ocean physics and chemistry. White areas = no data. (Figure courtesy of Toby Westberry (OSU) and Ivan Lima (WHOI), satellite data courtesy of NASA Ocean Biology Processing Group.)

derived NPP is insufficient to distinguish climate-change mediated shifts in NPP from those driven by natural climate variability (Henson et al., 2010; Beaulieu et al., 2013). Although multi-decadal, the available time series of oceanic NPP measurements are also not of sufficient duration relative to the time scales of longer-term climate variability modes as for example Atlantic Multi-decadal Oscillation (AMO), with periodicity of 60-70 years, Figure 6-1). Recent attempts to synthesize longer (i.e., centennial) records of chlorophyll as a proxy for phytoplankton stocks (e.g., Boyce et al., 2010) have been criticized for relying on questionable linkages between different proxies for chlorophyll over a century of records (e.g., Rykaczewski and Dunne, 2011).

Models in which projected climate change alters the environmental drivers of NPP provide estimates of spatial changes and of the rate of change of NPP. For example, four global coupled climate–ocean biogeochemical Earth System Models (WGI AR5 Chapter 6) projected an increase in NPP at high latitudes as a result of alleviation of light and temperature limitation of NPP, particularly in the high-latitude biomes (Steinacher et al., 2010). However, this regional increase in NPP was more than offset by decreases in NPP at lower latitudes and at mid-latitudes due to the reduced input of macronutrients into the photic zone. The reduced mixed-layer depth and reduced rate of circulation may cause a decrease in the flux of macronutrients to the euphotic zone (Figure 6-2). These changes to oceanic conditions result in a reduction in global mean NPP by 2 to 13% by 2100 relative to 2000 under a high emission scenario (Polovina et al., 2011; SRES (Special Report on Emission Scenarios) A2, between RCP6.0 and RCP8.5). This is consistent with a more recent analysis based on 10 Earth System Models (Bopp et al., 2013), which project decreases in global NPP by 8.6 (±7.9), 3.9 (±5.7), 3.6 (±5.7), and 2.0 (±4.1) % in the 2090s relative to the 1990s, under the scenarios RCP8.5, RCP6.0, RCP4.5, and RCP2.6, respectively. However, the magnitude of projected changes varies widely between models (e.g., from 0 to 20% decrease in NPP globally under RCP 8.5). The various models show very large differences in NPP at regional scales (i.e., provinces, see Figure PP-1b).

Model projections had predicted a range of changes in global NPP from an increase (relative to preindustrial rates) of up to 8.1% under an intermediate scenario (SRES A1B, similar to RCP6.0; Sarmiento et al., 2004; Schmittner et al., 2008) to a decrease of 2-20% under the SRES A2 emission scenario (Steinacher et al., 2010). These projections did not consider the potential contribution of primary production derived from atmospheric nitrogen fixation in tropical and subtropical regions, favoured by increasing stratification and reduced nutrient inputs from mixing. This mechanism is potentially important, although such episodic increases in nitrogen fixation are not sustainable without the presence of excess phosphate (e.g., Moore et al., 2009; Boyd et al., 2010). This may lead to an underestimation of NPP (Mohr et al., 2010; Mulholland et al., 2012; Wilson et al., 2012), however, the extent of such underestimation is unknown (Luo et al., 2012).

Care must be taken when comparing global, provincial (e.g., low-latitude waters, e.g., Behrenfeld et al., 2006) and regional trends in NPP derived from observations, as some regions have additional local environmental influences such as enhanced density stratification of the upper ocean from melting sea ice. For example, a longer phytoplankton growing season, due to more sea ice–free days, may have increased NPP (based on a regionally validated time-series of satellite NPP) in Arctic waters (Arrigo and van Dijken, 2011) by an average of 8.1×10^{12} g C yr^{-1} between 1998 and 2009. Other regional trends in NPP are reported in Sections 30.5.1 to 30.5.6. In addition, although future model projections of global NPP from different models (Steinacher et al., 2010; Bopp et al., 2013) are comparable, regional projections from each of the models differ substantially. This raises concerns as to which aspect(s) of the different model NPP parameterizations are responsible for driving regional differences in NPP, and moreover, how accurate model projections are of global NPP.

From a global perspective, open ocean NPP will decrease moderately by 2100 under both low- (SRES B1 or RCP4.5) and high-emission scenarios (*medium confidence*; SRES A2 or RCPs 6.0, 8.5, Sections 6.3.4, 6.5.1), paralleled by an increase in NPP at high latitudes and a decrease in the tropics (*medium confidence*). However, there is *limited evidence* and *low agreement* on the direction, magnitude and differences of a change of NPP in various ocean regions and coastal waters projected by 2100 (*low confidence*).

References

Arrigo, K.R. and G.L. van Dijken, 2011: Secular trends in Arctic Ocean net primary production. *Journal of Geophysical Research*, **116(C9)**, C09011, doi:10.1029/2011JC007151.

Beaulieu, C., S.A. Henson, J.L. Sarmiento, J.P. Dunne, S.C. Doney, R.R. Rykaczewski, and L. Bopp, 2013: Factors challenging our ability to detect long-term trends in ocean chlorophyll. *Biogeosciences*, **10(4)**, 2711-2724.

Behrenfeld, M.J., R.T. O'Malley, D.A. Siegel, C.R. McClain, J.L. Sarmiento, G.C. Feldman, A.J. Milligan, P.G. Falkowski, R.M. Letelier, and E.S. Boss, 2006: Climate-driven trends in contemporary ocean productivity. *Nature*, **444(7120)**, 752-755.

Bopp, L., L. Resplandy, J.C. Orr, S.C. Doney, J.P. Dunne, M. Gehlen, P. Halloran, C. Heinze, T. Ilyina, R. Séférian, J. Tijiputra, and M. Vichi, 2013: Multiple stressors of ocean ecosystems in the 21st century: projections with CMIP5 models. *Biogeosciences*, **10**, 6225-6245.

Boyce, D.G., M.R. Lewis, and B. Worm, 2010: Global phytoplankton decline over the past century. *Nature*, **466(7306)**, 591-596.

Boyd, P.W. and S.C. Doney, 2002: Modelling regional responses by marine pelagic ecosystems to global climate change. *Geophysical Research Letters*, **29(16)**, 53-1–53-4, doi:10.1029/2001GL014130.

Boyd, P.W., R. Strzepek, F.X. Fu, and D.A. Hutchins, 2010: Environmental control of open-ocean phytoplankton groups: now and in the future. *Limnology and Oceanography*, **55(3)**, 1353-1376.

Chavez, F.P., P.G. Strutton, C.E. Friederich, R.A. Feely, G.C. Feldman, D.C. Foley, and M.J. McPhaden, 1999: Biological and chemical response of the equatorial Pacific Ocean to the 1997-98 El Niño. *Science*, **286(5447)**, 2126-2131.

Chavez, F.P., M. Messié, and J.T. Pennington, 2011: Marine primary production in relation to climate variability and change. *Annual Review of Marine Science*, **3(1)**, 227-260.

Field, C.B., M.J. Behrenfeld, J.T. Randerson, and P. Falkowski, 1998: Primary production of the biosphere: integrating terrestrial and oceanic components. *Science*, **281(5374)**, 237-240.

Hamme, R.C., P.W. Webley, W.R. Crawford, F.A. Whitney, M.D. DeGrandpre, S.R. Emerson, C.C. Eriksen, K.E. Giesbrecht, J.F.R. Gower, M.T. Kavanaugh, M.A. Peña, C.L. Sabine, S.D. Batten, L.A. Coogan, D.S. Grundle, and D. Lockwood, 2010: Volcanic ash fuels anomalous plankton bloom in subarctic northeast Pacific. *Geophysical Research Letters*, **37(19)**, L19604, doi:10.1029/2010GL044629.

Henson, S.A., J.L. Sarmiento, J.P. Dunne, L. Bopp, I. Lima, S.C. Doney, J. John, and C. Beaulieu, 2010: Detection of anthropogenic climate change in satellite records of ocean chlorophyll and productivity. *Biogeosciences*, **7(2)**, 621-640.

Longhurst, A.R., 1998: *Ecological Geography of the Sea*. Academic Press, San Diego, CA, USA, 560 pp.

Luo, Y.-W., S.C. Doney, L.A. Anderson, M. Benavides, I. Berman-Frank, A. Bode, S. Bonnet, K.H. Boström, D. Böttjer, D.G. Capone, E.J. Carpenter, Y.L. Chen, M.J. Church, J.E. Dore, L.I. Falcón, A. Fernández, R.A. Foster, K. Furuya, F. Gómez, K. Gundersen, A.M. Hynes, D.M. Karl, S. Kitajima, R.J. Langlois, J. LaRoche, R.M. Letelier, E. Marañón, D.J. McGillicuddy Jr., P.H. Moisander, C.M. Moore, B. Mouriño-Carballido, M.R. Mulholland, J.A. Needoba, K.M. Orcutt, A.J. Poulton, E. Rahav, P. Raimbault, A.P. Rees, L. Riemann, T. Shiozaki, A. Subramaniam, T. Tyrrell, K.A. Turk-Kubo, M. Varela, T.A. Villareal, E.A. Webb, A.E. White, J. Wu, and J.P. Zehr, 2012: Database of diazotrophs in global ocean: abundances, biomass and nitrogen fixation rates. *Earth System Science Data*, **4**, 47-73, doi:10.5194/essd-4-47-2012.

Mohr, W., T. Großkopf, D.W.R. Wallace, and J. LaRoche, 2010: Methodological underestimation of oceanic nitrogen fixation rates. *PLoS ONE*, **5(9)**, e12583, doi:10.1371/journal.pone.0012583.

Moore, C.M., M.M. Mills, E.P. Achterberg, R.J. Geider, J. LaRoche, M.I. Lucas, E.L. McDonagh, X. Pan, A.J. Poulton, M.J.A. Rijkenberg, D.J. Suggett, S.J. Ussher, and E.M.S. Woodward, 2009: Large-scale distribution of Atlantic nitrogen fixation controlled by iron availability. *Nature Geoscience*, **2(12)**, 867-871.

Mulholland, M.R., P.W. Bernhardt, J.L. Blanco-Garcia, A. Mannino, K. Hyde, E. Mondragon, K. Turk, P.H. Moisander, and J.P. Zehr, 2012: Rates of dinitrogen fixation and the abundance of diazotrophs in North American coastal waters between Cape Hatteras and Georges Bank. *Limnology and Oceanography*, **57(4)**, 1067-1083.

Polovina, J.J., J.P. Dunne, P.A. Woodworth, and E.A. Howell, 2011: Projected expansion of the subtropical biome and contraction of the temperate and equatorial upwelling biomes in the North Pacific under global warming. *ICES Journal of Marine Science*, **68(6)**, 986-995.

Rykaczewski, R.R. and J.P. Dunne, 2011: A measured look at ocean chlorophyll trends. *Nature*, **472(7342)**, E5-E6, doi:10.1038/nature09952.

Saba, V.S., M.A.M. Friedrichs, M.-E. Carr, D. Antoine, R.A. Armstrong, I. Asanuma, O. Aumont, N.R. Bates, M.J. Behrenfeld, V. Bennington, L. Bopp, J. Bruggeman, E.T. Buitenhuis, M.J. Church, A.M. Ciotti, S.C. Doney, M. Dowell, J. Dunne, S. Dutkiewicz, W. Gregg, N. Hoepffner, K.J.W. Hyde, J. Ishizaka, T. Kameda, D.M. Karl, I. Lima, M.W. Lomas, J. Marra, G.A. McKinley, F. Mélin, J.K. Moore, A. Morel, J. O'Reilly, B. Salihoglu, M. Scardi, T.J. Smyth, S.L. Tang, J. Tjiputra, J. Uitz, M. Vichi, K. Waters, T.K. Westberry, and A. Yool, 2010: Challenges of modeling depth-integrated marine primary productivity over multiple decades: a case study at BATS and HOT. *Global Biogeochemical Cycles*, **24**, GB3020, doi:10.1029/2009GB003655.

Saba, V.S., M.A.M. Friedrichs, D. Antoine, R.A. Armstrong, I. Asanuma, M.J. Behrenfeld, A.M. Ciotti, M. Dowell, N. Hoepffner, K.J.W. Hyde, J. Ishizaka, T. Kameda, J. Marra, F. Mélin, A. Morel, J. O'Reilly, M. Scardi, W.O. Smith Jr., T.J. Smyth, S. Tang, J. Uitz, K. Waters, and T.K. Westberry, 2011: An evaluation of ocean color model estimates of marine primary productivity in coastal and pelagic regions across the globe. *Biogeosciences*, **8(2)**, 489-503.

Sarmiento, J.L., R. Slater, R. Barber, L. Bopp, S.C. Doney, A.C. Hirst, J. Kleypas, R. Matear, U. Mikolajewicz, P. Monfray, V. Soldatov, S.A. Spall, and R. Stouffer, 2004: Response of ocean ecosystems to climate warming. *Global Biogeochemical Cycles*, **18(3)**, GB3003, doi:10.1029/2003GB002134.

Schmittner, A., A. Oschlies, H.D. Matthews, and E.D. Galbraith, 2008: Future changes in climate, ocean circulation, ecosystems, and biogeochemical cycling simulated for a business-as-usual CO2 emission scenario until year 4000 AD. *Global Biogeochemical Cycles*, **22(1)**, GB1013, doi:10.1029/2007GB002953.

Steinacher, M., F. Joos, T.L. Frölicher, L. Bopp, P. Cadule, V. Cocco, S.C. Doney, M. Gehlen, K. Lindsay, J.K. Moore, B. Schneider, and J. Segschneider, 2010: Projected 21st century decrease in marine productivity: a multi-model analysis. *Biogeosciences*, **7(3)**, 979-1005.

Westberry, T., M.J. Behrenfeld, D.A Siegel, and E. Boss, 2008: Carbon-based primary productivity modeling with vertically resolved photoacclimation. *Global Biogeochemical Cycles*, **22(2)**, GB2024, doi:10.1029/2007GB003078.

Wilson, S.T., D. Böttjer, M.J. Church, and D.M. Karl, 2012: Comparative assessment of nitrogen fixation methodologies, conducted in the oligotrophic North Pacific Ocean. *Applied and Environmental Microbiology*, **78(18)**, 6516-6523.

This cross-chapter box should be cited as:

Boyd, P.W., S. Sundby, and H.-O. Pörtner, 2014: *Cross-chapter box on net primary production in the ocean. In: Climate Change 2014: Impacts, Adaptation, and Vulnerability. Summaries, Frequently Asked Questions, and Cross-Chapter Boxes. Contribution of Working Group II to the Fifth Assessment Report of the Intergovernmental Panel on Climate Change* [Field, C.B., V.R. Barros, D.J. Dokken, K.J. Mach, M.D. Mastrandrea, T.E. Bilir, M. Chatterjee, K.L. Ebi, Y.O. Estrada, R.C. Genova, B. Girma, E.S. Kissel, A.N. Levy, S. MacCracken, P.R. Mastrandrea, and L.L. White (eds.)]. World Meteorological Organization, Geneva, Switzerland, pp. 135-138. (in Arabic, Chinese, English, French, Russian, and Spanish)

Regional Climate Summary Figures

Noah Diffenbaugh (USA), Dáithí Stone (Canada/South Africa/USA), Peter Thorne (USA/Norway /UK, Filippo Giorgi (Italy), Bruce Hewitson (South Africa), Richard Jones (UK), Geert Jan van Oldenborgh (Netherlands)

Information about the likelihood of regional climate change, assessed by Working Group I (WGI), is foundational for the Working Group II assessment of climate-related risks. To help communicate this assessment, the regional chapters of WGII present a coordinated set of regional climate figures, which summarize observed and projected change in annual average temperature and precipitation during the near term and the longer term for RCP2.6 and RCP8.5. These WGII regional climate summary figures use the same temperature and precipitation fields that are assessed in WGI Chapter 2 and WGI Chapter 12, with spatial boundaries, uncertainty metrics, and data classes tuned to support the WGII assessment of climate-related risks and options for risk management. Additional details on regional climate and regional climate processes can be found in WGI Chapter 14 and WGI Annex 1.

The WGII maps of observed annual temperature and precipitation use the same source data, calculations of data sufficiency, and calculations of trend significance as WGI Chapter 2 and WGI Figures SPM.1 and SPM.2. (A full description of the observational data selection and significance testing can be found in WGI Box 2.2.) Observed trends are determined by linear regression over the 1901–2012 period of Merged Land–Ocean Surface Temperature (MLOST) for annual temperature, and over the 1951–2010 period of Global Precipitation Climatology Centre (GPCC) for annual precipitation. Data points on the maps are classified into three categories, reflecting the categories used in WGI Figures SPM.1 and SPM.2:

1) Solid colors indicate areas where (a) sufficient data exist to permit a robust estimate of the trend (i.e., only for grid boxes with greater than 70% complete records and more than 20% data availability in the first and last 10% of the time period), and (b) the trend is significant at the 10% level (after accounting for autocorrelation effects on significance testing).
2) Diagonal lines indicate areas where sufficient data exist to permit a robust estimate of the trend, but the trend is not significant at the 10% level.
3) White indicates areas where there are not sufficient data to permit a robust estimate of the trend.

The WGII maps of projected annual temperature and precipitation are based on the climate model simulations from the Coupled Model Intercomparison Project Phase 5 (CMIP5; Taylor et al., 2012), which also form the basis for the figures presented in WGI (including WGI Chapters 12, 14, and Annex I). The CMIP5 archive includes output from Atmosphere–Ocean General Circulation Models (AOGCMs), AOGCMs with coupled vegetation and/or carbon cycle components, and AOGCMs with coupled atmospheric chemistry components. The number of models from which output is available, and the number of realizations of each model, vary between the different CMIP5 experiments. The WGII regional climate maps use the same source data as WGI Chapter 12 (e.g., Box 12.1 Figure

1), including the WGI multi-model mean values; the WGI individual model values; the WGI measure of baseline ("internal") variability; and the WGI time periods for the reference (1986–2005), mid-21st century (2046–2065), and late-21st century (2081–2100) periods. The full description of the selection of models, the selection of realizations, the definition of internal variability, and the interpolation to a common grid can be found in WGI Chapter 12 and Annex I.

In contrast to the Coupled Model Intercomparison Project Phase 3 (CMIP3) (Meehl et al., 2007), which used the IPCC Special Report on Emission Scenarios (SRES) emission scenarios (IPCC, 2000), CMIP5 uses the Representative Concentration Pathways (RCPs) (van Vuuren et al., 2011) to characterize possible trajectories of climate forcing over the 21st century. The WGII regional climate projection maps include RCP2.6 and RCP8.5, which represent the high and low end of the RCP range at the end of the 21st century. Projected changes in global mean temperature are similar across the RCPs over the next few decades (Figure RC-1; WGI Figure 12.5). During this near-term era of committed climate change, risks will evolve as socioeconomic trends interact with the changing climate. In addition, societal responses, particularly adaptations, will influence near-term outcomes. In the second half of the 21st century and beyond, the magnitude of global temperature increase diverges across the RCPs (Figure RC-1; WGI Figure 12.5). For this longer-term era of climate options, near-term and longer-term mitigation and adaptation, as well as development pathways, will determine the risks of climate change. The benefits of mitigation and adaptation thereby occur over different but overlapping time frames, and present-day choices thus affect the risks of climate change throughout the 21st century.

The projection maps plot differences in annual average temperature and precipitation between the future and reference periods (Figures RC-2 and RC-3), categorized into four classes. The classes are constructed based on the IPCC uncertainty guidance, providing a quantitative basis for assigning likelihood (Mastrandrea et al., 2010), with *likely* defined as 66 to 100% and *very likely* defined as 90 to 100%.

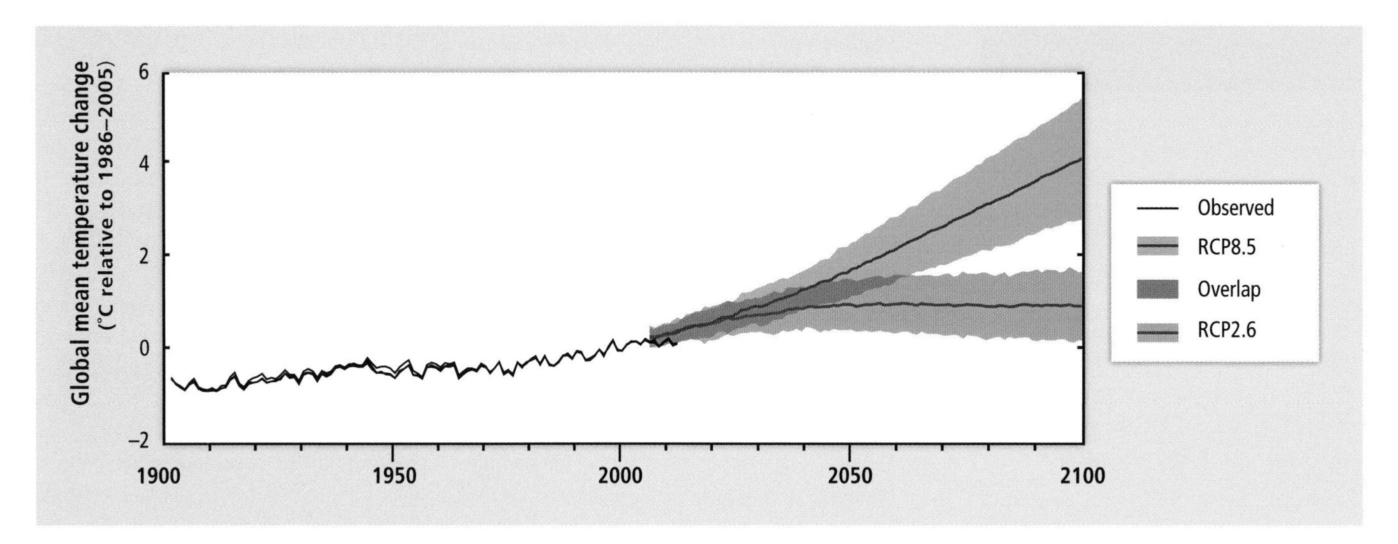

Figure RC-1 | Observed and projected changes in global annual average temperature. Values are expressed relative to 1986–2005. Black lines show the Goddard Institute for Space Studies Surface Temperature Analysis (GISTEMP), National Climate Data Center Merged Land–Ocean Surface Temperature (NCDC-MLOST), and Hadley Centre/Climatic Research Unit gridded surface temperature data set 4.2 (HadCRUT4.2) estimates from observational measurements. Blue and red lines and shading denote the ensemble mean and ±1.64 standard deviation range, based on Coupled Model Intercomparison Project Phase 5 (CMIP5) simulations from 32 models for Representative Concentration Pathway (RCP) 2.6 and 39 models for RCP8.5.

The classifications in the WGII regional climate projection figures are based on two aspects of likelihood (e.g., WGI Box 12.1 and Knutti et al., 2010). The first is the likelihood that projected changes exceed differences arising from internal climate variability (e.g., Tebaldi et al., 2011). The second is agreement among models on the sign of change (e.g., Christensen et al., 2007; and IPCC, 2012).

The four classifications of projected change depicted in the WGII regional climate maps are:

1) Solid colors indicate areas with very strong agreement, where the multi-model mean change is greater than twice the baseline variability (natural internal variability in 20-year means), and greater than or equal to 90% of models agree on sign of change. These criteria (and the areas that fall into this category) are identical to the highest confidence category in WGI Box 12.1. This category supersedes other categories in the WGII regional climate maps.
2) Colors with white dots indicate areas with strong agreement, where 66% or more of models show change greater than the baseline variability, and 66% or more of models agree on sign of change.
3) Gray indicates areas with divergent changes, where 66% or more of models show change greater than the baseline variability, but fewer than 66% agree on sign of change.
4) Colors with diagonal lines indicate areas with little or no change, where fewer than 66% of models show change greater than the baseline variability. It should be noted that areas that fall in this category for the annual average could still exhibit significant change at seasonal, monthly, and/or daily time scales.

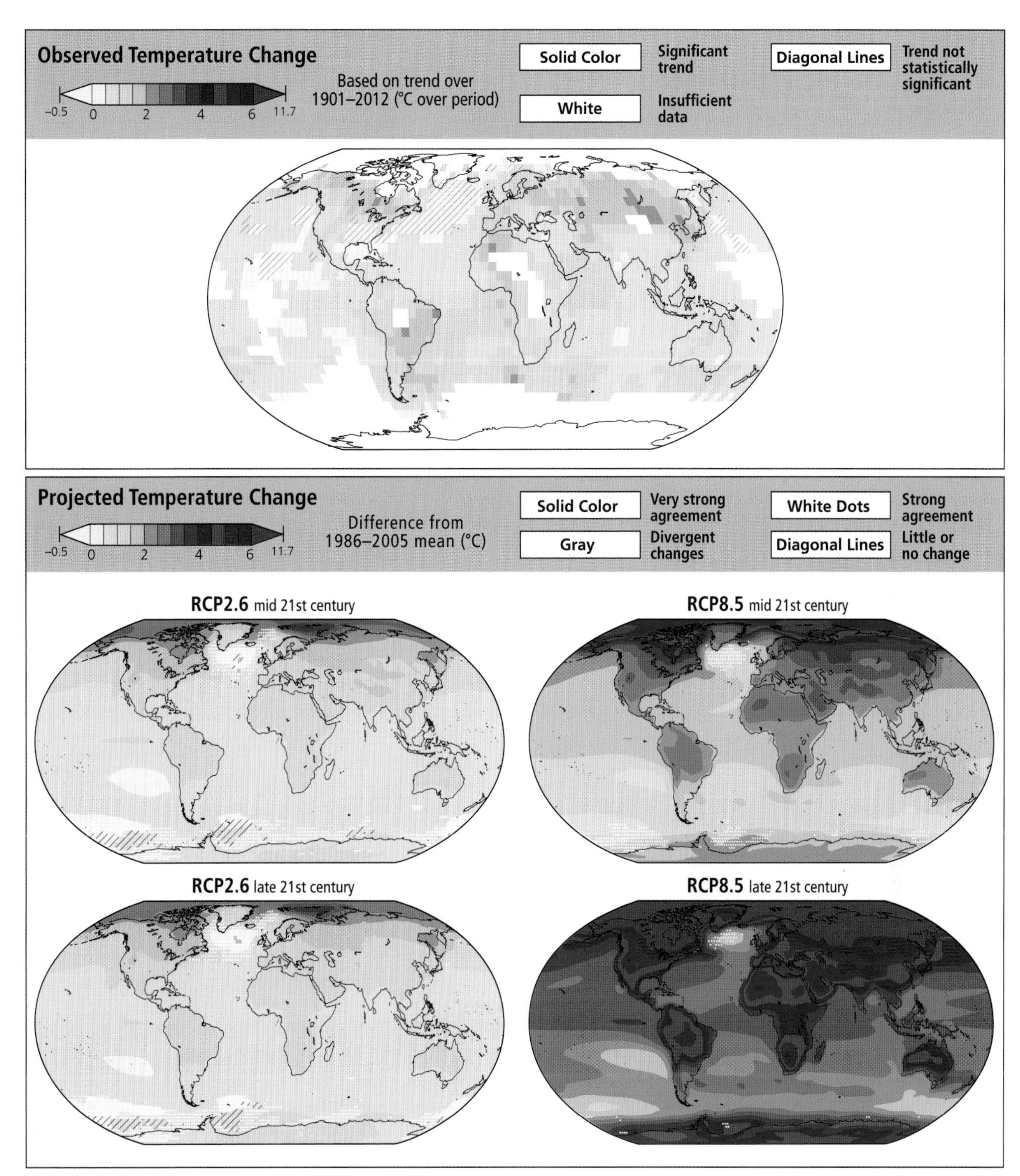

Figure RC-2 | Observed and projected changes in annual average surface temperature. (A) Map of observed annual average temperature change from 1901 to 2012, derived from a linear trend where sufficient data permit a robust estimate (i.e., only for grid boxes with greater than 70% complete records and more than 20% data availability in the first and last 10% of the time period); other areas are white. Solid colors indicate areas where trends are significant at the 10% level (after accounting for autocorrelation effects on significance testing). Diagonal lines indicate areas where trends are not significant. Observed data (range of grid-point values: –0.53 to +2.50°C over period) are from WGI AR5 Figures SPM.1 and 2.21. (B) Coupled Model Intercomparison Project Phase 5 (CMIP5) multi-model mean projections of annual average temperature changes for 2046–2065 and 2081–2100 under Representative Concentration Pathway (RCP) 2.6 and 8.5, relative to 1986–2005. Solid colors indicate areas with very strong agreement, where the multi-model mean change is greater than twice the baseline variability (natural internal variability in 20-year means) and ≥90% of models agree on sign of change. Colors with white dots indicate areas with strong agreement, where ≥66% of models show change greater than the baseline variability and ≥66% of models agree on sign of change. Gray indicates areas with divergent changes, where ≥66% of models show change greater than the baseline variability, but <66% agree on sign of change. Colors with diagonal lines indicate areas with little or no change, where <66% of models show change greater than the baseline variability, although there may be significant change at shorter timescales such as seasons, months, or days. Analysis uses model data from WGI AR5 Figure SPM.8, Box 12.1, and Annex I. The range of grid-point values for the multi-model mean is: +0.19 to +4.08°C for mid 21st century of RCP2.6; +0.06 to +3.85°C for late 21st century of RCP2.6; +0.70 to +7.04°C for mid 21st century of RCP8.5; and +1.38 to +11.71°C for late 21st century of RCP8.5.

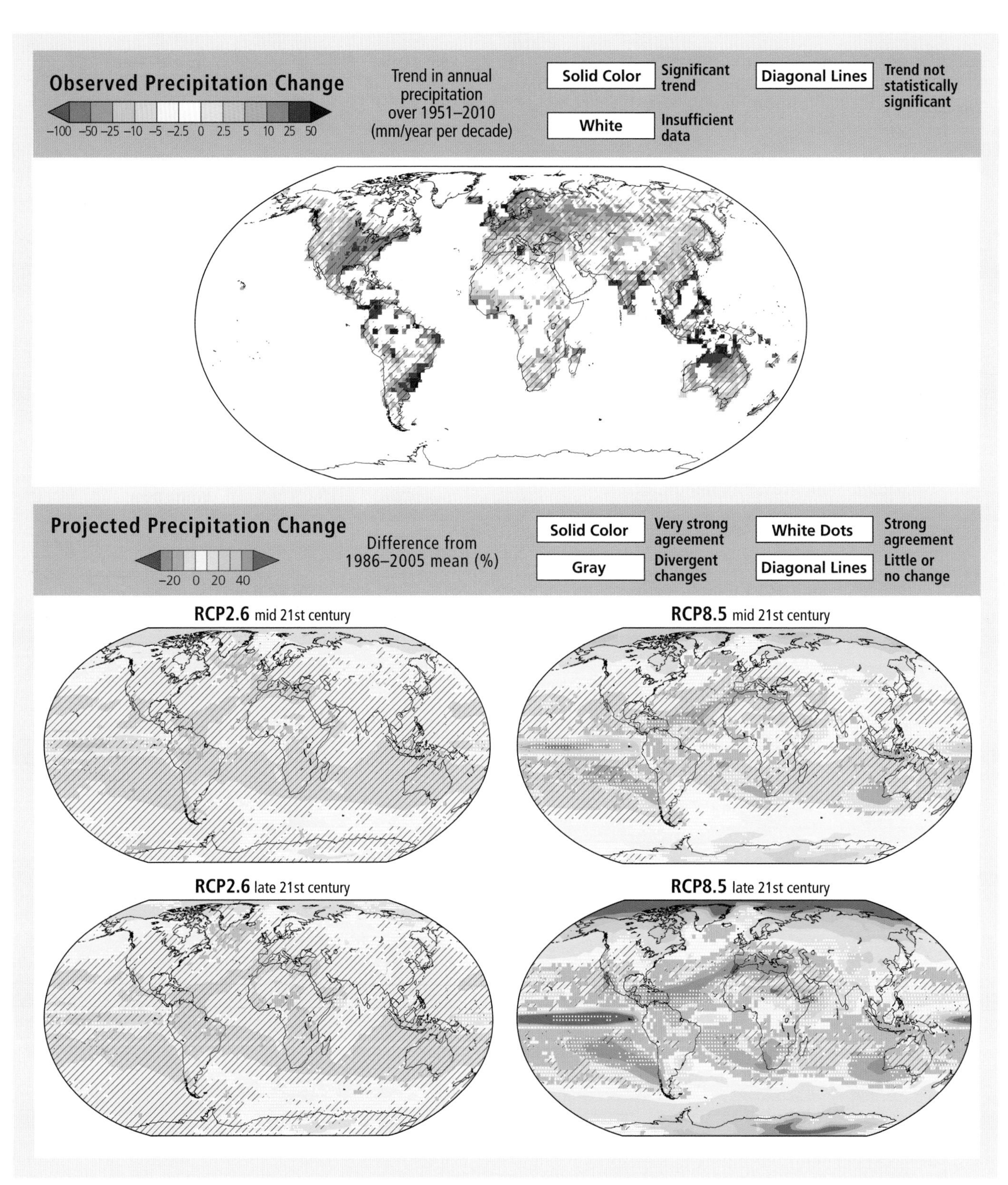

Figure RC-3 | Observed and projected changes in annual average precipitation. (A) Map of observed annual precipitation change from 1951–2010, derived from a linear trend where sufficient data permit a robust estimate (i.e., only for grid boxes with greater than 70% complete records and more than 20% data availability in the first and last 10% of the time period); other areas are white. Solid colors indicate areas where trends are significant at the 10% level (after accounting for autocorrelation effects on significance testing). Diagonal lines indicate areas where trends are not significant. Observed data (range of grid-point values: –185 to +111 mm/year per decade) are from WGI AR5 Figures SPM.2 and 2.29. (B) Coupled Model Intercomparison Project Phase 5 (CMIP5) multi-model average percent changes in annual mean precipitation for 2046–2065 and 2081–2100 under Representative Concentration Pathway (RCP) 2.6 and 8.5, relative to 1986–2005. Solid colors indicate areas with very strong agreement, where the multi-model mean change is greater than twice the baseline variability (natural internal variability in 20-yr means) and ≥90% of models agree on sign of change. Colors with white dots indicate areas with strong agreement, where ≥66% of models show change greater than the baseline variability and ≥66% of models agree on sign of change. Gray indicates areas with divergent changes, where ≥66% of models show change greater than the baseline variability, but <66% agree on sign of change. Colors with diagonal lines indicate areas with little or no change, where <66% of models show change greater than the baseline variability, although there may be significant change at shorter timescales such as seasons, months, or days. Analysis uses model data from WGI AR5 Figure SPM.8, Box 12.1, and Annex I. The range of grid-point values for the multi-model mean is: –10 to +24% for mid 21st century of RCP2.6; –9 to +22% for late 21st century of RCP2.6; –19 to +57% for mid 21st century of RCP8.5; and –34 to +112% for late 21st century of RCP8.5.

References

Christensen, J.H., B. Hewitson, A. Busuioc, A. Chen, X. Gao, I. Held, R. Jones, R.K. Kolli, W.-T. Kwon, R. Laprise, V. Magaña Rueda, L. Mearns, C.G. Menéndez, J. Räisänen, A. Rinke, A. Sarr, and P. Whetton, 2007: Regional climate projections. *Climate Change 2007: The Physical Science Basis. Contribution of Working Group I to the Fourth Assessment Report of the Intergovernmental Panel on Climate Change* [Solomon, S., D. Qin, M. Manning, Z. Chen, M. Marquis, K.B. Averyt, M. Tignor, and H.L. Miller (eds.)]. Cambridge University Press, Cambridge, UK and New York, NY, USA, pp. 847-940.

IPCC, 2000: *Special Report on Emissions Scenarios* [Nakicenovic, N. and R. Swart (eds.)]. Cambridge University Press, Cambridge, UK, 570 pp.

IPCC, 2012: *Managing the Risks of Extreme Events and Disasters to Advance Climate Change Adaptation. A Special Report of Working Groups I and II of the Intergovernmental Panel on Climate Change* [Field, C.B., V. Barros, T.F. Stocker, D. Qin, D.J. Dokken, K.L. Ebi, M.D. Mastrandrea, K.J. Mach, G.-K. Plattner, S.K. Allen, M. Tignor, and P.M. Midgley (eds.)]. Cambridge University Press, Cambridge, UK and New York, NY, USA, 582 pp.

Knutti, R., R. Furrer, C. Tebaldi, J. Cermak, and G.A. Meehl, 2010: Challenges in combining projections from multiple climate models. *Journal of Climate*, **23(10)**, 2739-2758.

Mastrandrea, M.D., C.B. Field, T.F. Stocker, O. Edenhofer, K.L. Ebi, D.J. Frame, H. Held, E. Kriegler, K.J. Mach, P.R. Matschoss, G.-K. Plattner, G.W. Yohe, and F.W. Zwiers, 2010: *Guidance Note for Lead Authors of the IPCC Fifth Assessment Report on Consistent Treatment of Uncertainties*. Intergovernmental Panel on Climate Change (IPCC), www.ipcc.ch/pdf/supporting-material/uncertainty-guidance-note.pdf.

Meehl, G.A., C. Covey, K.E. Taylor, T. Delworth, R.J. Stouffer, M. Latif, B. McAvaney, and J.F.B. Mitchell, 2007: The WCRP CMIP3 multimodel dataset – a new era in climate change research. *Bulletin of the American Meteorological Society*, **88(9)**, 1383-1394.

Taylor, K.E., R.J. Stouffer, and G.A. Meehl, 2012: An overview of CMIP5 and the experiment design. *Bulletin of the American Meteorological Society*, **93(4)**, 485-498.

Tebaldi, C., J.M. Arblaster, and Reto Knutti, 2011: Mapping model agreement on future climate projections. *Geophysical Research Letters*, **38(23)**, L23701, doi:10.1029/2011GL049863.

van Vuuren, D.P., J. Edmonds, M. Kainuma, K. Riahi, A. Thomson, K. Hibbard, G.C. Hurtt, T. Kram, V. Krey, J.-F. Lamarque, T. Masui, M. Meinshausen, N. Nakicenovic, S.J. Smith, and S.K. Rose 2011: The representative concentration pathways: an overview. *Climatic Change*, **109(1-2)**, 5-31.

This cross-chapter box should be cited as:

Diffenbaugh, N.S., D.A. Stone, P. Thorne, F. Giorgi, B.C. Hewitson, R.G. Jones, and G.J. van Oldenborgh, 2014: Cross-chapter box on the regional climate summary figures. In: *Climate Change 2014: Impacts, Adaptation, and Vulnerability. Summaries, Frequently Asked Questions, and Cross-Chapter Boxes. Contribution of Working Group II to the Fifth Assessment Report of the Intergovernmental Panel on Climate Change* [Field, C.B., V.R. Barros, D.J. Dokken, K.J. Mach, M.D. Mastrandrea, T.E. Bilir, M. Chatterjee, K.L. Ebi, Y.O. Estrada, R.C. Genova, B. Girma, E.S. Kissel, A.N. Levy, S. MacCracken, P.R. Mastrandrea, and L.L. White (eds.)]. World Meteorological Organization, Geneva, Switzerland, pp. 139-143. (in Arabic, Chinese, English, French, Russian, and Spanish)

Impact of Climate Change on Freshwater Ecosystems due to Altered River Flow Regimes

Petra Döll (Germany), Stuart E. Bunn (Australia)

It is widely acknowledged that the flow regime is a primary determinant of the structure and function of rivers and their associated floodplain wetlands, and flow alteration is considered to be a serious and continuing threat to freshwater ecosystems (Bunn and Arthington, 2002; Poff and Zimmerman, 2010; Poff et al., 2010). Most species distribution models do not consider the effect of changing flow regimes (i.e., changes to the frequency, magnitude, duration, and/or timing of key flow parameters) or they use precipitation as proxy for river flow (Heino et al., 2009).

There is growing evidence that climate change will significantly alter ecologically important attributes of hydrologic regimes in rivers and wetlands, and exacerbate impacts from human water use in developed river basins (*medium confidence*; Xenopoulos et al., 2005; Aldous et al., 2011). By the 2050s, climate change is projected to impact river flow characteristics such as long-term average discharge, seasonality, and statistical high flows (but not statistical low flows) more strongly than dam construction and water withdrawals have done up to around the year 2000 (Figure RF-1; Döll and Zhang, 2010). For one climate scenario (Special Report on Emission Scenarios (SRES) A2 emissions, Met Office Hadley Centre climate prediction model 3 (HadCM3)), 15% of the global land area may be negatively affected, by the 2050s, by a decrease of fish species in the upstream basin of more than 10%, as compared to only 10% of the land area that has already suffered from such decreases due to water withdrawals and dams (Döll and Zhang, 2010). Climate change may exacerbate the negative impacts of dams for freshwater ecosystems but may also provide opportunities for operating dams and power stations to the benefit of riverine ecosystems. This is the case if total runoff increases and, as occurs in Sweden, the annual hydrograph becomes more similar to variation in electricity demand, that is, with a lower spring flood and increased runoff during winter months (Renofalt et al., 2010).

Because biota are often adapted to a certain level of river flow variability, the projected larger variability of river flows that is due to increased climate variability is *likely* to select for generalist or invasive species (Ficke et al., 2007). The relatively stable habitats of groundwater-fed streams in snow-dominated or glacierized basins may be altered by reduced recharge by meltwater and as a result experience more variable (possibly intermittent) flows (Hannah et al., 2007). A high-impact change of flow variability is a flow regime shift from intermittent to perennial or vice versa. It is projected that until the 2050s, river flow regime shifts may occur on 5 to 7% of the global land area, mainly in semiarid areas (Döll and Müller Schmied, 2012; see Table 3-2 in Chapter 3).

In Africa, one third of fish species and one fifth of the endemic fish species occur in eco-regions that may experience a change in discharge or runoff of more than 40% by the 2050s (Thieme et al., 2010). Eco-regions containing more than 80% of Africa's freshwater fish species and several outstanding ecological and evolutionary phenomena are *likely* to experience hydrologic conditions substantially different from the present, with alterations in long-term average annual river discharge or runoff of more than 10% due to climate change and water use (Thieme et al., 2010).

As a result of increased winter temperatures, freshwater ecosystems in basins with significant snow storage are affected by higher river flows in winter, earlier spring peak flows, and possibly reduced summer low flows (Section 3.2.3). Strongly increased winter peak flows may lead to a decline in salmonid populations in the Pacific Northwest of the USA of 20 to 40% by the 2050s (depending on the climate model) due to scouring of the streambed during egg incubation, the relatively pristine high-elevation areas being affected most (Battin et al., 2007). Reductions in summer low flows will increase the competition for water between ecosystems and irrigation water users (Stewart et al., 2005). Ensuring environmental flows through purchasing or leasing water rights and altering reservoir release patterns will be an important adaptation strategy (Palmer et al., 2009).

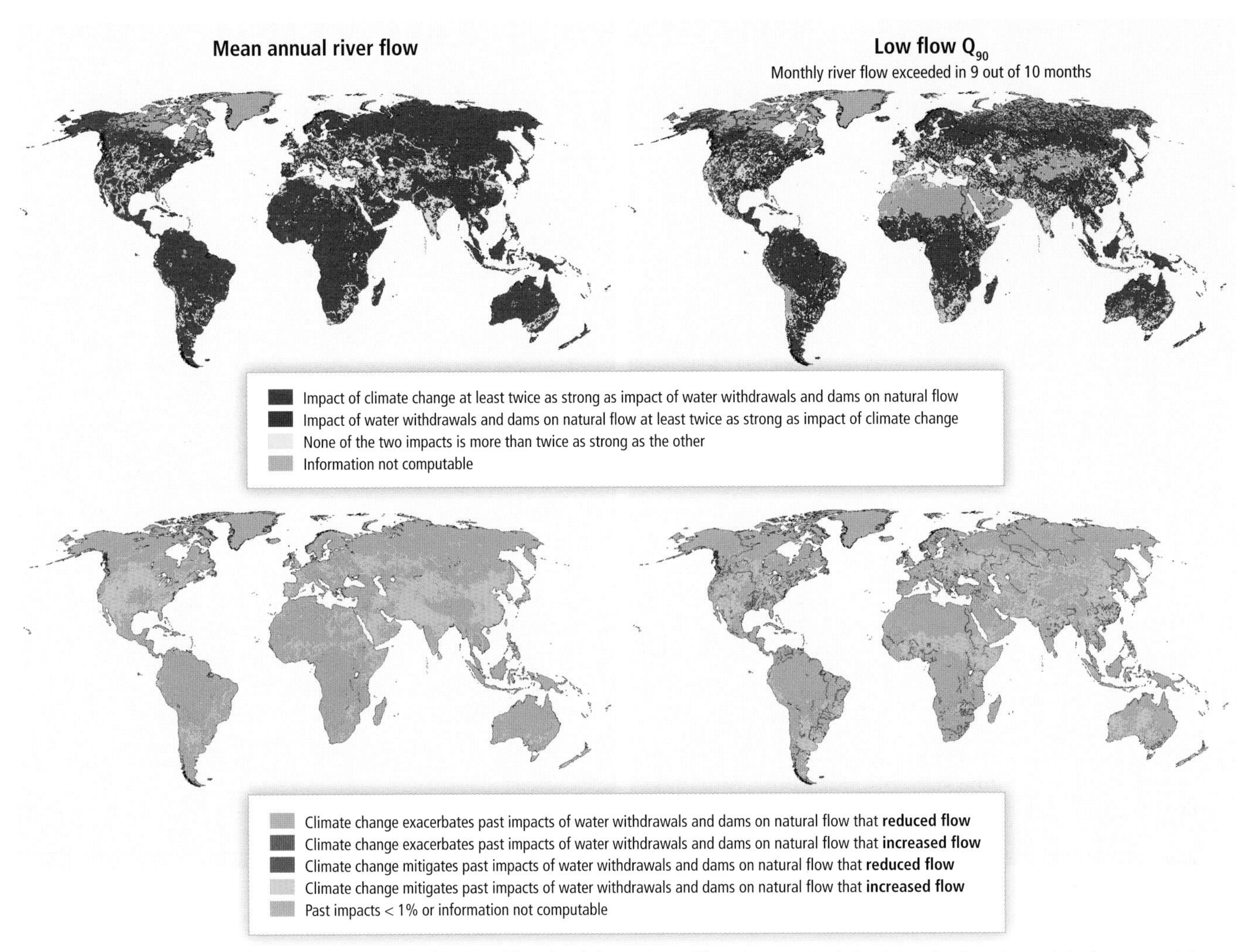

Figure RF-1 | Impact of climate change relative to the impact of water withdrawals and dams on natural flows for two ecologically relevant river flow characteristics (mean annual river flow and monthly low flow Q_{90}), computed by a global water model (Döll and Zhang, 2010). Impact of climate change is the percent change of flow between 1961–1990 and 2041–2070 according to the emissions scenario A2 as implemented by the global climate model Met Office Hadley Centre Coupled Model, version 3 (HadCM3). Impact of water withdrawals and reservoirs is computed by running the model with and without water withdrawals and dams that existed in 2002. Please note that the figure does not reflect spatial differences in the magnitude of change.

Observations and models suggest that global warming impacts on glacier and snow-fed streams and rivers will pass through two contrasting phases (Burkett et al., 2005; Vuille et al., 2008; Jacobsen et al., 2012). In the first phase, when river discharge is increased as a result of intensified melting, the overall diversity and abundance of species may increase. However, changes in water temperature and stream flow may have negative impacts on narrow range endemics (Jacobsen et al., 2012). In the second phase, when snowfields melt early and glaciers have shrunken to the point that late-summer stream flow is reduced, broad negative impacts are foreseen, with species diversity rapidly declining once a critical threshold of roughly 50% glacial cover is crossed (Figure RF-2).

River discharge also influences the response of river temperatures to increases of air temperature. Globally averaged, air temperature increases of 2°C, 4°C, and 6°C are estimated to lead to increases of annual mean river temperatures of 1.3°C, 2.6°C, and 3.8°C, respectively (van Vliet et al., 2011). Discharge decreases of 20% and 40% are computed to result in additional increases of river water temperature of 0.3° C and 0.8°C on average (van Vliet et al., 2011). Therefore, where rivers will experience drought more frequently in the future, freshwater-dependent biota will suffer not only directly by changed flow conditions but also by drought-induced river temperature increases, as well as by related decreased oxygen and increased pollutant concentrations.

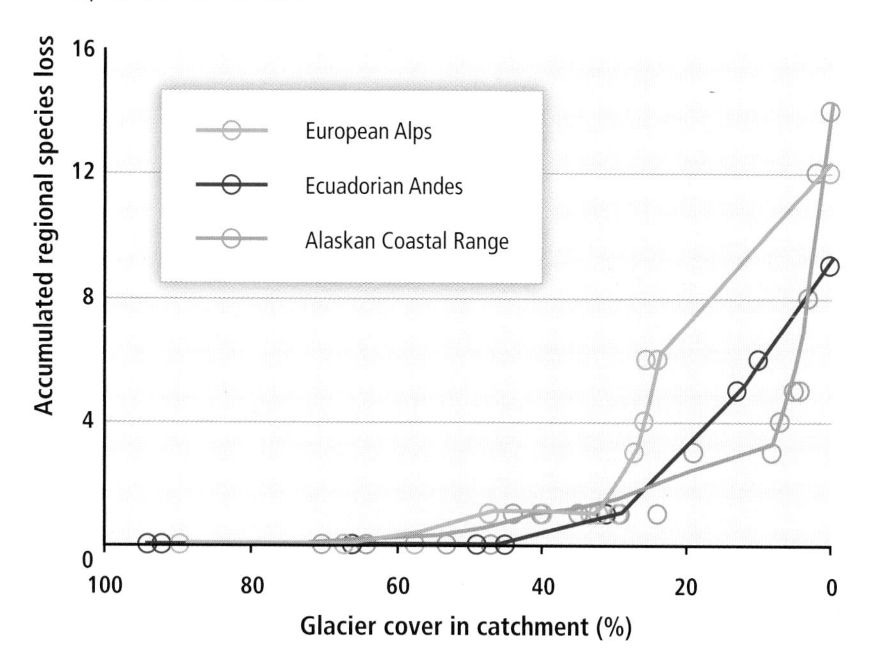

Figure RF-2 | Accumulated loss of regional species richness (gamma diversity) of macroinvertebrates as a function of glacial cover in catchment. Obligate glacial river macroinvertebrates begin to disappear from assemblages when glacial cover in the catchment drops below approximately 50%, and 9 to 14 species are predicted to be lost with the complete disappearance of glaciers in each region, corresponding to 11, 16, and 38% of the total species richness in the three study regions in Ecuador, Europe, and Alaska. Data are derived from multiple river sites from the Ecuadorian Andes and Swiss and Italian Alps, and a temporal study of a river in the Coastal Range Mountains of southeast Alaska over nearly three decades of glacial shrinkage. Each data point represents a river site (Europe or Ecuador) or date (Alaska), and lines are Lowess fits. (Adapted by permission from Jacobsen et al., 2012.)

RF

References

Aldous, A., J. Fitzsimons, B. Richter, and L. Bach, 2011: Droughts, floods and freshwater ecosystems: evaluating climate change impacts and developing adaptation strategies. *Marine and Freshwater Research*, **62(3)**, 223-231.

Battin, J., M.W. Wiley, M.H. Ruckelshaus, R.N. Palmer, E. Korb, K.K. Bartz, and H. Imaki, 2007: Projected impacts of climate change on salmon habitat restoration. *Proceedings of the National Academy of Sciences of the United States of America*, **104(16)**, 6720-6725.

Bunn, S.E. and A.H. Arthington, 2002: Basic principles and ecological consequences of altered flow regimes for aquatic biodiversity. *Environmental Management*, **30(4)**, 492-507.

Burkett, V., D. Wilcox, R. Stottlemyer, W. Barrow, D. Fagre, J. Baron, J. Price, J. Nielsen, C. Allen, D. Peterson, G. Ruggerone, and T. Doyle, 2005: Nonlinear dynamics in ecosystem response to climatic change: case studies and policy implications. *Ecological Complexity*, **2(4)**, 357-394.

Döll, P. and H. Müller Schmied, 2012: How is the impact of climate change on river flow regimes related to the impact on mean annual runoff? A global-scale analysis. *Environmental Research Letters*, **7(1)**, 014037, doi:10.1088/1748-9326/7/1/014037.

Döll, P. and J. Zhang, 2010: Impact of climate change on freshwater ecosystems: a global-scale analysis of ecologically relevant river flow alterations. *Hydrology and Earth System Sciences*, **14(5)**, 783-799.

Ficke, A.D., C.A. Myrick, and L.J. Hansen, 2007: Potential impacts of global climate change on freshwater fisheries. *Reviews in Fish Biology and Fisheries*, **17(4)**, 581-613.

Hannah, D.M., L.E. Brown, A.M. Milner, A.M. Gurnell, G.R. McGregord, G.E. Petts, B.P.G. Smith, and D.L. Snook, 2007: Integrating climate-hydrology-ecology for alpine river systems. *Aquatic Conservation: Marine and Freshwater Ecosystems*, **17(6)**, 636-656.

Heino, J., R. Virkalla, and H. Toivonen, 2009: Climate change and freshwater biodiversity: detected patterns, future trends and adaptations in northern regions. *Biological Reviews*, **84(1)**, 39-54.

Jacobsen, D., A.M. Milner, L.E. Brown, and O. Dangles, 2012: Biodiversity under threat in glacier-fed river systems. *Nature Climate Change*, **2(5)**, 361-364.

Palmer, M.A., D.P. Lettenmaier, N.L. Poff, S.L. Postel, B. Richter, and R. Warner, 2009: Climate change and river ecosystems: protection and adaptation options. *Environmental Management*, **44(6)**, 1053-1068.

Poff, N.L. and J.K.H. Zimmerman, 2010: Ecological responses to altered flow regimes: a literature review to inform the science and management of environmental flows. *Freshwater Biology*, **55(1)**, 194-205.

Poff, N.L., B.D. Richter, A.H. Arthington, S.E. Bunn, R.J. Naiman, E. Kendy, M. Acreman, C. Apse, B.P. Bledsoe, M.C. Freeman, J. Henriksen, R.B. Jacobson, J.G. Kennen, D.M. Merritt, J.H. O'Keeffe, J.D. Olden, K. Rogers, R.E. Tharme, and A. Warner, 2010: The ecological limits of hydrologic alteration (ELOHA): a new framework for developing regional environmental flow standards. *Freshwater Biology*, **55(1)**, 147-170.

Renofalt, B.M., R. Jansson, and C. Nilsson, 2010: Effects of hydropower generation and opportunities for environmental flow management in Swedish riverine ecosystems. *Freshwater Biology*, **55(1)**, 49-67.

Stewart, I., D. Cayan, and M. Dettinger, 2005: Changes toward earlier streamflow timing across western North America. *Journal of Climate*, **18(8)**, 1136-1155.

Thieme, M.L., B. Lehner, R. Abell, and J. Matthews, 2010: Exposure of Africa's freshwater biodiversity to a changing climate. *Conservation Letters*, **3(5)**, 324-331.

van Vliet, M.T.H., F. Ludwig, J.J.G. Zwolsman, G.P. Weedon, and P. Kabat, 2011: Global river temperatures and sensitivity to atmospheric warming and changes in river flow. *Water Resources Research*, **47(2)**, W02544, doi:10.1029/2010WR009198.

Vuille, M., B. Francou, P. Wagnon, I. Juen, G. Kaser, B.G. Mark, and R.S. Bradley, 2008: Climate change and tropical Andean glaciers: past, present and future. *Earth-Science Reviews*, **89(3-4)**, 79-96.

Xenopoulos, M., D. Lodge, J. Alcamo, M. Marker, K. Schulze, and D. Van Vuuren, 2005: Scenarios of freshwater fish extinctions from climate change and water withdrawal. *Global Change Biology*, **11(10)**, 1557-1564.

This cross-chapter box should be cited as:

Döll, P. and S.E. Bunn, 2014: *Cross-chapter box on the impact of climate change on freshwater ecosystems due to altered river flow regimes. In: Climate Change 2014: Impacts, Adaptation, and Vulnerability. Summaries, Frequently Asked Questions, and Cross-Chapter Boxes. Contribution of Working Group II to the Fifth Assessment Report of the Intergovernmental Panel on Climate Change* [Field, C.B., V.R. Barros, D.J. Dokken, K.J. Mach, M.D. Mastrandrea, T.E. Bilir, M. Chatterjee, K.L. Ebi, Y.O. Estrada, R.C. Genova, B. Girma, E.S. Kissel, A.N. Levy, S. MacCracken, P.R. Mastrandrea, and L.L. White (eds.)]. World Meteorological Organization, Geneva, Switzerland, pp. 145-148. (in Arabic, Chinese, English, French, Russian, and Spanish)

Building Long-Term Resilience from Tropical Cyclone Disasters

Yoshiki Saito (Japan), Kathleen McInnes (Australia)

Tropical cyclones (also referred to as hurricanes and typhoons in some regions) cause powerful winds, torrential rains, high waves, and storm surge, all of which can have major impacts on society and ecosystems. Bangladesh and India suffer 86% of mortality from tropical cyclones (Murray et al., 2012), which occurs mainly during the rarest and most severe storm categories (i.e., Categories 3, 4, and 5 on the Saffir–Simpson scale).

About 90 tropical cyclones occur globally each year (Seneviratne et al., 2012) although interannual variability is large. Changes in observing techniques, particularly after the introduction of satellites in the late 1970s, confounds the assessment of trends in tropical cyclone frequencies and intensities, which leads to *low confidence* that any observed long-term (i.e., 40 years or more) increases in tropical cyclone activity are robust, after accounting for past changes in observing capability (Seneviratne et al., 2012; Chapter 2). There is also *low confidence* in the detection and attribution of century scale trends in tropical cyclones. Future changes to tropical cyclones arising from climate change are *likely* to vary by region. This is because there is *medium confidence* that for certain regions, shorter-term forcing by natural and anthropogenic aerosols has had a measurable effect on tropical cyclones. Tropical cyclone frequency is *likely* to decrease or remain unchanged over the 21st century, while intensity (i.e., maximum wind speed and rainfall rates) is *likely* to increase (WGI AR5 Section 14.6). Regionally specific projections have *lower confidence* (see WGI AR5 Box 14.2).

Longer-term impacts from tropical cyclones include salinization of coastal soils and water supplies and subsequent food and water security issues from the associated storm surge and waves (Terry and Chui, 2012). However, preparation for extreme tropical cyclone events through improved governance and development to reduce their impacts provides an avenue for building resilience to longer-term changes associated with climate change.

Asian deltas are particularly vulnerable to tropical cyclones owing to their large population density in expanding urban areas (Nicholls et al., 2007). Extreme cyclones in Asia since 1970 caused more than 0.5 million fatalities (Murray et al., 2012), for example, cyclones Bhola in 1970, Gorky in 1991, Thelma in 1998, Gujarat in 1998, Orissa in 1999, Sidr in 2007, and Nargis in 2008. Tropical cyclone Nargis hit Myanmar on May 2, 2008 and caused more than 138,000 fatalities. Several-meter high storm surges widely flooded densely populated coastal areas of the Irrawaddy Delta and surrounding areas (Revenga et al., 2003; Brakenridge et al., 2013). The flooded areas were captured by a NASA Moderate Resolution Imaging Spectrometer (MODIS) image on May 5, 2008 (see Figure TC-1).

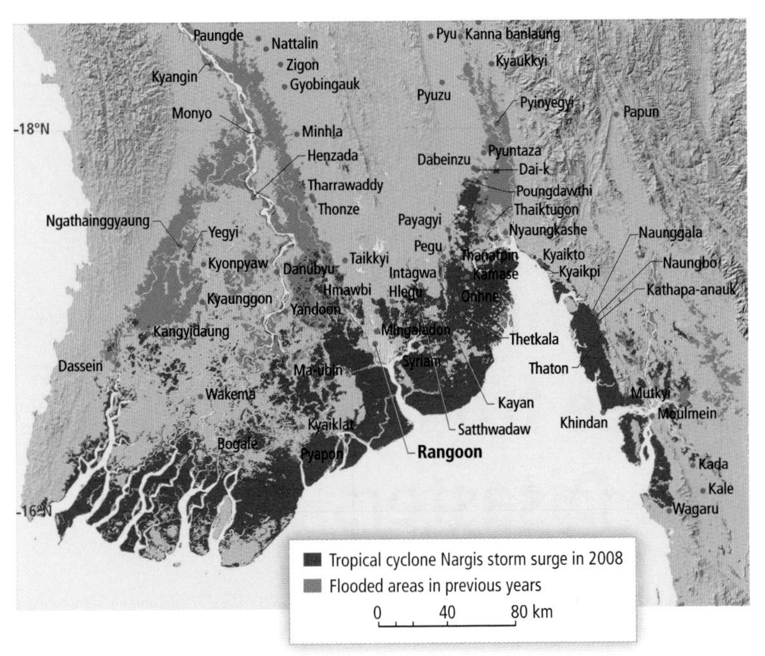

Figure TC-1 | The intersection of inland and storm surge flooding. Red shows May 5, 2008 Moderate Resolution Imaging Spectrometer (MODIS) mapping of the tropical cyclone Nargis storm surge along the Irrawaddy Delta and to the east, Myanmar. The purple areas to the north were flooded by the river in prior years. (Source: Brakenridge et al., 2013.)

Murray et al. (2012) compared the response to cyclone Sidr in Bangladesh in 2007 and Nargis in Myanmar in 2008 and demonstrated how disaster risk reduction methods could be successfully applied to climate change adaptation. Sidr, despite being of similar strength to Nargis, caused far fewer fatalities (3400 compared to more than 138,000) and this was attributed to advancement in preparedness and response in Bangladesh through experience in previous cyclones such as Bhola and Gorky. The responses included the construction of multistoried cyclone shelters, improvement of forecasting and warning capacity, establishing a coastal volunteer network, and coastal reforestation of mangroves. Disaster risk management strategies for tropical cyclones in coastal areas create protective measures, anticipate and plan for extreme events, and increase the resilience of potentially exposed communities. The integration of activities relating to education, training, and awareness-raising into relevant ongoing processes and practices is important for the long-term success of disaster risk reduction and management (Murray et al., 2012). However, Birkmann and Teichman (2010) caution that while the combination of risk reduction and climate change adaptation strategies may be desirable, different spatial and temporal scales, norm systems, and knowledge types and sources between the two goals can confound their effective combination.

References

Birkman, J. and K. von Teichman, 2010: Integrating disaster risk reduction and climate change adaptation: key challenges – scales, knowledge and norms. *Sustainability Science*, **5**, 171-184.

Brakenridge, G.R., J.P.M. Syvitski, I. Overeem, S.A. Higgins, A.J. Kettner, J.A. Stewart-Moore, and R. Westerhoff, 2013: Global mapping of storm surges and the assessment of delta vulnerability. *Natural Hazards*, **66**, 1295-1312, doi:10.1007/s11069-012-0317-z.

Murray V., G. McBean, M. Bhatt, S. Borsch, T.S. Cheong, W.F. Erian, S. Llosa, F. Nadim, M. Nunez, R. Oyun, and A.G. Suarez, 2012: Case studies. In: *Managing the Risks of Extreme Events and Disasters to Advance Climate Change Adaptation. A Special Report of Working Groups I and II of the Intergovernmental Panel on Climate Change* [Field, C.B., V. Barros, T.F. Stocker, D. Qin, D.J. Dokken, K.L. Ebi, M.D. Mastrandrea, K.J. Mach, G.-K. Plattner, S.K. Allen, M. Tignor, and P.M. Midgley (eds.)]. Cambridge University Press, Cambridge, UK and New York, NY, USA, pp. 487-542.

Nicholls, R.J., 2007: *Adaptation Options for Coastal Areas and Infrastructure: An Analysis for 2030*. Report to the United Nations Framework Convention on Climate Change (UNFCCC), UNFCCC Secretariat, Bonn, Germany, 35 pp.

Revenga, C., J. Nackoney, E. Hoshino, Y. Kura, and J. Maidens, 2003: Watersheds of Asia and Oceania: AS 12 Irrawaddy. In: *Water Resources eAtlas: Watersheds of the World*. A collaborative product of the International Union for Conservation of Nature (IUCN), the International Water Management Institute (IWMI), the Ramsar Convention Bureau, and the World Resources Institute (WRI), WRI, Washington, DC, USA, pdf.wri.org/watersheds_2003/as13.pdf.

Seneviratne, S.I., N. Nicholls, D. Easterling, C.M. Goodess, S. Kanae, J. Kossin, Y. Luo, J. Marengo, K. McInnes, M. Rahimi, M. Reichstein, A. Sorteberg, C. Vera, and X. Zhang, 2012: Changes in climate extremes and their impacts on the natural physical environment. In: *Managing the Risks of Extreme Events and Disasters to Advance Climate Change Adaptation. A Special Report of Working Groups I and II of the Intergovernmental Panel on Climate Change* [Field, C.B., V. Barros, T.F. Stocker, D. Qin, D.J. Dokken, K.L. Ebi, M.D. Mastrandrea, K.J. Mach, G.-K. Plattner, S.K. Allen, M. Tignor, and P.M. Midgley (eds.)]. Cambridge University Press, Cambridge, UK and New York, NY, USA, pp. 109-230.

Terry, J. and T.F.M. Chui, 2012: Evaluating the fate of freshwater lenses on atoll islands after eustatic sea level rise and cyclone driven inundation: a modelling approach. *Global and Planetary Change*, **88-89**, 76-84.

This cross-chapter box should be cited as:

Saito, Y. and K.L. McInnes, 2014: Cross-chapter box on building long-term resilience from tropical cyclone disasters. In: *Climate Change 2014: Impacts, Adaptation, and Vulnerability. Summaries, Frequently Asked Questions, and Cross-Chapter Boxes. Contribution of Working Group II to the Fifth Assessment Report of the Intergovernmental Panel on Climate Change* [Field, C.B., V.R. Barros, D.J. Dokken, K.J. Mach, M.D. Mastrandrea, T.E. Bilir, M. Chatterjee, K.L. Ebi, Y.O. Estrada, R.C. Genova, B. Girma, E.S. Kissel, A.N. Levy, S. MacCracken, P.R. Mastrandrea, and L.L. White (eds.)]. World Meteorological Organization, Geneva, Switzerland, pp. 149-150. (in Arabic, Chinese, English, French, Russian, and Spanish)

Uncertain Trends in Major Upwelling Ecosystems

Salvador E. Lluch-Cota (Mexico), Ove Hoegh-Guldberg (Australia), David Karl (USA), Hans O. Pörtner (Germany), Svein Sundby (Norway), Jean-Pierre Gattuso (France)

Upwelling is the vertical transport of cold, dense, nutrient-rich, relatively low-pH and often oxygen-poor waters to the euphotic zone where light is abundant. These conditions trigger high levels of primary production and a high biomass of benthic and pelagic organisms. The driving forces of upwelling include wind stress and the interaction of ocean currents with bottom topography. Upwelling intensity also depends on water column stratification. The major upwelling systems of the planet, the Equatorial Upwelling System (EUS; Section 30.5.2, Figure 30.1A) and the Eastern Boundary Upwelling Ecosystems (EBUE; Section 30.5.5, Figure 30.1A), represent only 10% of the ocean surface but contribute nearly 25% to global fish production (Figure 30.1B, Table SM30.1).

Marine ecosystems associated with upwelling systems can be influenced by a range of "bottom-up" trophic mechanisms, with upwelling, transport, and chlorophyll concentrations showing strong seasonal and interannual couplings and variability. These, in turn, influence trophic transfer up the food chain, affecting zooplankton, foraging fish, seabirds, and marine mammals.

There is considerable speculation as to how upwelling systems might change in a warming and acidifying ocean. Globally, the heat gain of the surface ocean has increased stratification by 4% (WGI Sections 3.2, 3.3, 3.8), which means that more wind energy is needed to bring deep waters to the surface. It is as yet unclear to what extent wind stress can offset the increased stratification, owing to the uncertainty in wind speed trends (WGI Section 3.4.4). In the tropics, observations of reductions in trade winds over several decades contrast more recent evidence indicating their strengthening since the late 1990s (WGI Section 3.4.4). Observations and modeling efforts in fact show diverging trends in coastal upwelling at the eastern boundaries of the Pacific and the Atlantic. Bakun (1990) proposed that the difference in rates of heat gain between land and ocean causes an increase in the pressure gradient, which results in increased alongshore winds and leads to intensified offshore transport of surface water through Ekman pumping and the upwelling of nutrient-rich, cold waters (Figure CC-UP). Some regional records support this hypothesis; others do not. There is considerable variability in warming and cooling trends over the past decades both within and among systems, making it difficult to predict changes in the intensity of all Eastern EBUEs (Section 30.5.5).

Understanding whether upwelling and climate change will impact resident biota in an additive, synergistic, or antagonistic manner is important for projections of how ecological goods and services provided for human society will change. Even though upwellings may prove more resilient to climate change than other ocean ecosystems because of their ability to function under extremely variable conditions (Capone and Hutchins, 2013), consequences of their shifts are highly relevant because these systems provide a significant portion of global primary productivity and fishery catch (Figure 30.1 A, B; Table SM30.1). Increased upwelling would enhance fisheries yields. However, the export of organic material from surface to deeper layers of the ocean may increase and stimulate its decomposition by microbial activity, thereby enhancing oxygen depletion and CO_2 enrichment in deeper water layers. Once this water returns to the surface through upwelling, benthic and pelagic coastal communities will be exposed to acidified and deoxygenated water which may combine with anthropogenic impact to negatively affect marine biota and ecosystem structure of the upper ocean (*high confidence*; Sections 6.3.2, 6.3.3, 30.3.2.2, 30.3.2.3). Extreme hypoxia may result in abnormal mortalities of fishes and invertebrates (Keller et al., 2010), reduce fisheries' catch potential, and impact aquaculture in coastal areas (Barton et al., 2012; see also Sections 5.4.3.3, 6.3.3, 6.4.1, 30.5.1.1.2, 30.5.5.1.3). Shifts in upwelling also coincide with an apparent increase in the frequency of submarine eruptions of methane and hydrogen sulfide gas, caused by enhanced formation and sinking of phytoplankton biomass to the hypoxic or anoxic sea floor. This combination of factors has been implicated in the extensive mortality of coastal fishes and invertebrates (Bakun and Weeks, 2004; Bakun et al., 2010), resulting in significant reductions in fishing productivity, such as Cape hake (*Merluccius capensis*), Namibia's most valuable fishery (Hamukuaya et al., 1998).

Reduced upwelling would also reduce the productivity of important pelagic fisheries, such as for sardines, anchovies and mackerel, with major consequences for the economies of several countries (Section 6.4.1, Chapter 7, Figure 30.1A, B, Table S30.1). However, under projected scenarios of reduced upward supply of nutrients due to stratification of the open ocean, upwelling of both nutrients and trace elements may become increasingly important to maintaining upper ocean nutrient and trace metal inventories. It has been suggested that upwelling areas may also increase nutrient content and productivity under enhanced stratification, and that upwelled and partially denitrified waters containing excess phosphate may select for N_2-fixing microorganisms (Deutsch et al., 2007; Deutsch and Weber, 2012), but field observations of N_2 fixation in these regions have not supported these predictions (Fernandez et al., 2011; Franz et al., 2012). The role of this process in global primary production thus needs to be validated (*low confidence*).

The central question therefore is whether or not upwelling will intensify, and if so, whether the effects of intensified upwelling on O_2 and CO_2 inventories will outweigh its benefits for primary production and associated fisheries and aquaculture (*low confidence*). In any case increasing atmospheric CO_2 concentrations will equilibrate with upwelling waters that may cause them to become more corrosive, depending on pCO_2 of the upwelled water, and potentially increasingly impact the biota of EBUEs.

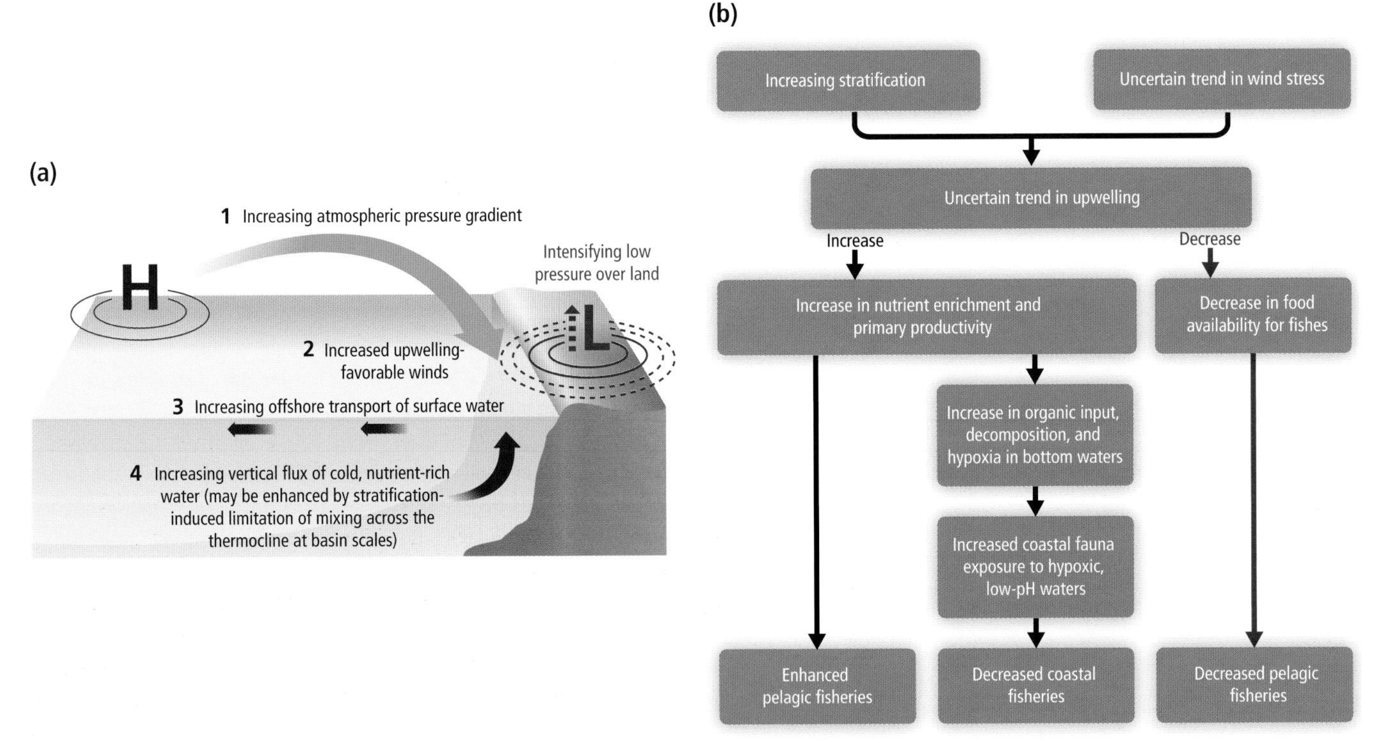

Figure UP-1 | (a) Hypothetic mechanism of increasing coastal wind–driven upwelling at Equatorial and Eastern Boundary upwelling systems (EUS, EBUE, Figure 30-1), where differential warming rates between land and ocean results in increased land–ocean (1) pressure gradients that produce (2) stronger alongshore winds and (3) offshore movement of surface water through Ekman transport, and (4) increased upwelling of deep cold nutrient rich waters to replace it. (b) Potential consequences of climate change in upwelling systems. Increasing stratification and uncertainty in wind stress trends result in uncertain trends in upwelling. Increasing upwelling may result in higher input of nutrients to the euphotic zone, and increased primary production, which in turn may enhance pelagic fisheries, but also decrease coastal fisheries due to an increased exposure of coastal fauna to hypoxic, low pH waters. Decreased upwelling may result in lower primary production in these systems with direct impacts on pelagic fisheries productivity.

References

Bakun, A., 1990: Global climate change and intensification of coastal ocean upwelling. *Science*, **247(4939)**, 198-201.

Bakun, A. and S.J. Weeks, 2004: Greenhouse gas buildup, sardines, submarine eruptions and the possibility of abrupt degradation of intense marine upwelling ecosystems. *Ecology Letters*, **7(11)**, 1015-1023.

Bakun, A., D. B. Field, A. N. A. Redondo-Rodriguez, and S. J. Weeks, 2010: Greenhouse gas, upwelling-favorable winds, and the future of coastal ocean upwelling ecosystems. *Global Change Biology* **16**:1213-1228.

Barton, A., B. Hales, G.G. Waldbusser, C. Langdon, and R.A. Feely, 2012: The Pacific oyster, *Crassostrea gigas*, shows negative correlation to naturally elevated carbon dioxide levels: implications for near-term ocean acidification effects. *Limnology and Oceanography*, **57(3)**, 698-710.

Capone, D.G. and D.A. Hutchins, 2013: Microbial biogeochemistry of coastal upwelling regimes in a changing ocean. *Nature Geoscience*, **6(9)** 711-717.

Deutsch, C. and T. Weber, 2012: Nutrient ratios as a tracer and driver of ocean biogeochemistry. *Annual Review of Marine Science*, **4**, 113-141.

Deutsch, C., J.L. Sarmiento, D.M. Sigman, N. Gruber, and J.P. Dunne, 2007: Spatial coupling of nitrogen inputs and losses in the ocean. *Nature*, **445(7124)**, 163-167.

Fernandez, C., L. Farías, and O. Ulloa, 2011: Nitrogen fixation in denitrified marine waters. *PLoS ONE*, **6(6)**, e20539, doi:10.1371/journal.pone.0020539.

Franz, J., G. Krahmann, G. Lavik, P. Grasse, T. Dittmar, and U. Riebesell, 2012: Dynamics and stoichiometry of nutrients and phytoplankton in waters influenced by the oxygen minimum zone in the eastern tropical Pacific. *Deep-Sea Research Part I: Oceanographic Research Papers*, **62**, 20-31.

Hamukuaya, H., M.J. O'Toole, and P.M.J. Woodhead, 1998: Observations of severe hypoxia and offshore displacement of Cape hake over the Namibian shelf in 1994. *South African Journal of Marine Science*, **19(1)**, 57-59.

Keller, A.A., V. Simon, F. Chan, W.W. Wakefield, M.E. Clarke, J.A. Barth, D. Kamikawa and E.L. Fruh, 2010: Demersal fish and invertebrate biomass in relation to an offshore hypoxic zone along the US West Coast. *Fisheries Oceanography*, **19**, 76-87.

This cross-chapter box should be cited as:

Lluch-Cota, S.E., O. Hoegh-Guldberg, D. Karl, H.-O. Pörtner, S. Sundby, and J.-P. Gattuso, 2014: *Cross-chapter box on uncertain trends in major upwelling ecosystems. In: Climate Change 2014: Impacts, Adaptation, and Vulnerability. Summaries, Frequently Asked Questions, and Cross-Chapter Boxes. Contribution of Working Group II to the Fifth Assessment Report of the Intergovernmental Panel on Climate Change* [Field, C.B., V.R. Barros, D.J. Dokken, K.J. Mach, M.D. Mastrandrea, T.E. Bilir, M. Chatterjee, K.L. Ebi, Y.O. Estrada, R.C. Genova, B. Girma, E.S. Kissel, A.N. Levy, S. MacCracken, P.R. Mastrandrea, and L.L. White (eds.)]. World Meteorological Organization, Geneva, Switzerland, pp. 151-153. (in Arabic, Chinese, English, French, Russian, and Spanish)

Urban–Rural Interactions – Context for Climate Change Vulnerability, Impacts, and Adaptation

John Morton (UK), William Solecki (USA), Purnamita Dasgupta (India), David Dodman (Jamaica), Marta G. Rivera-Ferre (Spain)

Rural areas and urban areas have always been interconnected and interdependent, but recent decades have seen new forms of these interconnections: a tendency for rural–urban boundaries to become less well defined, and new types of land use and economic activity on those boundaries. These conditions have important implications for understanding climate change impacts, vulnerabilities, and opportunities for adaptation. This box examines three critical implications of these interactions:

1) Climate extremes in rural areas resulting in urban impacts— teleconnections of resources and migration streams mean that climate extremes in non-urban locations with associated shifts in water supply, rural agricultural potential, and the habitability of rural areas will have downstream impacts in cities.
2) Events specific to the rural–urban interface— given the highly integrated nature of rural–urban interface areas and overarching demand to accommodate both rural and urban demands in these settings, there is a set of impacts, vulnerabilities, and opportunities for adaptation specific to these locations. These impacts include loss of local agricultural production, economic marginalization resulting from being neither rural or urban, and stress on human health.
3) Integrated infrastructure and service disruption—as urban demands often take preference, interdependent rural and urban resource systems place nearby rural areas at risk, because during conditions of climate stress, rural areas more often suffer resource shortages or other disruptions to sustain resources to cities. For example, under conditions of resource stress associated with climate risk (e.g., droughts) urban areas are at an advantage because of political, social, and economic requirements to maintain service supply to cities to the detriment of relatively marginal rural sites and settlements.

Urban areas historically have been dependent on the lands just beyond their boundaries for most of their critical resources including water, food, and energy. Although in many contexts, the connections between urban settlements and surrounding rural areas are still present, long distance, teleconnected, large-scale supply chains have been developed particularly with respect to energy resources and food supply (Güneralp et al., 2013). Extreme event disruptions in distant resource areas or to the supply chain and relevant infrastructure can negatively impact the urban areas dependent on these materials (Wilbanks et al., 2012). During the summer of 2012, for instance, an extended drought period in the central United States led to significantly reduced river levels on the Mississippi River that led to interruptions of barge traffic and delay of commodity flows to cities throughout the country. Urban water supply is also vulnerable to droughts in predominantly rural areas. In the case of Bulawayo, Zimbabwe, periodic urban water shortages over the last few decades have been triggered by rural droughts (Mkandla et al., 2005).

A further teleconnection between rural and urban areas is rural–urban migration. There have been cases where migration and urbanization patterns have been to attributed to climate change or its proxies such as in parts of Africa (Morton, 1989; Barrios et al., 2006). However, as recognized by Black et al. (2011), life in rural areas across the world typically involves complex patterns of rural–urban and rural–rural migration, subject to economic, political, social, and demographic drivers, patterns that are modified or exacerbated by climate events and trends rather than solely caused by them.

Globally, an increased blending of urban and rural qualities has occurred. Simon et al. (2006, p. 4) assert that the simple dichotomy between "rural" and "urban" has "long ceased to have much meaning in practice or for policy-making purposes in many parts of the global South." One approach to reconciling this is through the increasing application of the concept of "peri-urban areas" (Simon et al., 2006; Simon, 2008). These areas can be seen as rural locations that have "become more urban in character" (Webster, 2002, p. 5); as sites where households pursue a wider range of income-generating activities while still residing in what appear to be "largely rural landscapes" (Learner and Eakin, 2010, p. 1); or as locations in which rural and urban land uses coexist, whether in contiguous or fragmented units (Bowyer-Bower, 2006). The inhabitants of "core" urban areas within cities have also increasingly turned to agriculture, with production of staple foods, higher value crops and livestock (Bryld, 2003; Devendra et al., 2005; Lerner and Eakin, 2010; Lerner et al., 2013). Bryld (2003) sees this as driven by rural–urban migration and by structural adjustment (e.g., withdrawal of food price controls and food subsidies). Lerner and Eakin (2011; also Lerner et al., 2013) explored reasons why people produce food in urban environments, despite high opportunity costs of land and labor: buffering of risk from insecure urban labor markets; response to consumer demand; and the meeting of cultural needs.

Livelihoods and areas on the rural–urban interface suffer highly specific forms of vulnerability to disasters, including climate-related disasters. These may be summarized as specifically combining urban vulnerabilities of population concentration, dependence on infrastructure, and social diversity limiting social support with rural traits of distance, isolation, and invisibility to policymakers (Pelling and Mustafa, 2010). Increased connectivity can also encourage land expropriation to enable commercial land development (Pelling and Mustafa, 2010). Vulnerability may arise from the coexistence of rural and urban perspectives, which may give rise to conflicts between different social/interest groups and economic activities (Masuda and Garvin, 2008; Solona-Solona 2010; Darly and Torre, 2013).

Additional vulnerability of peri-urban areas is on account of the re-constituted institutional arrangements and their structural constraints (Iaquinta and Drescher, 2000). Rapid declines in traditional informal institutions and forms of collective action, and their imperfect replacement with formal state and market institutions, may also increase vulnerability (Pelling and Mustafa, 2010).

Peri-urban areas and livelihoods have low visibility to policymakers at both local and national levels, and may suffer from a lack of necessary services and inappropriate and uncoordinated policies. In Tanzania and Malawi, national policies of agricultural extension to farmer groups, for example, do not reach peri-urban farmers (Liwenga et al., 2012). In peri-urban areas around Mexico City (Eakin et al., 2013), management of the substantial risk of flooding is led *de facto* by agricultural and water agencies, in the absence of capacity within peri-urban municipalities and despite clear evidence that urban encroachment is a key driver of flood risk. In developed country contexts, suburban–exurban fringe areas often are overlooked in the policy arena that traditionally focuses on rural development and agricultural production, or urban growth and services (Hanlon et al., 2010). The environmental function of urban agriculture, in particular, in protection against flooding, will increase in the context of climate change (Aubry et al., 2012).

However, peri-urban areas and mixed livelihoods more generally on rural–urban interfaces, also exhibit specific factors that increase their resilience to climate shocks (Pelling and Mustafa, 2010). Increased transport connectivity in peri-urban areas can reduce disaster risk by providing a greater diversity of livelihood options and improving access to education. The expansion of local labor markets and wage labor in these areas can strengthen adaptive capacity through providing new livelihood opportunities (Pelling and Mustafa, 2010). Maintaining mixed portfolios of agricultural and non-agricultural livelihoods also spreads risk (Lerner et al., 2013).

In high-income countries, practices attempting to enhance the ecosystem services and localized agriculture more typically associated with lower density areas have been encouraged. In many situations these practices are focused increasingly on climate adaptation and mitigating the impacts of climate extremes such as those associated with heating and the urban heat island effect, or wetland restoration efforts to limit the impact of storm surge wave action (Verburg et al., 2012).

The dramatic growth of urban areas also implies that rural areas and communities are increasingly politically and economically marginalized within national contexts, resulting in potential infrastructure and service disruptions for such sites. Existing rural–urban conflicts for the management of natural resources (Castro and Nielsen, 2003) such as water (Celio et al., 2011) or land use conversion in rural areas, for example, wind farms in rural Catalonia (Zografos and Martínez-Alier, 2009); industrial coastal areas in Sweden (Stepanova and Bruckmeier, 2013); or conversion of rice land into industrial, residential, and recreational uses in the Philippines (Kelly, 1998) have been documented, and it is expected that stress from climate change impacts on land and natural resources will exacerbate these tensions. For instance, climate-induced reductions in water availability may be more of a concern than population growth or increased per capita use for securing continued supplies of water to large cities (Jenerette and Larsen, 2006), which requires an innovative approach to address such conflicts (Pearson et al., 2010).

References

Aubry, C., J. Ramamonjisoa, M.-H. Dabat, J. Rakotoarisoa, J. Rakotondraibe, and L. Rabeharisoa, 2012: Urban agriculture and land use in cities: an approach with the multi-functionality and sustainability concepts in the case of Antananarivo (Madagascar). *Land Use Policy*, **29**, 429-439.

Barrios, S., L. Bertinelli, and E. Strobl, 2006: Climatic change and rural-urban migration: the case of sub-Saharan Africa. *Journal of Urban Economics*, **60**, 357-371.

Black, R., W.N. Adger, N.W. Arnell, S. Dercon, A. Geddes, and D. Thomas, 2011: The effect of environmental change on human migration. *Global Environmental Change*, **21(Suppl. 1)**, S3-S11.

Bowyer-Bower, T., 2006: The inevitable illusiveness of 'sustainability' in the peri-urban interface: the case of Harare. In: *The Peri-Urban Interface: Approaches to Sustainable Natural and Human Resource Use* [McGregor, D., D. Simon, and D. Thompson (eds.)]. Earthscan, London, UK and Sterling, VA, USA, pp. 151-164.

Bryld, E., 2003: Potentials, problems, and policy implications for urban agriculture in developing countries. *Agriculture and Human Values*, **20**, 79-86.

Castro, A.P. and E. Nielsen, 2003: *Natural Resource Conflict Management Case Studies: An Analysis of Power, Participation and Protected Areas*. Food and Agriculture Organization of the United Nations (FAO), Rome, Italy, 268 pp.

Darly, S. and A. Torre, 2013: Conflicts over farmland uses and the dynamics of "agri-urban" localities in the Greater Paris Region: an empirical analysis based on daily regional press and field interviews. *Land Use Policy*, **30**, 90-99.

Devendra, C., J. Morton, B. Rischowsky, and D. Thomas, 2005: Livestock systems. In: *Livestock and Wealth Creation: Improving the Husbandry of Livestock Kept by the Poor in Developing Countries* [Owen, E., A. Kitalyi, N. Jayasuriya, and T. Smith (eds.)]. Nottingham University Press, Nottingham, UK, pp. 29-52.

Dixon, J.M., K.J. Donati, L.L. Pike, and L. Hattersley, 2009: Functional foods and urban agriculture: two responses to climate change-related food insecurity. *New South Wales Public Health Bulletin*, **20(2)**, 14-18.

Eakin, H., A. Lerner, and F. Murtinho, 2013: Adaptive capacity in evolving peri-urban spaces; responses to flood risk in the Upper Lerma River Valley, Mexico. *Global Environmental Change*, **20(1)**, 14-22.

Güneralp, B., K.C. Seto, and M. Ramachandran, 2013: Evidence of urban land teleconnections and impacts on hinterlands. *Current Opinion in Environmental Sustainability*, **5(5)**, 445-451.

Hanlon, B., J.R. Short, and T.J. Vicino, 2010: *Cities and Suburbs: New Metropolitan Realities in the US*. Routledge, Oxford, UK and New York, NY, USA, 304 pp.

Hoggart, K., 2005: *The City's Hinterland: Dynamism and Divergence in Europe's Peri-Urban Territories*. Ashgate Publishing, Ltd., Aldershot, UK and Ashgate Publishing Co., Burlington, VT, USA, 186 pp.

Iaquinta, D.L. and A.W. Drescher, 2000: Defining the peri-urban: rural-urban linkages and institutional connections. *Land Reform: Land Settlement and Cooperatives*, **2000(2)**, 8-26, www.fao.org/docrep/003/X8050T/X8050T00.HTM.

Jenerette, GD and L Larsen, 2006: A global perspective on changing sustainable urban water supplies. *Global and Planetary Change*, **50(3-4)**, 202-211.

Kelly, P.F., 1998: The politics of urban-rural relations: land use conversion in the Philippines. *Environment and Urbanization*, **10(1)**, 35-54, doi:10.1177/095624789801000116.

Lerner, A.M. and H. Eakin, 2010: An obsolete dichotomy? Rethinking the rural-urban interface in terms of food security and production in the global south. *Geographical Journal*, **177(4)**, 311-320.

Lerner, A.M., H. Eakin, and S. Sweeney, 2013: Understanding peri-urban maize production through an examination of household livelihoods in the Toluca Metropolitan Area, Mexico. *Journal of Rural Studies*, **30**, 52-63.

Liwenga, E., E. Swai, L. Nsemwa, A. Katunzi, B. Gwambene, M. Joshua, F. Chipungu, T. Stathers, and R. Lamboll, 2012: *Exploring Urban Rural Interdependence and the Impact of Climate Change in Tanzania and Malawi: Final Narrative Report*. Project Report, International Development Research Centre (IDRC), Ottawa, ON, Canada.

Masuda, J. and T. Garvin, 2008: Whose heartland? The politics of place at the rural-urban interface. *Journal of Rural Studies*, **24**, 118-123.

Mattia, C., C.A. Scott, and M. Giordano, 2010: Urban-agricultural water appropriation: the Hyderabad, India case. *Geographical Journal*, **176(1)**, 39-57.

Mkandla, N., P. Van der Zaag, and P. Sibanda, 2005: Bulawayo water supplies: sustainable alternatives for the next decade. *Physics and Chemistry of the Earth, Parts A/B/C*, **30(11-16)**, 935-942.

Morton, J., 1989: Ethnicity and politics in Red Sea Province, Sudan. *African Affairs*, **88(350)**, 63-76.

Pearson, L.J., A. Coggan, W. Proctor, and T.F. Smith, 2010: A sustainable decision support framework for urban water management. *Water Resources Management*, **24(2)**, 363-376.

Pelling, M. and D. Mustafa, 2010: *Vulnerability, Disasters and Poverty in Desakota Systems*. Political and Development Working Paper Series, No. 31, King's College London, London, UK, 26 pp.

Simon, D., 2008: Urban environments: issues on the peri-urban fringe. *Annual Review of Environmental Resources*, **33**, 167-185.

Simon, D., D. McGregor, and D. Thompson, 2006: Contemporary perspectives on the peri-urban zones of cities in developing countries. In: *The Peri-Urban Interface: Approaches to Sustainable Natural and Human Resource Use* [McGregor, D., D. Simon, and D. Thompson (eds.)]. Earthscan, London, UK and Sterling, VA, USA, pp. 3-17.

Solana-Solana, M., 2010: Rural gentrification in Catalonia, Spain: a case study of migration, social change and conflicts in the Empordanet area. *Geoforum*, **41(3)**, 508-517.

Stepanova, O. and K. Bruckmeier, 2013: Resource use conflicts and urban-rural resource use dynamics in Swedish coastal landscapes: comparison and synthesis. *Journal of Environmental Policy & Planning*, **15(4)**, 467-492, doi:10.1080/1523908X.2013.778173.

Verburg, P.H., E. Koomen, M. Hilferink, M. Perez-Soba, and J.P. Lesschen, 2012: An assessment of the impact of climate adaptation measures to reduce flood risk on ecosystem services. *Landscape Ecology*, **27**, 473-486.

Webster, D., 2002: *On the Edge: Shaping the Future of Peri-Urban East Asia*. Asia/Pacific Research Center (A/PARC), Stanford, CA, USA, 49 pp.

Wilbanks, T., S. Fernandez, G. Backus, P. Garcia, K. Jonietz, P. Kirshen, M. Savonis, W. Solecki, and T. Toole, 2012: *Climate Change and Infrastructure, Urban systems and Vulnerabilities*. Technical Report prepared by the Oak Ridge National Laboratory (ORNL) for the US Department of Energy in support of the National Climate Assessment, ORNL, Oak Ridge, TN, 19 pp., www.esd.ornl.gov/eess/Infrastructure.pdf.

Zasada, I., 2011: Multifunctional peri-urban agriculture – a review of societal demands and the provision of goods and services by farming. *Land Use Policy*, **28(4)**, 639-648.

Zografos, C. and J. Martínez-Alier, 2009: The politics of landscape value: a case study of wind farm conflict in rural Catalonia. *Environment and Planning A*, **41(7)**, 1726-1744.

This cross-chapter box should be cited as:

Morton, J.F., W. Solecki, P. Dasgupta, D. Dodman, and M.G. Rivera-Ferre, 2014: Cross-chapter box on urban–rural interactions—context for climate change vulnerability, impacts, and adaptation. In: *Climate Change 2014: Impacts, Adaptation, and Vulnerability. Summaries, Frequently Asked Questions, and Cross-Chapter Boxes. Contribution of Working Group II to the Fifth Assessment Report of the Intergovernmental Panel on Climate Change* [Field, C.B., V.R. Barros, D.J. Dokken, K.J. Mach, M.D. Mastrandrea, T.E. Bilir, M. Chatterjee, K.L. Ebi, Y.O. Estrada, R.C. Genova, B. Girma, E.S. Kissel, A.N. Levy, S. MacCracken, P.R. Mastrandrea, and L.L. White (eds.)]. World Meteorological Organization, Geneva, Switzerland, pp. 155-157. (in Arabic, Chinese, English, French, Russian, and Spanish)

Active Role of Vegetation in Altering Water Flows under Climate Change

Dieter Gerten (Germany), Richard Betts (UK), Petra Döll (Germany)

Climate, vegetation, and carbon and water cycles are intimately coupled, in particular via the simultaneous transpiration and CO_2 uptake through plant stomata in the process of photosynthesis. Hence, water flows such as runoff and evapotranspiration are affected not only directly by anthropogenic climate change as such (i.e., by changes in climate variables such as temperature and precipitation), but also indirectly by plant responses to increased atmospheric CO_2 concentrations. In addition, effects of climate change (e.g., higher temperature or altered precipitation) on vegetation structure, biomass production, and plant distribution have an indirect influence on water flows. Rising CO_2 concentration affects vegetation and associated water flows in two contrasting ways, as suggested by ample evidence from Free Air CO_2 Enrichment (FACE), laboratory and modeling experiments (e.g., Leakey et al., 2009; Reddy et al., 2010; de Boer et al., 2011). On the one hand, a *physiological* effect leads to reduced opening of stomatal apertures, which is associated with lower water flow through the stomata, that is, lower leaf-level transpiration. On the other hand, a *structural* effect ("fertilization effect") stimulates photosynthesis and biomass production of C_3 plants including all tree species, which eventually leads to higher transpiration at regional scales. A key question is to what extent the climate- and CO_2-induced changes in vegetation and transpiration translate into changes in regional and global runoff.

The physiological effect of CO_2 is associated with an increased intrinsic water use efficiency (WUE) of plants, which means that less water is transpired per unit of carbon assimilated. Records of stable carbon isotopes in woody plants (Peñuelas et al., 2011) verify this finding, suggesting an increase in WUE of mature trees by 20.5% between the early 1960s and the early 2000s. Increases since pre-industrial times have also been found for several forest sites (Andreu-Hayles et al., 2011; Gagen et al., 2011; Loader et al., 2011; Nock et al., 2011) and in a temperate semi-natural grassland (Koehler et al., 2010), although in one boreal tree species WUE ceased to increase after 1970 (Gagen et al., 2011). Analysis of long-term whole-ecosystem carbon and water flux measurements from 21 sites in North American temperate and boreal forests corroborates a notable increase in WUE over the two past decades (Keenan et al., 2013). An increase in global WUE over the past century is supported by ecosystem model results (Ito and Inatomi, 2012).

A key influence on the significance of increased WUE for large-scale transpiration is whether vegetation structure and production has remained approximately constant (as assumed in the global modeling study by Gedney et al., 2006) or has increased in some regions due to the structural CO_2 effect (as assumed in models by Piao et al., 2007; Gerten et al., 2008). While field-based results vary considerably among sites, tree ring studies suggest that tree growth did not increase globally since the 1970s in response to climate and CO_2 change (Andreu-Hayles et al.,

2011; Peñuelas et al., 2011). However, basal area measurements at more than 150 plots across the tropics suggest that biomass and growth rates in intact tropical forests have increased in recent decades (Lewis et al., 2009). This is also confirmed for 55 temperate forest plots, with a suspected contribution of CO_2 effects (McMahon et al., 2010). Satellite observations analyzed in Donohue et al. (2013) suggest that an increase in vegetation cover by 11% in warm drylands (1982–2010 period) is attributable to CO_2 fertilization. Owing to the interplay of physiological and structural effects, the net impact of CO_2 increase on global-scale transpiration and runoff remains rather poorly constrained. This is also true because nutrient limitation, often omitted in modeling studies, can suppress the CO_2 fertilization effect (see Rosenthal and Tomeo, 2013).

Therefore, there are conflicting views on whether the direct CO_2 effects on plants already have a significant influence on evapotranspiration and runoff at global scale. AR4 reported work by Gedney et al. (2006) that suggested that the physiological CO_2 effect (lower transpiration) contributed to a supposed increase in global runoff seen in reconstructions by Labat et al. (2004). However, a more recent analysis based on a more complete data set (Dai et al., 2009) suggested that river basins with decreasing runoff outnumber basins with increasing runoff, such that a small decline in global runoff is *likely* for the period 1948–2004. Hence, detection of vegetation contributions to changes in water flows critically depends on the availability and quality of hydrometeorological observations (Haddeland et al., 2011; Lorenz and Kunstmann, 2012). Overall, the evidence since AR4 suggests that climatic variations and trends have been the main driver of global runoff change in the past decades; both CO_2 increase and land use change have contributed less (Piao et al., 2007; Gerten et al., 2008; Alkama et al., 2011; Sterling et al., 2013). Oliveira et al. (2011) furthermore pointed to the importance of changes in incident solar radiation and the mediating role of vegetation; according to their global simulations, a higher diffuse radiation fraction during 1960–1990 may have increased evapotranspiration in the tropics by 3% due to higher photosynthesis from shaded leaves.

It is uncertain how vegetation responses to future increases in CO_2 and to climate change will modulate the impacts of climate change on freshwater flows. Twenty-first century continental- and basin-scale runoff is projected by some models to either increase more or decrease less when the physiological CO_2 effect is included in addition to climate change effects (Betts et al., 2007; Murray et al., 2012). This could somewhat ease the increase in water scarcity anticipated in response to future climate change and population growth (Gerten et al., 2011; Wiltshire et al., 2013). In absolute terms, the isolated effect of CO_2 has been modeled to increase future global runoff by 4 to 5% (Gerten et al., 2008) up to 13% (Nugent and Matthews, 2012) compared to the present, depending on the assumed CO_2 trajectory and whether feedbacks of changes in vegetation structure and distribution to the atmosphere are accounted for (they were in Nugent and Matthews, 2012). In a global model intercomparison study (Davie et al., 2013), two out of four models projected stronger increases and, respectively, weaker decreases in runoff when considering CO_2 effects compared to simulations with constant CO_2 concentration (consistent with the above findings, though magnitudes differed between the models), but two other models showed the reverse. Thus, the choice of models and the way they represent the coupling between CO_2, stomatal closure, and plant growth is a source of uncertainty, as also suggested by Cao et al. (2009). Lower transpiration due to rising CO_2 concentration may also affect future regional climate change itself (Boucher et al., 2009) and enhance the contrast between land and ocean surface warming (Joshi et al., 2008). Overall, although physiological and structural effects will influence water flows in many regions, precipitation and temperature effects are *likely* to remain the prime influence on global runoff (Alkama et al., 2010).

An application of a soil–vegetation–atmosphere–transfer model indicates complex responses of groundwater recharge to vegetation-mediated changes in climate, with computed groundwater recharge being always larger than would be expected from just accounting for changes in rainfall (McCallum et al., 2010). Another study found that even if precipitation slightly decreased, groundwater recharge might increase as a net effect of vegetation responses to climate change and CO_2 rise, that is, increasing WUE and either increasing or decreasing leaf area (Crosbie et al., 2010). Depending on the type of grass in Australia, the same change in climate is suggested to lead to either increasing or decreasing groundwater recharge in this location (Green et al., 2007). For a site in the Netherlands, a biomass decrease was computed for each of eight climate scenarios indicating drier summers and wetter winters (A2 emissions scenario), using a fully coupled vegetation and variably saturated hydrological model. The resulting increase in groundwater recharge up-slope was simulated to lead to higher water tables and an extended habitat for down-slope moisture-adapted vegetation (Brolsma et al., 2010).

Using a large ensemble of climate change projections, Konzmann et al. (2013) put hydrological changes into an agricultural perspective and suggested that the net result of physiological and structural CO_2 effects on crop irrigation requirements would be a global reduction (Figure VW-1). Thus, adverse climate change impacts on irrigation requirements and crop yields might be partly buffered as WUE and crop production improve (Fader et al., 2010). However, substantial CO_2-driven improvements will be realized only if proper management abates limitation of plant growth by nutrient availability or other factors.

Changes in vegetation coverage and structure due to long-term climate change or shorter-term extreme events such as droughts (Anderegg et al., 2013) also affect the partitioning of precipitation into evapotranspiration and runoff, sometimes involving complex feedbacks with the atmosphere such as in the Amazon region (Port et al., 2012; Saatchi et al., 2013). One model in the study by Davie et al. (2013) showed regionally diverse climate change effects on vegetation distribution and structure, which had a much weaker effect on global runoff than the structural and physiological CO_2 effects. As water, carbon, and vegetation dynamics evolve synchronously and interactively under climate change (Heyder et al., 2011; Gerten et al., 2013), it remains a challenge to disentangle the individual effects of climate, CO_2, and land cover change on the water cycle.

(a) Impact of climate change including physiological and structural crop responses to increased atmospheric CO_2

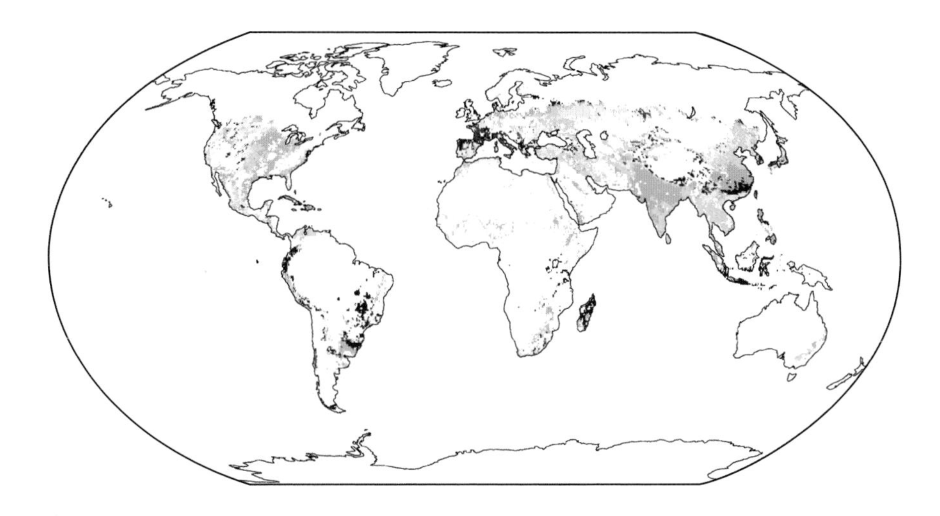

(b) Impact of climate change only

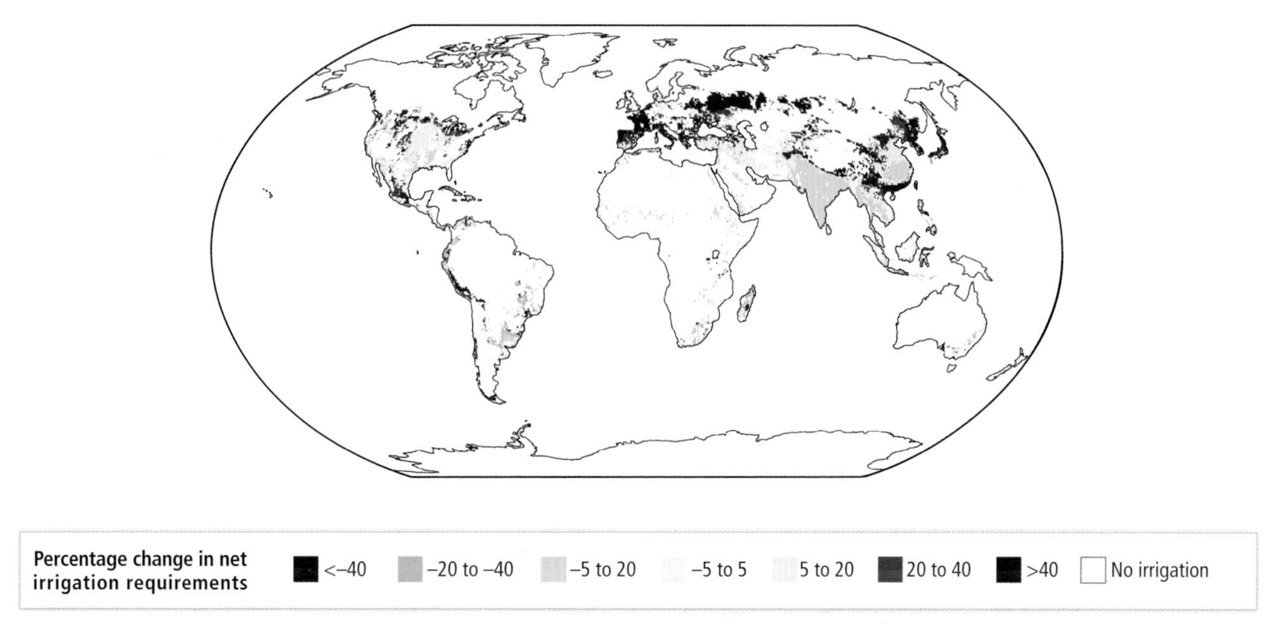

Figure VW-1 | Percentage change in net irrigation requirements of 11 major crops from 1971–2000 to 2070–2099 on areas currently equipped for irrigation, assuming current management practices. (a) Impact of climate change including physiological and structural crop responses to increased atmospheric CO_2 concentration (co-limitation by nutrients not considered). (b) Impact of climate change only. Shown is the median change derived from climate change projections by 19 General Circulation Models (GCMs; based on the Special Report on Emission Scenarios (SRES) A2 emissions scenario) used to force a vegetation and hydrology model. (Modified after Konzmann et al., 2013.)

References

Alkama, R., M. Kageyama, and G. Ramstein, 2010: Relative contributions of climate change, stomatal closure, and leaf area index changes to 20th and 21st century runoff change: a modelling approach using the Organizing Carbon and Hydrology in Dynamic Ecosystems (ORCHIDEE) land surface model. *Journal of Geophysical Research: Atmospheres*, **115(D17)**, D17112, doi:10.1029/2009JD013408.

Alkama, R., B. Decharme, H. Douville, and A. Ribes, 2011: Trends in global and basin-scale runoff over the late twentieth century: methodological issues and sources of uncertainty. *Journal of Climate*, **24(12)**, 3000-3014.

Anderegg, W.R.L., J.M. Kane, and L.D.L. Anderegg, 2013: Consequences of widespread tree mortality triggered by drought and temperature stress. *Nature Climate Change*, **3**, 30-36.

Andreu-Hayles, L., O. Planells, E. Gutierrez, E. Muntan, G. Helle, K.J. Anchukaitis, and G.H. Schleser, 2011: Long tree-ring chronologies reveal 20th century increases in water-use efficiency but no enhancement of tree growth at five Iberian pine forests. *Global Change Biology*, **17(6)**, 2095-2112.

Betts, R.A., O. Boucher, M. Collins, P.M. Cox, P.D. Falloon, N. Gedney, D.L. Hemming, C. Huntingford, C.D. Jones, D.M.H. Sexton, and M.J. Webb, 2007: Projected increase in continental runoff due to plant responses to increasing carbon dioxide. *Nature*, **448(7157)**, 1037-1041.

Boucher, O., A. Jones, and R.A. Betts, 2009: Climate response to the physiological impact of carbon dioxide on plants in the Met Office Unified Model HadCM3. *Climate Dynamics*, **32(2-3)**, 237-249.

Brolsma, R.J., M.T.H. van Vliet, and M.F.P. Bierkens, 2010: Climate change impact on a groundwater-influenced hillslope ecosystem. *Water Resources Research*, **46(11)**, W11503, doi:10.1029/2009WR008782.

Cao, L., G. Bala, K. Caldeira, R. Nemani, and G. Ban-Weiss, 2009: Climate response to physiological forcing of carbon dioxide simulated by the coupled Community Atmosphere Model (CAM3.1) and Community Land Model (CLM3.0). *Geophysical Research Letters*, **36(10)**, L10402, doi:10.1029/2009GL037724.

Crosbie, R.S., J.L. McCallum, G.R. Walker, and F.H.S. Chiew, 2010: Modelling climate-change impacts on groundwater recharge in the Murray-Darling Basin, Australia. *Hydrogeology Journal*, **18(7)**, 1639-1656.

Dai, A., T. Qian, K.E. Trenberth, and J.D. Milliman, 2009: Changes in continental freshwater discharge from 1948 to 2004. *Journal of Climate*, **22(10)**, 2773-2792.

Davie, J.C.S., P.D. Falloon, R. Kahana, R. Dankers, R. Betts, F.T. Portmann, D.B. Clark, A. Itoh, Y. Masaki, K. Nishina, B. Fekete, Z. Tessler, X. Liu, Q. Tang, S. Hagemann, T. Stacke, R. Pavlick, S. Schaphoff, S.N. Gosling, W. Franssen, and N. Arnell, 2013: Comparing projections of future changes in runoff and water resources from hydrological and ecosystem models in ISI-MIP. *Earth System Dynamics*, **4**, 359-374.

de Boer, H.J., E.I. Lammertsma, F. Wagner-Cremer, D.L. Dilcher, M.J. Wassen, and S.C. Dekker, 2011: Climate forcing due to optimization of maximal leaf conductance in subtropical vegetation under rising CO_2. *Proceedings of the National Academy of Sciences of the United States of America*, **108(10)**, 4041-4046.

Donohue, R.J., M.L. Roderick, T.R. McVicar, and G.D. Farquhar, 2013: Impact of CO_2 fertilization on maximum foliage cover across the globe's warm, arid environments. *Geophysical Research Letters*, **40(12)**, 3031-3035.

Fader, M., S. Rost, C. Müller, A. Bondeau, and D. Gerten, 2010: Virtual water content of temperate cereals and maize: present and potential future patterns. *Journal of Hydrology*, **384(3-4)**, 218-231.

Gagen, M., W. Finsinger, F. Wagner-Cremer, D. McCarroll, N.J. Loader, I. Robertson, R. Jalkanen, G. Young, and A. Kirchhefer, 2011: Evidence of changing intrinsic water-use efficiency under rising atmospheric CO_2 concentrations in Boreal Fennoscandia from subfossil leaves and tree ring $\partial^{13}C$ ratios. *Global Change Biology*, **17(2)**, 1064-1072.

Gedney, N., P.M. Cox, R.A. Betts, O. Boucher, C. Huntingford, and P.A. Stott, 2006: Detection of a direct carbon dioxide effect in continental river runoff records. *Nature*, **439(7078)**, 835-838.

Gerten, D., S. Rost, W. von Bloh, and W. Lucht, 2008: Causes of change in 20th century global river discharge. *Geophysical Research Letters*, **35(20)**, L20405, doi:10.1029/2008GL035258.

Gerten, D., J. Heinke, H. Hoff, H. Biemans, M. Fader, and K. Waha, 2011: Global water availability and requirements for future food production. *Journal of Hydrometeorology*, **12(5)**, 885-899.

Gerten, D., W. Lucht, S. Ostberg, J. Heinke, M. Kowarsch, H. Kreft, Z.W. Kundzewicz, J. Rastgooy, R. Warren, and H.J. Schellnhuber, 2013: Asynchronous exposure to global warming: freshwater resources and terrestrial ecosystems. *Environmental Research Letters*, **8**, 034032, doi:10.1088/1748-9326/8/3/034032.

Green, T.R., B.C. Bates, S.P. Charles, and P.M. Fleming, 2007: Physically based simulation of potential effects of carbon dioxide-altered climates on groundwater recharge. *Vadose Zone Journal*, **6(3)**, 597-609.

Haddeland, I., D.B. Clark, W. Franssen, F. Ludwig, F. Voss, N.W. Arnell, N. Bertrand, M. Best, S. Folwell, D. Gerten, S. Gomes, S.N. Gosling, S. Hagemann, N. Hanasaki, R. Harding, J. Heinke, P. Kabat, S. Koirala, T. Oki, J. Polcher, T. Stacke, P. Viterbo, G.P. Weedon, and P. Yeh, 2011: Multimodel estimate of the global terrestrial water balance: setup and first results. *Journal of Hydrometeorology*, **12(5)**, 869-884.

Heyder, U., S. Schaphoff, D. Gerten, and W. Lucht, 2011: Risk of severe climate change impact on the terrestrial biosphere. *Environmental Research Letters*, **6(3)**, 034036, doi:10.1088/1748-9326/6/3/034036.

Ito, A. and M. Inatomi, 2012: Water-use efficiency of the terrestrial biosphere: a model analysis focusing on interactions between the global carbon and water cycles. *Journal of Hydrometeorology*, **13(2)**, 681-694.

Joshi, M.M., J.M. Gregory, M.J. Webb, D.M.H. Sexton, and T.C. Johns, 2008: Mechanisms for the land/sea warming contrast exhibited by simulations of climate change. *Climate Dynamics*, **30(5)**, 455-465.

Keenan, T.F., D.Y. Hollinger, G. Bohrer, D. Dragoni, J.W. Munger, H.P. Schmid, and A.D. Richardson, 2013: Increase in forest water-use efficiency as atmospheric carbon dioxide concentrations rise. *Nature*, **499(7458)**, 324-327.

Koehler, I.H., P.R. Poulton, K. Auerswald, and H. Schnyder, 2010: Intrinsic water-use efficiency of temperate seminatural grassland has increased since 1857: an analysis of carbon isotope discrimination of herbage from the Park Grass Experiment. *Global Change Biology*, **16(5)**, 1531-1541.

Konzmann, M., D. Gerten, and J. Heinke, 2013: Climate impacts on global irrigation requirements under 19 GCMs, simulated with a vegetation and hydrology model. *Hydrological Sciences Journal*, **58(1)**, 88-105.

Labat, D., Y. Godderis, J. Probst, and J. Guyot, 2004: Evidence for global runoff increase related to climate warming. *Advances in Water Resources*, **27(6)**, 631-642.

Leakey, A.D.B., E.A. Ainsworth, C.J. Bernacchi, A. Rogers, S.P. Long, and D.R. Ort, 2009: Elevated CO_2 effects on plant carbon, nitrogen, and water relations: six important lessons from FACE. *Journal of Experimental Botany*, **60(10)**, 2859-2876.

Lewis, S.L., J. Lloyd, S. Sitch, E.T.A. Mitchard, and W.F. Laurance, 2009: Changing ecology of tropical forests: evidence and drivers. *Annual Review of Ecology Evolution and Systematics*, **40**, 529-549.

Loader, N.J., R.P.D. Walsh, I. Robertson, K. Bidin, R.C. Ong, G. Reynolds, D. McCarroll, M. Gagen, and G.H.F. Young, 2011: Recent trends in the intrinsic water-use efficiency of ringless rainforest trees in Borneo. *Philosophical Transactions of the Royal Society B*, **366(1582)**, 3330-3339.

Lorenz, C. and H. Kunstmann, 2012: The hydrological cycle in three state-of-the-art reanalyses: intercomparison and performance analysis. *Journal of Hydrometeorology*, **13(5)**, 1397-1420.

McCallum, J.L., R.S. Crosbie, G.R. Walker, and W.R. Dawes, 2010: Impacts of climate change on groundwater in Australia: a sensitivity analysis of recharge. *Hydrogeology Journal*, **18(7)**, 1625-1638.

McMahon, S.M., G.G. Parker, and D.R. Miller, 2010: Evidence for a recent increase in forest growth. *Proceedings of the National Academy of Sciences of the United States of America*, **107(8)**, 3611-3615.

Murray, S.J., P.N. Foster, and I.C. Prentice, 2012: Future global water resources with respect to climate change and water withdrawals as estimated by a dynamic global vegetation model. *Journal of Hydrology*, **448-449**, 14-29.

Nock, C.A., P.J. Baker, W. Wanek, A. Leis, M. Grabner, S. Bunyavejchewin, and P. Hietz, 2011: Long-term increases in intrinsic water-use efficiency do not lead to increased stem growth in a tropical monsoon forest in western Thailand. *Global Change Biology*, **17(2)**, 1049-1063.

Nugent, K.A. and H.D. Matthews, 2012: Drivers of future northern latitude runoff change. *Atmosphere-Ocean*, **50(2)**, 197-206.

Oliveira, P.J.C., E.L. Davin, S. Levis, and S.I. Seneviratne, 2011: Vegetation-mediated impacts of trends in global radiation on land hydrology: a global sensitivity study. *Global Change Biology*, **17(11)**, 3453-3467.

Peñuelas, J., J.G. Canadell, and R. Ogaya, 2011: Increased water-use efficiency during the 20th century did not translate into enhanced tree growth. *Global Ecology and Biogeography*, **20(4)**, 597-608.

Piao, S., P. Friedlingstein, P. Ciais, N. de Noblet-Ducoudre, D. Labat, and S. Zaehle, 2007: Changes in climate and land use have a larger direct impact than rising CO_2 on global river runoff trends. *Proceedings of the National Academy of Sciences of the United States of America*, **104(39)**, 15242-15247.

Port, U., V. Brovkin, and M. Claussen, 2012: The influence of vegetation dynamics on anthropogenic climate change. *Earth System Dynamics*, **3**, 233-243.

Reddy, A.R., G.K. Rasineni, and A.S. Raghavendra, 2010: The impact of global elevated CO_2 concentration on photosynthesis and plant productivity. *Current Science*, **99(1)**, 46-57.

Rosenthal, D.M. and N.J. Tomeo, 2013: Climate, crops and lacking data underlie regional disparities in the CO_2 fertilization effect. *Environmental Research Letters*, **8(3)**, 031001, doi:10.1088/1748-9326/8/3/031001.

Saatchi, S., S. Asefi-Najafabady, Y. Malhi, L.E.O.C. Aragão, L.O. Anderson, R.B. Myneni, and R. Nemani, 2013: Persistent effects of a severe drought on Amazonian forest canopy. *Proceedings of the National Academy of Sciences of the United States of America*, **110(2)**, 565-570.

Sterling, S.M., A. Ducharne, and J. Polcher, 2013: The impact of global land-cover change on the terrestrial water cycle. *Nature Climate Change*, **3**, 385-390.

Wiltshire, A., J. Gornall, B. Booth, E. Dennis, P. Falloon, G. Kay, D. McNeall, C. McSweeney, and R. Betts, 2013 : The importance of population, climate change and CO_2 plant physiological forcing in determining future global water stress. *Global Environmental Change*, **23(5)**, 1083-1097.

This cross-chapter box should be cited as:

Gerten, D., R. Betts, and P. Döll, 2014: Cross-chapter box on the active role of vegetation in altering water flows under climate change. In: *Climate Change 2014: Impacts, Adaptation, and Vulnerability. Summaries, Frequently Asked Questions, and Cross-Chapter Boxes. Contribution of Working Group II to the Fifth Assessment Report of the Intergovernmental Panel on Climate Change* [Field, C.B., V.R. Barros, D.J. Dokken, K.J. Mach, M.D. Mastrandrea, T.E. Bilir, M. Chatterjee, K.L. Ebi, Y.O. Estrada, R.C. Genova, B. Girma, E.S. Kissel, A.N. Levy, S. MacCracken, P.R. Mastrandrea, and L.L. White (eds.)]. World Meteorological Organization, Geneva, Switzerland, pp. 159-163. (in Arabic, Chinese, English, French, Russian, and Spanish).

The Water–Energy–Food/ Feed/Fiber Nexus as Linked to Climate Change

Douglas J. Arent (USA), Petra Döll (Germany), Kenneth M. Strzepek (USA), Blanca Elena Jiménez Cisneros (Mexico), Andy Reisinger (New Zealand), Ferenc Toth (Hungary), Taikan Oki (Japan)

Water, energy, and food/feed/fiber are linked through numerous interactive pathways and subject to a changing climate, as depicted in Figure CC-WE-1. The depth and intensity of those linkages vary enormously among countries, regions, and production systems. Energy technologies (e.g., biofuels, hydropower, thermal power plants), transportation fuels and modes, and food products (from irrigated crops, in particular animal protein produced by feeding irrigated crops and forages) may require significant amounts of water (Sections 3.7.2, 7.3.2, 10.2,10.3.4, 22.3.3, 25.7.2; Allan, 2003; King and Weber, 2008; McMahon and Price, 2011; Macknick et al., 2012a). In irrigated agriculture, climate, irrigating procedure, crop choice, and yields determine water requirements per unit of produced crop. In areas where water (and wastewater) must be pumped and/or treated, energy must be provided (Metcalf & Eddy, Inc. et al., 2007; Khan and Hanjra, 2009; EPA, 2010; Gerten et al., 2011). While food production, refrigeration, transport, and processing require large amounts of energy (Pelletier et al., 2011), a major link between food and energy as related to climate change is the competition of bioenergy and food production for land and water (*robust evidence, high agreement*; Section 7.3.2, Box 25-10; Diffenbaugh et al., 2012; Skaggs et al., 2012). Food and crop wastes, and wastewater, may be used as sources of energy, saving not only the consumption of conventional nonrenewable fuels used in their traditional processes, but also the consumption of the water and energy employed for processing or treatment and disposal (Schievano et al., 2009; Oh et al., 2010; Olson, 2012). Examples of this can be found in several countries across all income ranges. For example, sugar cane byproducts are increasingly used to produce electricity or for cogeneration (McKendry, 2002; Kim and Dale, 2004) for economic benefits, and increasingly as an option for greenhouse gas mitigation.

Most energy production methods require significant amounts of water, either directly (e.g., crop-based energy sources and hydropower) or indirectly (e.g., cooling for thermal energy sources or other operations) (*robust evidence, high agreement*; Sections 10.2.2, 10.3.4, 25.7.4; and van Vliet et al., 2012; Davies et al., 2013. Water for biofuels, for example, under the International Energy Agency (IEA) Alternative Policy Scenario, which has biofuels production increasing to 71 EJ in 2030, has been reported by Gerbens-Leenes et al. (2012) to drive global consumptive irrigation water use from 0.5% of global renewable water resources in 2005 to 5.5% in 2030, resulting in increased pressure on freshwater resources, with potential negative impacts on freshwater ecosystems. Water is also required for mining (Section 25.7.3), processing, and residue disposal of fossil and nuclear fuels or their byproducts. Water for energy currently ranges from a few percent in most developing countries to more than 50% of freshwater withdrawals in some developed countries, depending on the country (Kenny et al., 2009; WEC, 2010). Future water requirements will depend on electricity demand growth, the portfolio of generation technologies and water management options employed (*medium evidence, high agreement*; WEC, 2010; Sattler et al.,

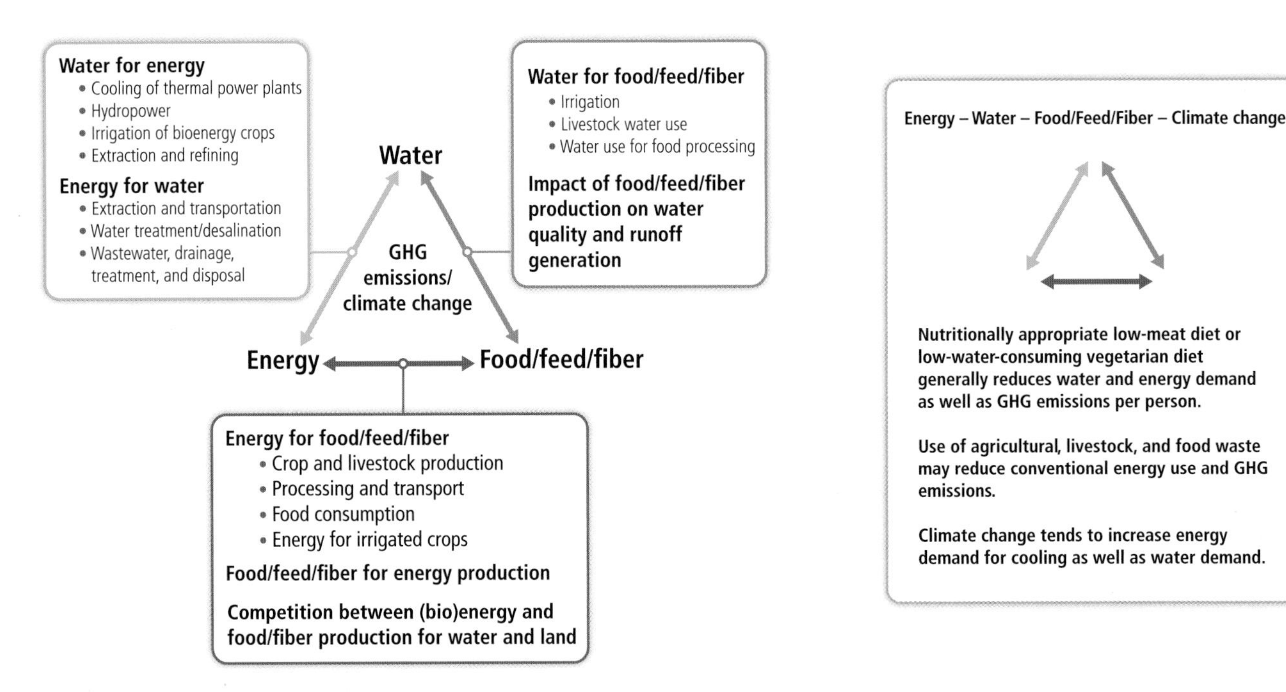

Figure WE-1 | The water–energy–food nexus as related to climate change. The interlinkages of supply/demand, quality and quantity of water, and energy and food/feed/fiber with changing climatic conditions have implications for both adaptation and mitigation strategies.

2012). Future water availability for energy production will change due to climate change (*robust evidence, high agreement*; Sections 3.4, 3.5.1, 3.5.2.2.

Water may require significant amounts of energy for lifting, transport, and distribution and for its treatment either to use it or to depollute it. Wastewater and even excess rainfall in cities requires energy to be treated or disposed. Some non-conventional water sources (wastewater or seawater) are often highly energy intensive. Energy intensities per m^3 of water vary by about a factor of 10 between different sources, for example, locally produced potable water from ground/surface water sources versus desalinated seawater (Box 25-2, Tables 25-6, 25-7; Macknick et al., 2012b; Plappally and Lienhard, 2012). Groundwater (35% of total global water withdrawals, with irrigated food production being the largest user; Döll et al., 2012) is generally more energy intensive than surface water. In India, for example, 19% of total electricity use in 2012 was for agricultural purposes (Central Statistics Office, 2013), with a large share for groundwater pumping. Pumping from greater depth increases energy demand significantly—electricity use (kWh m^{-3} of water) increases by a factor of 3 when going from 35 to 120 m depth (Plappally and Lienhard, 2012). The reuse of appropriate wastewater for irrigation (reclaiming both water and energy-intense nutrients) may increase agricultural yields, save energy, and prevent soil erosion (*medium confidence*; Smit and Nasr, 1992; Jiménez-Cisneros, 1996; Qadir et al., 2007; Raschid-Sally and Jayakody, 2008). More energy efficient treatment methods enable poor quality ("black") wastewater to be treated to quality levels suitable for discharge into water courses, avoiding additional freshwater and associated energy demands (Keraita et al., 2008). If properly treated to retain nutrients, such treated water may increase soil productivity, contributing to increased crop yields/food security in regions unable to afford high power bills or expensive fertilizer (*high confidence*; Oron, 1996; Lazarova and Bahri, 2005; Redwood and Huibers, 2008; Jiménez-Cisneros, 2009).

Linkages among water, energy, food/feed/fiber, and climate are also strongly related to land use and management (*robust evidence, high agreement*; Section 4.4.4, Box 25-10). Land degradation often reduces efficiency of water and energy use (e.g., resulting in higher fertilizer demand and surface runoff), and compromises food security (Sections 3.7.2, 4.4.4). On the other hand, afforestation activities to sequester carbon have important co-benefits of reducing soil erosion and providing additional (even if only temporary) habitat (see Box 25-10) but may reduce renewable water resources. Water abstraction for energy, food, or biofuel production or carbon sequestration can also compete with minimal environmental flows needed to maintain riverine habitats and wetlands, implying a potential conflict between economic and other valuations and uses of water (*medium evidence, high agreement*; Sections 25.4.3, 25.6.2, Box 25-10). Only a few reports have begun to evaluate the multiple interactions among energy, food, land, and water and climate (McCornick et al., 2008; Bazilian et al., 2011; Bierbaum and Matson, 2013), addressing the issues from a security standpoint and describing early integrated modeling approaches. The interaction among each of these factors is influenced by the changing climate, which in turn impacts energy and water demand, bioproductivity, and other factors (see Figure CC-WE-1 and Wise et al., 2009), and has implications for security of supplies of energy, food, and water; adaptation and mitigation pathways; and air pollution reduction, as well as the implications for health and economic impacts as described throughout this Assessment Report.

The interconnectivity of food/fiber, water, land use, energy, and climate change, including the perhaps not yet well understood cross-sector impacts, are increasingly important in assessing the implications for adaptation/mitigation policy decisions. Fuel–food–land use–water–greenhouse gas (GHG) mitigation strategy interactions, particularly related to bioresources for food/feed, power, or fuel, suggest that combined assessment of water, land type, and use requirements, energy requirements, and potential uses and GHG impacts often epitomize the interlinkages. For example, mitigation scenarios described in the IPCC Special Report on Renewable Energy Sources and Climate Change Mitigation (IPCC, 2011) indicate up to 300 EJ of biomass primary energy by 2050 under increasingly stringent mitigation scenarios. Such high levels of biomass production, in the absence of technology and process/management/operations change, would have significant implications for land use, water, and energy, as well as food production and pricing. Consideration of the interlinkages of energy, food/feed/fiber, water, land use, and climate change is increasingly recognized as critical to effective climate resilient pathway decision making (*medium evidence, high agreement*), although tools to support local- and regional-scale assessments and decision support remain very limited.

References

Allan, T., 2003: Virtual water – the water, food, and trade nexus: useful concept or misleading metaphor? *Water International*, **28(1)**, 4-10.

Bazilian, M., H. Rogner, M. Howells, S. Hermann, D. Arent, D. Gielen, P. Steduto, A. Mueller, P. Komor, R.S.J. Tol, and K. Yumkella, 2011: Considering the energy, water and food nexus: towards an integrated modelling approach. *Energy Policy*, **39(12)**, 7896-7906.

Bierbaum, R. and P. Matson, 2013: Energy in the context of sustainability. *Daedalus*, **142(1)**, 146-161.

Davies, E., K. Page, and J.A. Edmonds, 2013: An integrated assessment of global and regional water demands for electricity generation to 2095. *Advances in Water Resources*, **52**, 296-313, doi:10.1016/j.advwatres.2012.11.020.

Diffenbaugh, N., T. Hertel, M. Scherer, and M. Verma, 2012: Response of corn markets to climate volatility under alternative energy futures. *Nature Climate Change*, **2**, 514-518.

Döll, P., H. Hoffmann-Dobrev, F.T. Portmann, S. Siebert, A. Eicker, M. Rodell, G. Strassberg, and B. Scanlon, 2012: Impact of water withdrawals from groundwater and surface water on continental water storage variations. *Journal of Geodynamics*, **59-60**, 143-156, doi:10.1016/j.jog.2011.05.001.

EPA, 2010: *Evaluation of Energy Conservation Measures for Wastewater Treatment Facilities*. EPA 832-R-10-005, U.S. Environmental Protection Agency (EPA), Office of Wastewater Management, Washington, DC, USA, 222 pp., water.epa.gov/scitech/wastetech/upload/Evaluation-of-Energy-Conservation-Measures-for-Wastewater-Treatment-Facilities.pdf.

Gerbens-Leenes, P.W., A.R. van Lienden, A.Y. Hoekstra, and Th.H. van der Meer, 2012: Biofuel scenarios in a water perspective: the global blue and green water footprint of road transport in 2030. *Global Environmental Change*, **22(3)**, 764-775.

Gerber, N., M. van Eckert, and T. Breuer, 2008: *The Impacts of Biofuel Production on Food Prices: A Review*. ZEF – Discussion Papers on Development Policy, No. 127, Center for Development Research [Zentrum für Entwicklungsforschung (ZEF)], Bonn, Germany, 19 pp.

Gerten, D., H. Heinke, H. Hoff, H. Biemans, M. Fader, and K. Waha, 2011: Global water availability and requirements for future food production. *Journal of Hydrometeorology*, **12**, 885-899.

IPCC, 2011: Summary for Policymakers. In: *IPCC Special Report on Renewable Energy Sources and Climate Change Mitigation. Special Report of Working Group III of the Intergovernmental Panel on Climate Change* [Edenhofer, O., R. Pichs-Madruga, Y. Sokona, K. Seyboth, P. Matschoss, S. Kadner, T. Zwickel, P. Eickemeier, G. Hansen, S. Schlmer, and C. von Stechow (eds.)]. Cambridge University Press, Cambridge, UK and New York, NY, USA, pp. 3-26.

Jiménez-Cisneros, B., 1996: Wastewater reuse to increase soil productivity. *Water Science and Technology*, **32(12)**, 173-180.

Jiménez-Cisneros, B., 2009: 4.06 – Safe sanitation in low economic development areas. In: *Treatise on Water Science, Volume 4: Water-Quality Engineering* [Wilderer, P.A. (ed.)]. Reference Module in Earth Systems and Environmental Sciences, Academic Press, Oxford, UK, pp.147-200.

Kenny, J.F., N.L. Barber, S.S. Hutson, K.S. Linsey, J.K. Lovelace, and M.A. Maupin, 2009: *Estimated Use of Water in the United States in 2005*. U.S. Department of the Interior, U.S. Geological Survey (USGS) Circular 1344, USGS, Reston, VA, USA, 53 pp.

Keraita, B., B. Jiménez, and P. Drechsel, 2008: Extent and implications of agricultural reuse of untreated, partly treated and diluted wastewater in developing countries. *CAB Reviews: Perspectives in Agriculture, Veterinary Science, Nutrition and Natural Resources*, **3(58)**, 15-27.

Khan, S. and M.A. Hanjra, 2009: Footprints of water and energy inputs in food production – global perspectives. *Food Policy*, **34**, 130-140.

Kim, S. and B. Dale, 2004: Global potential bioethanol production from wasted crops and crop residues. *Biomass and Bioenergy*, **26(4)**, 361-375.

King, C. and M.E. Webber, 2008: Water intensity of transportation. *Environmental Science and Technology*, **42(21)**, 7866-7872.

Lazarova, V. and A. Bahri, 2005: *Water Reuse for Irrigation: Agriculture, Landscapes, and Turf Grass*. CRC Press, Boca Raton, FL, USA, 408 pp.

McKendry, P., 2002: Energy production from biomass (part 1): overview of biomass. *Bioresource Technology*, **83(1)**, 37-46.

Macknick, J., R. Newmark, G. Heath, K.C. Hallett, J. Meldrum, and S. Nettles-Anderson, 2012a: Operational water consumption and withdrawal factors for electricity generating technologies: a review of existing literature. *Environmental Research Letters*, **7(4)**, 045802, doi:10.1088/1748-9326/7/4/045802.

Macknick, J., S. Sattler, K. Averyt, S. Clemmer, and J. Rogers, 2012b: Water implications of generating electricity: water use across the United States based on different electricity pathways through 2050. *Environmental Research Letters*, **7(4)**, 045803, doi:10.1088/1748-9326/7/4/045803.

McCornick, P.G., S.B. Awulachew, and M. Abebe, 2008: Water-food-energy-environment synergies and tradeoffs: major issues and case studies. *Water Policy*, **10**, 23-36.

McMahon, J.E. and S.K. Price, 2011: Water and energy interactions. *Annual Review of Environment and Resources*, **36**, 163-191.

Metcalf & Eddy, Inc. an AECOM Company, T. Asano, F. Burton, H. Leverenz, R. Tsuchihashi, and G. Tchobanoglous, 2007: *Water Reuse: Issues, Technologies, and Applications*. McGraw-Hill Professional, New York, NY, USA, 1570 pp.

Oh, S.T., J.R. Kim, G.C. Premier, T.H. Lee, C. Kim, and W.T. Sloan, 2010: Sustainable wastewater treatment: how might microbial fuel cells contribute. *Biotechnology Advances*, **28(6)**, 871-881.

Olson, G., 2012: *Water and Energy Nexus: Threats and Opportunities*. IWA Publishing, London, UK, 294 pp.

Oron, G., 1996: Soil as a complementary treatment component for simultaneous wastewater disposal and reuse. *Water Science and Technology*, **34(11)**, 243-252.

Pelletier, N., E. Audsley, S. Brodt, T. Garnett, P. Henriksson, A. Kendall, K.J. Kramer, D. Murphy, T. Nemeck, and M. Troell, 2011: Energy intensity of agriculture and food systems. *Annual Review of Environment and Resources*, **36**, 223-246.

Plappally, A.K. and J.H. Lienhard V, 2012: Energy requirements for water production, treatment, end use, reclamation, and disposal. *Renewable and Sustainable Energy Reviews*, **16(7)**, 4818-4848.

Qadir, M., D. Wichelns, L. Raschid-Sally, P. Singh Minhas, P. Drechsel, A. Bahri, P. McCornick, R. Abaidoo, F. Attia, S. El-Guindy, J.H.J. Ensink, B. Jiménez, J.W. Kijne, S. Koo-Oshima, J.D. Oster, L. Oyebande, J.A. Sagardoy, and W. van der Hoek, 2007: Agricultural use of marginal-quality water – opportunities and challenges. In: *Water for Food, Water for Life: A Comprehensive Assessment of Water Management in Agriculture* [Molden, D. (ed.)]. Earthscan Publications, Ltd., London, UK, pp. 425-458.

Raschid-Sally, L. and P. Jayakody, 2008: *Drivers and Characteristics of Wastewater Agriculture in Developing Countries: Results from a Global Assessment*. IWMI Research Report 127, International Water Management Institute (IWMI), Colombo, Sri Lanka, 29 pp.

Redwood, M. and F. Huibers, 2008: Wastewater irrigation in urban agriculture. In: *Water Reuse: An International Survey of Current Practice, Issues and Needs* [Jiménez, B. and T. Asano (ed.)]. IWA Publishing, London, UK, pp. 228-240.

Sattler, S., J. Macknick, D. Yates, F. Flores-Lopez, A. Lopez, and J. Rogers, 2012: Linking electricity and water models to assess electricity choices at water-relevant scales. *Environmental Research Letters*, **7(4)**, 045804, doi:10.1088/1748-9326/7/4/045804.

Schievano A., G. D'Imporzano, and F. Adani, 2009: Substituting energy crops with organic wastes and agro-industrial residues for biogas production. *Journal of Environmental Management*, **90(8)**, 2537-2541.

Skaggs, R., K. Hibbard, P. Frumhoff, T. Lowry, R. Middleton, R. Pate, V. Tidwell, J. Arnold, K. Averyt, A. Janetos, C. Izaurralde, J. Rice, and S. Rose, 2012: *Climate and Energy-Water-Land System Interactions*. PNNL 21185, Technical Report to the US Department of Energy in support of the National Climate Assessment, Pacific Northwest National Laboratory (PNNL), Richland, WA, USA, 152 pp.

Smit, J. and J. Nasr, 1992: Urban agriculture for sustainable cities: using wastes and idle land and water bodies as resources. *Environment and Urbanization*, **4(2)**, 141-152.

van Vliet, M.T.H., J.R. Yearsley, F. Ludwig, S. Vögele, D.P. Lettenmaier, and P. Kabat, 2012: Vulnerability of US and European electricity supply to climate change. *Nature Climate Change*, **2**, 676-681.

Wise, M., K. Calvin, A. Thomson, L. Clarke, B. Bond-Lamberty, R. Sands, S.J. Smith, A. Janetos, and J. Edmonds, 2009: Implications of limiting CO_2 concentrations for land use and energy. *Science*, **324**, 1183-1186.

WEC, 2010: *Water for Energy*. World Energy Council (WEC), London, UK, 51 pp.

This cross-chapter box should be cited as:

Arent, D.J., P. Döll, K.M. Strzepek, B.E. Jiménez Cisneros, A. Reisinger, F.L. Tóth, and T. Oki, 2014: Cross-chapter box on the water–energy–food/feed/fiber nexus as linked to climate change. In: *Climate Change 2014: Impacts, Adaptation, and Vulnerability. Summaries, Frequently Asked Questions, and Cross-Chapter Boxes. Contribution of Working Group II to the Fifth Assessment Report of the Intergovernmental Panel on Climate Change* [Field, C.B., V.R. Barros, D.J. Dokken, K.J. Mach, M.D. Mastrandrea, T.E. Bilir, M. Chatterjee, K.L. Ebi, Y.O. Estrada, R.C. Genova, B. Girma, E.S. Kissel, A.N. Levy, S. MacCracken, P.R. Mastrandrea, and L.L. White (eds.)]. World Meteorological Organization, Geneva, Switzerland, pp. 165-168. (in Arabic, Chinese, English, French, Russian, and Spanish)

Glossary

Glossary

Editorial Board Co-Chairs:
John Agard (Trinidad and Tobago), E. Lisa F. Schipper (Sweden)

Editorial Board:
Joern Birkmann (Germany), Maximiliano Campos (Costa Rica), Carolina Dubeux (Brazil), Yukihiro Nojiri (Japan), Lennart Olsson (Sweden), Balgis Osman-Elasha (Sudan), Mark Pelling (UK), Michael J. Prather (USA), Marta G. Rivera-Ferre (Spain), Oliver C. Ruppel (Namibia), Asbury Sallenger (USA), Kirk R. Smith (USA), Asuncion L. St. Clair (Norway)

TSU Facilitation:
Katharine J. Mach (USA), Michael D. Mastrandrea (USA), T. Eren Bilir (USA)

Abrupt climate change
A large-scale change in the climate system that takes place over a few decades or less, persists (or is anticipated to persist) for at least a few decades, and causes substantial disruptions in human and natural systems.

Access to food
One of the three components underpinning food security, the other two being availability and utilization. Access to food is dependent on (1) the affordability of food (i.e., people have income or other resources to exchange for food); (2) satisfactory allocation within the household or society; and (3) preference (i.e., it is what people want to eat, influenced by socio-cultural norms). See also Food security.

Acclimatization
A change in functional or morphological traits occurring once or repeatedly (e.g., seasonally) during the lifetime of an individual organism in its natural environment. Through acclimatization the individual maintains performance across a range of environmental conditions. For a clear differentiation between findings in laboratory and field studies, the term *acclimation* is used in ecophysiology for the respective phenomena when observed in well-defined experimental settings. The term *(adaptive) plasticity* characterizes the generally limited scope of changes in phenotype that an individual can reach through the process of acclimatization.

Adaptability
See Adaptive capacity.

Adaptation[1]
The process of adjustment to actual or expected climate and its effects. In human systems, adaptation seeks to moderate or avoid harm or exploit beneficial opportunities. In some natural systems, human intervention may facilitate adjustment to expected climate and its effects.

Incremental adaptation Adaptation actions where the central aim is to maintain the essence and integrity of a system or process at a given scale.[2]

Transformational adaptation Adaptation that changes the fundamental attributes of a system in response to climate and its effects.

See also Autonomous adaptation, Evolutionary adaptation, and Transformation.

Adaptation assessment
The practice of identifying options to adapt to climate change and evaluating them in terms of criteria such as availability, benefits, costs, effectiveness, efficiency, and feasibility.

Adaptation constraint
Factors that make it harder to plan and implement adaptation actions or that restrict options.

Adaptation deficit
The gap between the current state of a system and a state that minimizes adverse impacts from existing climate conditions and variability.

Adaptation limit
The point at which an actor's objectives (or system needs) cannot be secured from intolerable risks through adaptive actions.

Hard adaptation limit No adaptive actions are possible to avoid intolerable risks.

Soft adaptation limit Options are currently not available to avoid intolerable risks through adaptive action.

Adaptation needs
The circumstances requiring action to ensure safety of populations and security of assets in response to climate impacts.

Adaptation opportunity
Factors that make it easier to plan and implement adaptation actions, that expand adaptation options, or that provide ancillary co-benefits.

Adaptation options
The array of strategies and measures that are available and appropriate for addressing adaptation needs. They include a wide range of actions that can be categorized as structural, institutional, or social.

Adaptive capacity
The ability of systems, institutions, humans, and other organisms to adjust to potential damage, to take advantage of opportunities, or to respond to consequences.[3]

Adaptive management
A process of iteratively planning, implementing, and modifying strategies for managing resources in the face of uncertainty and change. Adaptive management involves adjusting approaches in response to observations of their effect and changes in the system brought on by resulting feedback effects and other variables.

Aggregate impacts
Total impacts integrated across sectors and/or regions. The aggregation of impacts requires knowledge of (or assumptions about) the relative importance of different impacts. Measures of aggregate impacts include, for example, the total number of people affected, or the total economic costs, and are usually bound by time, place, and/or sector.

Ancillary benefits
See Co-benefits.

Anomaly
The deviation of a variable from its value averaged over a reference period.

---

[1] Reflecting progress in science, this glossary entry differs in breadth and focus from the entry used in the Fourth Assessment Report and other IPCC reports.
[2] This definition builds from the definition used in Park et al. (2012).
[3] This glossary entry builds from definitions used in previous IPCC reports and the Millennium Ecosystem Assessment (MEA, 2005).

Anthropogenic
Resulting from or produced by human activities.

Anthropogenic emissions
Emissions of greenhouse gases, greenhouse gas precursors, and aerosols caused by human activities. These activities include the burning of fossil fuels, deforestation, land use changes, livestock production, fertilization, waste management, and industrial processes.

Arid zone
Areas where vegetation growth is severely constrained due to limited water availability. For the most part, the native vegetation of arid zones is sparse. There is high rainfall variability, with annual averages below 300 mm. Crop farming in arid zones requires irrigation.

Atlantic Multi-decadal Oscillation/Variability (AMO/AMV)
A multi-decadal (65- to 75-year) fluctuation in the North Atlantic, in which sea surface temperatures showed warm phases during roughly 1860 to 1880 and 1930 to 1960 and cool phases during 1905 to 1925 and 1970 to 1990 with a range of approximately 0.4°C. See AMO Index in WGI AR5 Box 2.5.

Atmosphere-Ocean General Circulation Model (AOGCM)
See Climate model.

Attribution
See Detection and attribution.

Autonomous adaptation
Adaptation in response to experienced climate and its effects, without planning explicitly or consciously focused on addressing climate change. Also referred to as spontaneous adaptation.

Baseline/reference
The baseline (or reference) is the state against which change is measured. A baseline period is the period relative to which anomalies are computed. The baseline concentration of a trace gas is that measured at a location not influenced by local anthropogenic emissions.

Biodiversity
The variability among living organisms from terrestrial, marine, and other ecosystems. Biodiversity includes variability at the genetic, species, and ecosystem levels.[4]

Bioenergy
Energy derived from any form of biomass such as recently living organisms or their metabolic by-products.

Biofuel
A fuel, generally in liquid form, developed from organic matter or combustible oils produced by living or recently living plants. Examples of biofuel include alcohol (bioethanol), black liquor from the paper-manufacturing process, and soybean oil.

First-generation manufactured biofuel First-generation manufactured biofuel is derived from grains, oilseeds, animal fats, and waste vegetable oils with mature conversion technologies.

Second-generation biofuel Second-generation biofuel uses non-traditional biochemical and thermochemical conversion processes and feedstock mostly derived from the lignocellulosic fractions of, for example, agricultural and forestry residues, municipal solid waste, etc.

Third-generation biofuel Third-generation biofuel would be derived from feedstocks such as algae and energy crops by advanced processes still under development.

These second- and third-generation biofuels produced through new processes are also referred to as next-generation or advanced biofuels, or advanced biofuel technologies.

Biomass
The total mass of living organisms in a given area or volume; dead plant material can be included as dead biomass. Biomass burning is the burning of living and dead vegetation.

Biome
A biome is a major and distinct regional element of the biosphere, typically consisting of several ecosystems (e.g., forests, rivers, ponds, swamps within a region). Biomes are characterized by typical communities of plants and animals.

Biosphere
The part of the Earth system comprising all ecosystems and living organisms, in the atmosphere, on land (terrestrial biosphere), or in the oceans (marine biosphere), including derived dead organic matter, such as litter, soil organic matter, and oceanic detritus.

Boundary organization
A bridging institution, social arrangement, or network that acts as an intermediary between science and policy.

Business As Usual (BAU)
Business as usual projections are based on the assumption that operating practices and policies remain as they are at present. Although baseline scenarios could incorporate some specific features of BAU scenarios (e.g., a ban on a specific technology), BAU scenarios imply that no practices or policies other than the current ones are in place. See also Baseline/reference, Climate scenario, Emission scenario, Representative Concentration Pathways, Scenario, Socioeconomic scenario, and SRES scenarios.

Capacity building
The practice of enhancing the strengths and attributes of, and resources available to, an individual, community, society, or organization to respond to change.

[4] This glossary entry builds from definitions used in the Global Biodiversity Assessment (Heywood, 1995) and the Millennium Ecosystem Assessment (MEA, 2005).

Carbon cycle
The term used to describe the flow of carbon (in various forms, e.g., as carbon dioxide) through the atmosphere, ocean, terrestrial and marine biosphere, and lithosphere. In this report, the reference unit for the global carbon cycle is GtC or equivalently PgC (10^{15}g).

Carbon dioxide (CO_2)
A naturally occurring gas, also a by-product of burning fossil fuels from fossil carbon deposits, such as oil, gas, and coal, of burning biomass, of land use changes, and of industrial processes (e.g., cement production). It is the principal anthropogenic greenhouse gas that affects the Earth's radiative balance. It is the reference gas against which other greenhouse gases are measured and therefore has a Global Warming Potential of 1.

Carbon dioxide (CO_2) fertilization
The enhancement of the growth of plants as a result of increased atmospheric carbon dioxide (CO_2) concentration.

Carbon sequestration
See Uptake.

Clean Development Mechanism (CDM)
A mechanism defined under Article 12 of the Kyoto Protocol through which investors (governments or companies) from developed (Annex B) countries may finance greenhouse gas emission reduction or removal projects in developing (Non-Annex B) countries, and receive Certified Emission Reduction Units for doing so, which can be credited towards the commitments of the respective developed countries. The CDM is intended to facilitate the two objectives of promoting sustainable development in developing countries and of helping industrialized countries to reach their emissions commitments in a cost-effective way.

Climate
Climate in a narrow sense is usually defined as the average weather, or more rigorously, as the statistical description in terms of the mean and variability of relevant quantities over a period of time ranging from months to thousands or millions of years. The classical period for averaging these variables is 30 years, as defined by the World Meteorological Organization. The relevant quantities are most often surface variables such as temperature, precipitation, and wind. Climate in a wider sense is the state, including a statistical description, of the climate system.

Climate-altering pollutants (CAPs)
Gases and particles released from human activities that affect the climate either directly, through mechanisms such as radiative forcing from changes in greenhouse gas concentrations, or indirectly, by, for example, affecting cloud formation or the lifetime of greenhouse gases in the atmosphere. CAPs include both those pollutants that have a warming effect on the atmosphere, such as CO_2, and those with cooling effects, such as sulfates.

Climate change
Climate change refers to a change in the state of the climate that can be identified (e.g., by using statistical tests) by changes in the mean and/or the variability of its properties, and that persists for an extended period, typically decades or longer. Climate change may be due to natural internal processes or external forcings such as modulations of the solar cycles, volcanic eruptions, and persistent anthropogenic changes in the composition of the atmosphere or in land use. Note that the Framework Convention on Climate Change (UNFCCC), in its Article 1, defines climate change as: "a change of climate which is attributed directly or indirectly to human activity that alters the composition of the global atmosphere and which is in addition to natural climate variability observed over comparable time periods." The UNFCCC thus makes a distinction between climate change attributable to human activities altering the atmospheric composition, and climate variability attributable to natural causes. See also Climate change commitment and Detection and Attribution.

Climate change commitment
Due to the thermal inertia of the ocean and slow processes in the cryosphere and land surfaces, the climate would continue to change even if the atmospheric composition were held fixed at today's values. Past change in atmospheric composition leads to a committed climate change, which continues for as long as a radiative imbalance persists and until all components of the climate system have adjusted to a new state. The further change in temperature after the composition of the atmosphere is held constant is referred to as the constant composition temperature commitment or simply committed warming or warming commitment. Climate change commitment includes other future changes, for example, in the hydrological cycle, in extreme weather events, in extreme climate events, and in sea level change. The constant emission commitment is the committed climate change that would result from keeping anthropogenic emissions constant and the zero emission commitment is the climate change commitment when emissions are set to zero. See also Climate change.

Climate extreme (Extreme weather or climate event)
See Extreme weather event.

Climate feedback
An interaction in which a perturbation in one climate quantity causes a change in a second, and the change in the second quantity ultimately leads to an additional change in the first. A negative feedback is one in which the initial perturbation is weakened by the changes it causes; a positive feedback is one in which the initial perturbation is enhanced. In this Assessment Report, a somewhat narrower definition is often used in which the climate quantity that is perturbed is the global mean surface temperature, which in turn causes changes in the global radiation budget. In either case, the initial perturbation can either be externally forced or arise as part of internal variability.

Climate governance
Purposeful mechanisms and measures aimed at steering social systems towards preventing, mitigating, or adapting to the risks posed by climate change (Jagers and Stripple, 2003).

Climate model (spectrum or hierarchy)
A numerical representation of the climate system based on the physical, chemical, and biological properties of its components, their interactions, and feedback processes, and accounting for some of its known properties. The climate system can be represented by models of varying complexity; that is, for any one component or combination of components, a spectrum or hierarchy of models can be identified, differing in such

aspects as the number of spatial dimensions, the extent to which physical, chemical, or biological processes are explicitly represented, or the level at which empirical parameterizations are involved. Coupled Atmosphere-Ocean General Circulation Models (AOGCMs) provide a representation of the climate system that is near or at the most comprehensive end of the spectrum currently available. There is an evolution towards more complex models with interactive chemistry and biology. Climate models are applied as a research tool to study and simulate the climate, and for operational purposes, including monthly, seasonal, and interannual climate predictions. See also Earth System Model.

Climate prediction

A climate prediction or climate forecast is the result of an attempt to produce (starting from a particular state of the climate system) an estimate of the actual evolution of the climate in the future, for example, at seasonal, interannual, or decadal time scales. Because the future evolution of the climate system may be highly sensitive to initial conditions, such predictions are usually probabilistic in nature. See also Climate projection, Climate scenario, and Predictability.

Climate projection

A climate projection is the simulated response of the climate system to a scenario of future emission or concentration of greenhouse gases and aerosols, generally derived using climate models. Climate projections are distinguished from climate predictions by their dependence on the emission/concentration/radiative-forcing scenario used, which is in turn based on assumptions concerning, for example, future socioeconomic and technological developments that may or may not be realized. See also Climate scenario.

Climate-resilient pathways

Iterative processes for managing change within complex systems in order to reduce disruptions and enhance opportunities associated with climate change.

Climate scenario

A plausible and often simplified representation of the future climate, based on an internally consistent set of climatological relationships that has been constructed for explicit use in investigating the potential consequences of anthropogenic climate change, often serving as input to impact models. Climate projections often serve as the raw material for constructing climate scenarios, but climate scenarios usually require additional information such as the observed current climate. See also Emission scenario and Scenario.

Climate sensitivity

In IPCC reports, equilibrium climate sensitivity (units: °C) refers to the equilibrium (steady state) change in the annual global mean surface temperature following a doubling of the atmospheric equivalent carbon dioxide concentration. Owing to computational constraints, the equilibrium climate sensitivity in a climate model is sometimes estimated by running an atmospheric general circulation model coupled to a mixed-layer ocean model, because equilibrium climate sensitivity is largely determined by atmospheric processes. Efficient models can be run to equilibrium with a dynamic ocean. The climate sensitivity parameter (units: °C $(W\ m^{-2})^{-1}$) refers to the equilibrium change in the annual global mean surface temperature following a unit change in radiative forcing.

The effective climate sensitivity (units: °C) is an estimate of the global mean surface temperature response to doubled carbon dioxide concentration that is evaluated from model output or observations for evolving non-equilibrium conditions. It is a measure of the strengths of the climate feedbacks at a particular time and may vary with forcing history and climate state, and therefore may differ from equilibrium climate sensitivity.

The transient climate response (units: °C) is the change in the global mean surface temperature, averaged over a 20-year period, centered at the time of atmospheric carbon dioxide doubling, in a climate model simulation in which CO_2 increases at 1% yr^{-1}. It is a measure of the strength and rapidity of the surface temperature response to greenhouse gas forcing.

Climate system

The climate system is the highly complex system consisting of five major components: the atmosphere, the hydrosphere, the cryosphere, the lithosphere, and the biosphere, and the interactions among them. The climate system evolves in time under the influence of its own internal dynamics and because of external forcings such as volcanic eruptions, solar variations, and anthropogenic forcings such as the changing composition of the atmosphere and land use change.

Climate variability

Climate variability refers to variations in the mean state and other statistics (such as standard deviations, the occurrence of extremes, etc.) of the climate on all spatial and temporal scales beyond that of individual weather events. Variability may be due to natural internal processes within the climate system (internal variability), or to variations in natural or anthropogenic external forcing (external variability). See also Climate change.

Climate velocity

The speed at which isolines of a specified climate variable travel across landscapes or seascapes due to changing climate. For example, climate velocity for temperature is the speed at which isotherms move due to changing climate (km yr^{-1}) and is calculated as the temporal change in temperature (°C yr^{-1}) divided by the current spatial gradient in temperature (°C km^{-1}). It can be calculated using additional climate variables such as precipitation or can be based on the climatic niche of organisms.

Climatic driver (Climate driver)

A changing aspect of the climate system that influences a component of a human or natural system.

CMIP3 and CMIP5

Phases three and five of the Coupled Model Intercomparison Project (CMIP3 and CMIP5), coordinating and archiving climate model simulations based on shared model inputs by modeling groups from around the world. The CMIP3 multi-model data set includes projections using SRES scenarios. The CMIP5 data set includes projections using the Representative Concentration Pathways.

Coastal squeeze
A narrowing of coastal ecosystems and amenities (e.g., beaches, salt marshes, mangroves, and mud and sand flats) confined between landward-retreating shorelines (from sea level rise and/or erosion) and naturally or artificially fixed shorelines including engineering defenses (e.g., seawalls), potentially making the ecosystems or amenities vanish.

Co-benefits
The positive effects that a policy or measure aimed at one objective might have on other objectives, irrespective of the net effect on overall social welfare. Co-benefits are often subject to uncertainty and depend on local circumstances and implementation practices, among other factors. Co-benefits are also referred to as ancillary benefits.

Community-based adaptation
Local, community-driven adaptation. Community-based adaptation focuses attention on empowering and promoting the adaptive capacity of communities. It is an approach that takes context, culture, knowledge, agency, and preferences of communities as strengths.

Confidence
The validity of a finding based on the type, amount, quality, and consistency of evidence (e.g., mechanistic understanding, theory, data, models, expert judgment) and on the degree of agreement. Confidence is expressed qualitatively (Mastrandrea et al., 2010). See Box 1-1. See also Uncertainty.

Contextual vulnerability (Starting-point vulnerability)
A present inability to cope with external pressures or changes, such as changing climate conditions. Contextual vulnerability is a characteristic of social and ecological systems generated by multiple factors and processes (O'Brien et al., 2007).

Convection
Vertical motion driven by buoyancy forces arising from static instability, usually caused by near-surface cooling or increases in salinity in the case of the ocean and near-surface warming or cloud-top radiative cooling in the case of the atmosphere. In the atmosphere, convection gives rise to cumulus clouds and precipitation and is effective at both scavenging and vertically transporting chemical species. In the ocean, convection can carry surface waters to deep within the ocean.

Coping
The use of available skills, resources, and opportunities to address, manage, and overcome adverse conditions, with the aim of achieving basic functioning of people, institutions, organizations, and systems in the short to medium term.[5]

Coping capacity
The ability of people, institutions, organizations, and systems, using available skills, values, beliefs, resources, and opportunities, to address, manage, and overcome adverse conditions in the short to medium term.[6]

Coral bleaching
Loss of coral pigmentation through the loss of intracellular symbiotic algae (known as zooxanthellae) and/or loss of their pigments.

Cryosphere
All regions on and beneath the surface of the Earth and ocean where water is in solid form, including sea ice, lake ice, river ice, snow cover, glaciers and ice sheets, and frozen ground (which includes permafrost).

Cultural impacts
Impacts on material and ecological aspects of culture and the lived experience of culture, including dimensions such as identity, community cohesion and belonging, sense of place, worldview, values, perceptions, and tradition. Cultural impacts are closely related to ecological impacts, especially for iconic and representational dimensions of species and landscapes. Culture and cultural practices frame the importance and value of the impacts of change, shape the feasibility and acceptability of adaptation options, and provide the skills and practices that enable adaptation.

Dead zones
Extremely hypoxic (i.e., low-oxygen) areas in oceans and lakes, caused by excessive nutrient input from human activities coupled with other factors that deplete the oxygen required to support many marine organisms in bottom and near-bottom water. See also Eutrophication and Hypoxic events.

Decarbonization
The process by which countries or other entities aim to achieve a low-carbon economy, or by which individuals aim to reduce their consumption of carbon.

Deforestation
Conversion of forest to non-forest. For a discussion of the term *forest* and related terms such as *afforestation*, *reforestation*, and *deforestation* see the IPCC Special Report on Land Use, Land-Use Change, and Forestry (IPCC, 2000). See also the report on Definitions and Methodological Options to Inventory Emissions from Direct Human-induced Degradation of Forests and Devegetation of Other Vegetation Types (IPCC, 2003).

Desertification
Land degradation in arid, semi-arid, and dry sub-humid areas resulting from various factors, including climatic variations and human activities. Land degradation in arid, semi-arid, and dry sub-humid areas is reduction or loss of the biological or economic productivity and complexity of rainfed cropland, irrigated cropland, or range, pasture, forest, and woodlands resulting from land uses or from a process or combination of processes, including processes arising from human activities and habitation patterns, such as (1) soil erosion caused by wind and/or water; (2) deterioration of the physical, chemical, biological, or economic properties of soil; and (3) long-term loss of natural vegetation (UNCCD, 1994).

---

[5] This glossary entry builds from the definition used in UNISDR (2009) and IPCC (2012a).
[6] This glossary entry builds from the definition used in UNISDR (2009) and IPCC (2012a).

Detection and attribution
Detection of change is defined as the process of demonstrating that climate or a system affected by climate has changed in some defined statistical sense, without providing a reason for that change. An identified change is detected in observations if its likelihood of occurrence by chance due to internal variability alone is determined to be small, for example, <10%. Attribution is defined as the process of evaluating the relative contributions of multiple causal factors to a change or event with an assignment of statistical confidence (Hegerl et al., 2010).

Detection of impacts of climate change
For a natural, human, or managed system, identification of a change from a specified baseline. The baseline characterizes behavior in the absence of climate change and may be stationary or non-stationary (e.g., due to land use change).

Disadvantaged populations
Sectors of a society that are marginalized, often because of low socioeconomic status, low income, lack of access to basic services such as health or education, lack of power, race, gender, religion, or poor access to communication technologies.

Disaster
Severe alterations in the normal functioning of a community or a society due to hazardous physical events interacting with vulnerable social conditions, leading to widespread adverse human, material, economic, or environmental effects that require immediate emergency response to satisfy critical human needs and that may require external support for recovery.

Disaster management
Social processes for designing, implementing, and evaluating strategies, policies, and measures that promote and improve disaster preparedness, response, and recovery practices at different organizational and societal levels.

Disaster risk
The likelihood within a specific time period of disaster. See Disaster.

Disaster Risk Management (DRM)
Processes for designing, implementing, and evaluating strategies, policies, and measures to improve the understanding of disaster risk, foster disaster risk reduction and transfer, and promote continuous improvement in disaster preparedness, response, and recovery practices, with the explicit purpose of increasing human security, well-being, quality of life, and sustainable development.

Disaster Risk Reduction (DRR)
Denotes both a policy goal or objective, and the strategic and instrumental measures employed for anticipating future disaster risk; reducing existing exposure, hazard, or vulnerability; and improving resilience.

Discounting
A mathematical operation making monetary (or other) amounts received or expended at different times (years) comparable across time. The discounter uses a fixed or possibly time-varying discount rate (>0) from year to year that makes future value worth less today.

Disturbance regime
Frequency, intensity, and types of disturbances of ecological systems, such as fires, insect or pest outbreaks, floods, and droughts.

Diurnal temperature range
The difference between the maximum and minimum temperature during a 24-hour period.

Downscaling
Downscaling is a method that derives local- to regional-scale (10 to 100 km) information from larger-scale models or data analyses. Two main methods exist: dynamical downscaling and empirical/statistical downscaling. The dynamical method uses the output of regional climate models, global models with variable spatial resolution, or high-resolution global models. The empirical/statistical methods develop statistical relationships that link the large-scale atmospheric variables with local/regional climate variables. In all cases, the quality of the driving model remains an important limitation on quality of the downscaled information.

Drought
A period of abnormally dry weather long enough to cause a serious hydrological imbalance. Drought is a relative term; therefore any discussion in terms of precipitation deficit must refer to the particular precipitation-related activity that is under discussion. For example, shortage of precipitation during the growing season impinges on crop production or ecosystem function in general (due to soil moisture drought, also termed agricultural drought), and during the runoff and percolation season primarily affects water supplies (hydrological drought). Storage changes in soil moisture and groundwater are also affected by increases in actual evapotranspiration in addition to reductions in precipitation. A period with an abnormal precipitation deficit is defined as a meteorological drought. A megadrought is a very lengthy and pervasive drought, lasting much longer than normal, usually a decade or more. For the corresponding indices, see WGI AR5 Box 2.4.

Dynamic Global Vegetation Model (DGVM)
A model that simulates vegetation development and dynamics through space and time, as driven by climate and other environmental changes.

Early warning system
The set of capacities needed to generate and disseminate timely and meaningful warning information to enable individuals, communities, and organizations threatened by a hazard to prepare to act promptly and appropriately to reduce the possibility of harm or loss.[7]

Earth System Model (ESM)
A coupled atmosphere-ocean general circulation model in which a representation of the carbon cycle is included, allowing for interactive calculation of atmospheric CO_2 or compatible emissions. Additional components (e.g., atmospheric chemistry, ice sheets, dynamic vegetation, nitrogen cycle, but also urban or crop models) may be included. See also Climate model.

[7] This glossary entry builds from the definition used in UNISDR (2009) and IPCC (2012a).

Ecophysiological process
Processes in which individual organisms respond continuously to environmental variability or change, such as climate change, generally at a microscopic or sub-organ scale. Ecophysiological mechanisms underpin individual organisms' tolerance to environmental stress, and comprise a broad range of responses defining the absolute tolerances by individuals of environmental conditions. Ecophysiological responses may scale up to control species' geographic ranges.

Ecosystem
A functional unit consisting of living organisms, their non-living environment, and the interactions within and between them. The components included in a given ecosystem and its spatial boundaries depend on the purpose for which the ecosystem is defined: in some cases they are relatively sharp, while in others they are diffuse. Ecosystem boundaries can change over time. Ecosystems are nested within other ecosystems, and their scale can range from very small to the entire biosphere. In the current era, most ecosystems either contain people as key organisms, or are influenced by the effects of human activities in their environment.

Ecosystem approach
A strategy for the integrated management of land, water, and living resources that promotes conservation and sustainable use in an equitable way. An ecosystem approach is based on the application of scientific methodologies focused on levels of biological organization, which encompass the essential structure, processes, functions, and interactions of organisms and their environment. It recognizes that humans, with their cultural diversity, are an integral component of many ecosystems. The ecosystem approach requires adaptive management to deal with the complex and dynamic nature of ecosystems and the absence of complete knowledge or understanding of their functioning. Priority targets are conservation of biodiversity and of the ecosystem structure and functioning, in order to maintain ecosystem services.[8]

Ecosystem-based adaptation
The use of biodiversity and ecosystem services as part of an overall adaptation strategy to help people to adapt to the adverse effects of climate change. Ecosystem-based adaptation uses the range of opportunities for the sustainable management, conservation, and restoration of ecosystems to provide services that enable people to adapt to the impacts of climate change. It aims to maintain and increase the resilience and reduce the vulnerability of ecosystems and people in the face of the adverse effects of climate change. Ecosystem-based adaptation is most appropriately integrated into broader adaptation and development strategies (CBD, 2009).

Ecosystem services
Ecological processes or functions having monetary or non-monetary value to individuals or society at large. These are frequently classified as (1) supporting services such as productivity or biodiversity maintenance, (2) provisioning services such as food, fiber, or fish, (3) regulating services such as climate regulation or carbon sequestration, and (4) cultural services such as tourism or spiritual and aesthetic appreciation.

El Niño-Southern Oscillation (ENSO)
The term El Niño was initially used to describe a warm-water current that periodically flows along the coast of Ecuador and Peru, disrupting the local fishery. It has since become identified with a basin-wide warming of the tropical Pacific Ocean east of the dateline. This oceanic event is associated with a fluctuation of a global-scale tropical and subtropical surface pressure pattern called the Southern Oscillation. This coupled atmosphere-ocean phenomenon, with preferred time scales of 2 to about 7 years, is known as the El Niño-Southern Oscillation (ENSO). It is often measured by the surface pressure anomaly difference between Tahiti and Darwin or the sea surface temperatures in the central and eastern equatorial Pacific. During an ENSO event, the prevailing trade winds weaken, reducing upwelling and altering ocean currents such that the sea surface temperatures warm, further weakening the trade winds. This event has a great impact on the wind, sea surface temperature, and precipitation patterns in the tropical Pacific. It has climatic effects throughout the Pacific region and in many other parts of the world, through global teleconnections. The cold phase of ENSO is called La Niña. For the corresponding indices, see WGI AR5 Box 2.5.

Emergent risk
A risk that arises from the interaction of phenomena in a complex system, for example, the risk caused when geographic shifts in human population in response to climate change lead to increased vulnerability and exposure of populations in the receiving region.

Emission scenario
A plausible representation of the future development of emissions of substances that are potentially radiatively active (e.g., greenhouse gases, aerosols) based on a coherent and internally consistent set of assumptions about driving forces (such as demographic and socioeconomic development, technological change) and their key relationships. Concentration scenarios, derived from emission scenarios, are used as input to a climate model to compute climate projections. In IPCC (1992) a set of emission scenarios was presented, which were used as a basis for the climate projections in IPCC (1996). These emission scenarios are referred to as the IS92 scenarios. In the IPCC Special Report on Emissions Scenarios (Nakićenović and Swart, 2000) emission scenarios, the so-called SRES scenarios, were published, some of which were used, among others, as a basis for the climate projections presented in Chapters 9 to 11 of IPCC (2001) and Chapters 10 and 11 of IPCC (2007). New emission scenarios for climate change, the four Representative Concentration Pathways, were developed for, but independently of, the present IPCC assessment. See also Climate scenario and Scenario.

Ensemble
A collection of model simulations characterizing a climate prediction or projection. Differences in initial conditions and model formulation result in different evolutions of the modeled system and may give information on uncertainty associated with model error and error in initial conditions in the case of climate forecasts and on uncertainty associated with model error and with internally generated climate variability in the case of climate projections.

[8] This glossary entry builds from definitions used in CBD (2000), MEA (2005), and the Fourth Assessment Report.

Environmental migration
Human migration involves movement over a significant distance and duration. Environmental migration refers to human migration where environmental risks or environmental change plays a significant role in influencing the migration decision and destination. Migration may involve distinct categories such as direct, involuntary, and temporary displacement due to weather-related disasters; voluntary relocation as settlements and economies become less viable; or planned resettlement encouraged by government actions or incentives. All migration decisions are multi-causal, and hence it is not meaningful to describe any migrant flow as being solely for environmental reasons.

Environmental services
See Ecosystem services.

Eutrophication
Over-enrichment of water by nutrients such as nitrogen and phosphorus. It is one of the leading causes of water quality impairment. The two most acute symptoms of eutrophication are hypoxia (or oxygen depletion) and harmful algal blooms. See also Dead zones.

Evolutionary adaptation
For a population or species, change in functional characteristics as a result of selection acting on heritable traits. The rate of evolutionary adaptation depends on factors such as strength of selection, generation turnover time, and degree of outcrossing (as opposed to inbreeding). See also Adaptation.

Exposure
The presence of people, livelihoods, species or ecosystems, environmental functions, services, and resources, infrastructure, or economic, social, or cultural assets in places and settings that could be adversely affected.

External forcing
External forcing refers to a forcing agent outside the climate system causing a change in the climate system. Volcanic eruptions, solar variations, and anthropogenic changes in the composition of the atmosphere and land use change are external forcings. Orbital forcing is also an external forcing as the insolation changes with orbital parameters eccentricity, tilt, and precession of the equinox.

Externalities/external costs/external benefits
Externalities arise from a human activity when agents responsible for the activity do not take full account of the activity's impacts on others' production and consumption possibilities, and no compensation exists for such impacts. When the impacts are negative, they are external costs. When the impacts are positive, they are external benefits.

Extratropical cyclone
A large-scale (of order 1000 km) storm in the middle or high latitudes having low central pressure and fronts with strong horizontal gradients in temperature and humidity. A major cause of extreme wind speeds and heavy precipitation especially in wintertime.

Extreme climate event
See Extreme weather event.

Extreme sea level
See Storm surge.

Extreme weather event
An extreme weather event is an event that is rare at a particular place and time of year. Definitions of rare vary, but an extreme weather event would normally be as rare as or rarer than the 10th or 90th percentile of a probability density function estimated from observations. By definition, the characteristics of what is called extreme weather may vary from place to place in an absolute sense. When a pattern of extreme weather persists for some time, such as a season, it may be classed as an extreme climate event, especially if it yields an average or total that is itself extreme (e.g., drought or heavy rainfall over a season).

Famine
Scarcity of food over an extended period and over a large geographical area, such as a country, or lack of access to food for socioeconomic, political, or cultural reasons. Famines may be caused by climate-related extreme events such as droughts or floods and by disease, war, or other factors.

Feedback
See Climate feedback.

Fire weather
Weather conditions conducive to triggering and sustaining wild fires, usually based on a set of indicators and combinations of indicators including temperature, soil moisture, humidity, and wind. Fire weather does not include the presence or absence of fuel load.

Fitness (Darwinian)
Fitness is the relative capacity of an individual or genotype to both survive and reproduce, quantified as the average contribution of the genotype to the gene pool of the next generations. During evolution, natural selection favors functions providing greater fitness such that the functions become more common over generations.

Flood
The overflowing of the normal confines of a stream or other body of water, or the accumulation of water over areas not normally submerged. Floods include river (fluvial) floods, flash floods, urban floods, pluvial floods, sewer floods, coastal floods, and glacial lake outburst floods.

Food security
A state that prevails when people have secure access to sufficient amounts of safe and nutritious food for normal growth, development, and an active and healthy life.[9] See also Access to food.

Food system
A food system includes the suite of activities and actors in the food chain (i.e., producing, processing and packaging, storing and transporting, trading and retailing, and preparing and consuming food); and the outcome of these activities relating to the three components underpinning food security (i.e., access to food, utilization of food, and food availability), all of which need to be stable over time. Food security is therefore

[9] This glossary entry builds from definitions used in FAO (2000) and previous IPCC reports.

underpinned by food systems, and is an emergent property of the behavior of the whole food system. Food insecurity arises when any aspect of the food system is stressed.

Forecast
See Climate prediction and Climate projection.

General Circulation Model (GCM)
See Climate model.

Geoengineering
Geoengineering refers to a broad set of methods and technologies that aim to deliberately alter the climate system in order to alleviate the impacts of climate change. Most, but not all, methods seek to either (1) reduce the amount of absorbed solar energy in the climate system (Solar Radiation Management) or (2) increase net carbon sinks from the atmosphere at a scale sufficiently large to alter climate (Carbon Dioxide Removal). Scale and intent are of central importance. Two key characteristics of geoengineering methods of particular concern are that they use or affect the climate system (e.g., atmosphere, land, or ocean) globally or regionally and/or could have substantive unintended effects that cross national boundaries. Geoengineering is different from weather modification and ecological engineering, but the boundary can be fuzzy (IPCC, 2012b, p. 2).

Global change
A generic term to describe global scale changes in systems, including the climate system, ecosystems, and social-ecological systems.

Global Climate Model (also referred to as General Circulation Model, both abbreviated as GCM)
See Climate model.

Global mean surface temperature
An estimate of the global mean surface air temperature. However, for changes over time, only anomalies, as departures from a climatology, are used, most commonly based on the area-weighted global average of the sea surface temperature anomaly and land surface air temperature anomaly.

Greenhouse effect
The infrared radiative effect of all infrared-absorbing constituents in the atmosphere. Greenhouse gases, clouds, and (to a small extent) aerosols absorb terrestrial radiation emitted by the Earth's surface and elsewhere in the atmosphere. These substances emit infrared radiation in all directions, but, everything else being equal, the net amount emitted to space is normally less than would have been emitted in the absence of these absorbers because of the decline of temperature with altitude in the troposphere and the consequent weakening of emission. An increase in the concentration of greenhouse gases increases the magnitude of this effect; the difference is sometimes called the enhanced greenhouse effect. The change in a greenhouse gas concentration because of anthropogenic emissions contributes to an instantaneous radiative forcing. Surface temperature and troposphere warm in response to this forcing, gradually restoring the radiative balance at the top of the atmosphere.

Greenhouse gas (GHG)
Greenhouse gases are those gaseous constituents of the atmosphere, both natural and anthropogenic, that absorb and emit radiation at specific wavelengths within the spectrum of terrestrial radiation emitted by the Earth's surface, the atmosphere itself, and clouds. This property causes the greenhouse effect. Water vapor (H_2O), carbon dioxide (CO_2), nitrous oxide (N_2O), methane (CH_4), and ozone (O_3) are the primary greenhouse gases in the Earth's atmosphere. Moreover, there are a number of entirely human-made greenhouse gases in the atmosphere, such as the halocarbons and other chlorine- and bromine-containing substances, dealt with under the Montreal Protocol. Beside CO_2, N_2O, and CH_4, the Kyoto Protocol deals with the greenhouse gases sulfur hexafluoride (SF_6), hydrofluorocarbons (HFCs), and perfluorocarbons (PFCs). For a list of well-mixed greenhouse gases, see WGI AR5 Table 2.SM.1.

Ground-level ozone
Atmospheric ozone formed naturally or from human-emitted precursors near Earth's surface, thus affecting human health, agriculture, and ecosystems. Ozone is a greenhouse gas, but ground-level ozone, unlike stratospheric ozone, also directly affects organisms at the surface. Ground-level ozone is sometimes referred to as tropospheric ozone, although much of the troposphere is well above the surface and thus does not directly expose organisms at the surface. See also Ozone.

Groundwater recharge
The process by which external water is added to the zone of saturation of an aquifer, either directly into a geologic formation that traps the water or indirectly by way of another formation.

Hazard
The potential occurrence of a natural or human-induced physical event or trend or physical impact that may cause loss of life, injury, or other health impacts, as well as damage and loss to property, infrastructure, livelihoods, service provision, ecosystems, and environmental resources. In this report, the term hazard usually refers to climate-related physical events or trends or their physical impacts.

Heat wave
A period of abnormally and uncomfortably hot weather.

Hotspot
A geographical area characterized by high vulnerability and exposure to climate change.

Human security
A condition that is met when the vital core of human lives is protected, and when people have the freedom and capacity to live with dignity. In the context of climate change, the vital core of human lives includes the universal and culturally specific, material and non-material elements necessary for people to act on behalf of their interests and to live with dignity.

Human system
Any system in which human organizations and institutions play a major role. Often, but not always, the term is synonymous with society or

social system. Systems such as agricultural systems, political systems, technological systems, and economic systems are all human systems in the sense applied in this report.

Hydrological cycle
The cycle in which water evaporates from the oceans and the land surface, is carried over the Earth in atmospheric circulation as water vapor, condenses to form clouds, precipitates over ocean and land as rain or snow, which on land can be intercepted by trees and vegetation, provides runoff on the land surface, infiltrates into soils, recharges groundwater, discharges into streams, and ultimately, flows out into the oceans, from which it will eventually evaporate again. The various systems involved in the hydrological cycle are usually referred to as hydrological systems.

Hypoxic events
Events that lead to deficiencies of oxygen in water bodies. See also Dead zones and Eutrophication.

Ice cap
A dome-shaped ice mass that is considerably smaller in extent than an ice sheet.

Ice sheet
A mass of land ice of continental size that is sufficiently thick to cover most of the underlying bed, so that its shape is mainly determined by its dynamics (the flow of the ice as it deforms internally and/or slides at its base). An ice sheet flows outward from a high central ice plateau with a small average surface slope. The margins usually slope more steeply, and most ice is discharged through fast flowing ice streams or outlet glaciers, in some cases into the sea or into ice shelves floating on the sea. There are only two ice sheets in the modern world, one on Greenland and one on Antarctica. During glacial periods there were others.

Ice shelf
A floating slab of ice of considerable thickness extending from the coast (usually of great horizontal extent with a very gently sloping surface), often filling embayments in the coastline of an ice sheet. Nearly all ice shelves are in Antarctica, where most of the ice discharged into the ocean flows via ice shelves.

(climate change) Impact assessment
The practice of identifying and evaluating, in monetary and/or non-monetary terms, the effects of climate change on natural and human systems.

Impacts (Consequences, Outcomes)[10]
Effects on natural and human systems. In this report, the term impacts is used primarily to refer to the effects on natural and human systems of extreme weather and climate events and of climate change. Impacts generally refer to effects on lives, livelihoods, health, ecosystems, economies, societies, cultures, services, and infrastructure due to the interaction of climate changes or hazardous climate events occurring within a specific time period and the vulnerability of an exposed society or system. Impacts are also referred to as consequences and outcomes. The impacts of climate change on geophysical systems, including floods, droughts, and sea level rise, are a subset of impacts called physical impacts.

Income
The maximum amount that a household, or other unit, can consume without reducing its real net worth. Total income is the broadest measure of income and refers to regular receipts such as wages and salaries, income from self-employment, interest and dividends from invested funds, pensions or other benefits from social insurance, and other current transfers receivable.[11]

Indian Ocean Dipole (IOD)
Large-scale mode of interannual variability of sea surface temperature in the Indian Ocean. This pattern manifests through a zonal gradient of tropical sea surface temperature, which in one extreme phase in boreal autumn shows cooling off Sumatra and warming off Somalia in the west, combined with anomalous easterlies along the equator.

Indigenous peoples
Indigenous peoples and nations are those that, having a historical continuity with pre-invasion and pre-colonial societies that developed on their territories, consider themselves distinct from other sectors of the societies now prevailing on those territories, or parts of them. They form at present principally non-dominant sectors of society and are often determined to preserve, develop, and transmit to future generations their ancestral territories, and their ethnic identity, as the basis of their continued existence as peoples, in accordance with their own cultural patterns, social institutions, and common law system.[12]

Industrial Revolution
A period of rapid industrial growth with far-reaching social and economic consequences, beginning in Britain during the second half of the 18th century and spreading to Europe and later to other countries including the United States. The invention of the steam engine was an important trigger of this development. The industrial revolution marks the beginning of a strong increase in the use of fossil fuels and emission of, in particular, fossil carbon dioxide. In this report the terms *preindustrial* and *industrial* refer, somewhat arbitrarily, to the periods before and after 1750, respectively.

Industrialized/developed/developing countries
There are a diversity of approaches for categorizing countries on the basis of their level of development, and for defining terms such as industrialized, developed, or developing. Several categorizations are used in this report. In the United Nations system, there is no established convention for the designation of developed and developing countries or areas. The United Nations Statistics Division specifies developed and developing regions based on common practice. In addition, specific countries are designated as least developed countries, landlocked

[10] Reflecting progress in science, this glossary entry differs in breadth and focus from the entry used in the Fourth Assessment Report and other IPCC reports.
[11] This glossary entry builds from the definition used in OECD (2003).
[12] This glossary entry builds from the definitions used in Cobo (1987) and previous IPCC reports.

developing countries, small island developing states, and transition economies. Many countries appear in more than one of these categories. The World Bank uses income as the main criterion for classifying countries as low, lower middle, upper middle, and high income. The UNDP aggregates indicators for life expectancy, educational attainment, and income into a single composite human development index (HDI) to classify countries as low, medium, high, or very high human development. See Box 1-2.

Informal sector
Commercial enterprises (mostly small) that are not registered or that otherwise fall outside official rules and regulations. Among the businesses that make up the informal sector, there is great diversity in the value of the goods or services produced, the numbers employed, the extent of illegality, and the connection to the formal sector. Many informal enterprises have some characteristics of formal-sector enterprises, and some people are in informal employment in the formal sector as they lack legal protection or employment benefits.

Informal settlement
A term given to settlements or residential areas that by at least one criterion fall outside official rules and regulations. Most informal settlements have poor housing (with widespread use of temporary materials) and are developed on land that is occupied illegally with high levels of overcrowding. In most such settlements, provision for safe water, sanitation, drainage, paved roads, and basic services is inadequate or lacking. The term *slum* is often used for informal settlements, although it is misleading as many informal settlements develop into good quality residential areas, especially where governments support such development.

Institutions
Institutions are rules and norms held in common by social actors that guide, constrain, and shape human interaction. Institutions can be formal, such as laws and policies, or informal, such as norms and conventions. Organizations—such as parliaments, regulatory agencies, private firms, and community bodies—develop and act in response to institutional frameworks and the incentives they frame. Institutions can guide, constrain, and shape human interaction through direct control, through incentives, and through processes of socialization.

Insurance/reinsurance
A family of financial instruments for sharing and transferring risk among a pool of at-risk households, businesses, and/or governments. See also Risk transfer.

Integrated assessment
A method of analysis that combines results and models from the physical, biological, economic, and social sciences, and the interactions among these components, in a consistent framework to evaluate the status and the consequences of environmental change and the policy responses to it.

Integrated Coastal Zone Management (ICZM)
An integrated approach for sustainably managing coastal areas, taking into account all coastal habitats and uses.

Invasive species/Invasive Alien Species (IAS)
A species introduced outside its natural past or present distribution (i.e., an alien species) that becomes established in natural or semi-natural ecosystems or habitat, is an agent of change, and threatens native biological diversity (IUCN, 2000; CBD, 2002).

Key vulnerability, Key risk, Key impact
A vulnerability, risk, or impact relevant to the definition and elaboration of "dangerous anthropogenic interference (DAI) with the climate system," in the terminology of United Nations Framework Convention on Climate Change (UNFCCC) Article 2, meriting particular attention by policy makers in that context.

Key risks are potentially severe adverse consequences for humans and social-ecological systems resulting from the interaction of climate-related hazards with vulnerabilities of societies and systems exposed. Risks are considered "key" due to high hazard or high vulnerability of societies and systems exposed, or both.

Vulnerabilities are considered "key" if they have the potential to combine with hazardous events or trends to result in key risks. Vulnerabilities that have little influence on climate-related risk, for instance, due to lack of exposure to hazards, would not be considered key.

Key impacts are severe consequences for humans and social-ecological systems.

Land grabbing
Large acquisitions of land or water rights for industrial agriculture, mitigation projects, or biofuels that have negative consequences on local and marginalized communities.

Land surface air temperature
The surface air temperature as measured in well-ventilated screens over land at 1.5 m above the ground.

Land use and Land use change
Land use refers to the total of arrangements, activities, and inputs undertaken in a certain land cover type (a set of human actions). The term *land use* is also used in the sense of the social and economic purposes for which land is managed (e.g., grazing, timber extraction, and conservation). Land use change refers to a change in the use or management of land by humans, which may lead to a change in land cover. Land cover and land use change may have an impact on the surface albedo, evapotranspiration, sources and sinks of greenhouse gases, or other properties of the climate system and may thus give rise to radiative forcing and/or other impacts on climate, locally or globally. See also the IPCC Special Report on Land Use, Land-Use Change, and Forestry (IPCC, 2000).

La Niña
See El Niño-Southern Oscillation.

Last Glacial Maximum (LGM)
The period during the last ice age when the glaciers and ice sheets reached their maximum extent, approximately 21 ka ago. This period

has been widely studied because the radiative forcings and boundary conditions are relatively well known.

Likelihood
The chance of a specific outcome occurring, where this might be estimated probabilistically. Likelihood is expressed in this report using a standard terminology (Mastrandrea et al., 2010), defined in Box 1-1. See also Confidence and Uncertainty.

Livelihood
The resources used and the activities undertaken in order to live. Livelihoods are usually determined by the entitlements and assets to which people have access. Such assets can be categorized as human, social, natural, physical, or financial.

Low regrets policy
A policy that would generate net social and/or economic benefits under current climate and a range of future climate change scenarios.

Maladaptive actions (Maladaptation)
Actions that may lead to increased risk of adverse climate-related outcomes, increased vulnerability to climate change, or diminished welfare, now or in the future.

Mean sea level
The surface level of the ocean at a particular point averaged over an extended period of time such as a month or year. Mean sea level is often used as a national datum to which heights on land are referred.

Meridional Overturning Circulation (MOC)
Meridional (north-south) overturning circulation in the ocean quantified by zonal (east-west) sums of mass transports in depth or density layers. In the North Atlantic, away from the subpolar regions, the MOC (which is in principle an observable quantity) is often identified with the thermohaline circulation (THC), which is a conceptual and incomplete interpretation. It must be borne in mind that the MOC is also driven by wind, and can also include shallower overturning cells such as occur in the upper ocean in the tropics and subtropics, in which warm (light) waters moving poleward are transformed to slightly denser waters and subducted equatorward at deeper levels. See also Thermohaline circulation.

Microclimate
Local climate at or near the Earth's surface. See also Climate.

Mitigation (of climate change)
A human intervention to reduce the sources or enhance the sinks of greenhouse gases.

Mitigation (of disaster risk and disaster)
The lessening of the potential adverse impacts of physical hazards (including those that are human-induced) through actions that reduce hazard, exposure, and vulnerability.

Mode of climate variability
Underlying space-time structure with preferred spatial pattern and temporal variation that helps account for the gross features in variance and for teleconnections. A mode of variability is often considered to be the product of a spatial climate pattern and an associated climate index time series.

Monsoon
A monsoon is a tropical and subtropical seasonal reversal in both the surface winds and associated precipitation, caused by differential heating between a continental-scale land mass and the adjacent ocean. Monsoon rains occur mainly over land in summer.

Non-climatic driver (Non-climate driver)
An agent or process outside the climate system that influences a human or natural system.

Nonlinearity
A process is called nonlinear when there is no simple proportional relation between cause and effect. The climate system contains many such nonlinear processes, resulting in a system with potentially very complex behavior. Such complexity may lead to abrupt climate change. See also Predictability.

North Atlantic Oscillation (NAO)
The North Atlantic Oscillation consists of opposing variations of surface pressure near Iceland and near the Azores. It therefore corresponds to fluctuations in the strength of the main westerly winds across the Atlantic into Europe, and thus to fluctuations in the embedded extratropical cyclones with their associated frontal systems. See NAO Index in WGI AR5 Box 2.5.

Ocean acidification
Ocean acidification refers to a reduction in the pH of the ocean over an extended period, typically decades or longer, which is caused primarily by uptake of carbon dioxide from the atmosphere, but can also be caused by other chemical additions or subtractions from the ocean. Anthropogenic ocean acidification refers to the component of pH reduction that is caused by human activity (IPCC, 2011, p. 37).

Opportunity costs
The benefits of an activity forgone through the choice of another activity.

Outcome vulnerability (End-point vulnerability)
Vulnerability as the end point of a sequence of analyses beginning with projections of future emission trends, moving on to the development of climate scenarios, and concluding with biophysical impact studies and the identification of adaptive options. Any residual consequences that remain after adaptation has taken place define the levels of vulnerability (Kelly and Adger, 2000; O'Brien et al., 2007).

Oxygen Minimum Zone (OMZ)
The midwater layer (200 to 1000 m) in the open ocean in which oxygen saturation is the lowest in the ocean. The degree of oxygen depletion depends on the largely bacterial consumption of organic matter, and the distribution of the OMZs is influenced by large-scale ocean circulation. In coastal oceans, OMZs extend to the shelves and may also affect benthic ecosystems.

Ozone
Ozone, the triatomic form of oxygen (O_3), is a gaseous atmospheric constituent. In the troposphere, it is created both naturally and by photochemical reactions involving gases resulting from human activities (smog). Tropospheric ozone acts as a greenhouse gas. In the stratosphere, it is created by the interaction between solar ultraviolet radiation and molecular oxygen (O_2). Stratospheric ozone plays a dominant role in the stratospheric radiative balance. Its concentration is highest in the ozone layer.

Pacific Decadal Oscillation (PDO)
The pattern and time series of the first empirical orthogonal function of sea surface temperature over the North Pacific north of 20°N. The PDO broadened to cover the whole Pacific Basin is known as the Inter-decadal Pacific Oscillation (IPO). The PDO and IPO exhibit similar temporal evolution.

Parameterization
In climate models, this term refers to the technique of representing processes that cannot be explicitly resolved at the spatial or temporal resolution of the model (sub-grid scale processes) by relationships between model-resolved larger-scale variables and the area- or time-averaged effect of such sub-grid scale processes.

Particulates
Very small solid particles emitted during the combustion of fossil and biomass fuels. Particulates may consist of a wide variety of substances. Of greatest concern for health are particulates of diameter less than or equal to 10 nm, usually designated as PM_{10}.

Pastoralism
A livelihood strategy based on moving livestock to seasonal pastures primarily in order to convert grasses, forbs, tree leaves, or crop residues into human food. The search for feed is however not the only reason for mobility; people and livestock may move to avoid various natural and/or social hazards, to avoid competition with others, or to seek more favorable conditions. Pastoralism can also be thought of as a strategy that is shaped by both social and ecological factors concerning uncertainty and variability of precipitation, and low and unpredictable productivity of terrestrial ecosystems.

Path dependence
The generic situation where decisions, events, or outcomes at one point in time constrain adaptation, mitigation, or other actions or options at a later point in time.

Permafrost
Ground (soil or rock and included ice and organic material) that remains at or below 0°C for at least 2 consecutive years.

Persistent Organic Pollutants (POPs)
Toxic organic chemical substances that persist in the environment for long periods of time, are transported and deposited in locations distant from their sources of release, bioaccumulate, and can have adverse effects on human health and ecosystems.[13]

Phenology
The relationship between biological phenomena that recur periodically (e.g., development stages, migration) and climate and seasonal changes.

Photochemical smog
A mix of oxidizing air pollutants produced by the reaction of sunlight with primary air pollutants, especially hydrocarbons.

Poverty
Poverty is a complex concept with several definitions stemming from different schools of thought. It can refer to material circumstances (such as need, pattern of deprivation, or limited resources), economic conditions (such as standard of living, inequality, or economic position), and/or social relationships (such as social class, dependency, exclusion, lack of basic security, or lack of entitlement).

Poverty trap
Poverty trap is understood differently across disciplines. In the social sciences, the concept, primarily employed at the individual, household, or community level, describes a situation in which escaping poverty becomes impossible due to unproductive or inflexible resources. A poverty trap can also be seen as a critical minimum asset threshold, below which families are unable to successfully educate their children, build up their productive assets, and get out of poverty. Extreme poverty is itself a poverty trap, since poor persons lack the means to participate meaningfully in society. In economics, the term *poverty trap* is often used at national scales, referring to a self-perpetuating condition where an economy, caught in a vicious cycle, suffers from persistent underdevelopment (Matsuyama, 2008). Many proposed models of poverty traps are found in the literature.

Predictability
The extent to which future states of a system may be predicted based on knowledge of current and past states of the system. Because knowledge of the climate system's past and current states is generally imperfect, as are the models that utilize this knowledge to produce a climate prediction, and because the climate system is inherently nonlinear and chaotic, predictability of the climate system is inherently limited. Even with arbitrarily accurate models and observations, there may still be limits to the predictability of such a nonlinear system (AMS, 2000).

Preindustrial
See Industrial Revolution.

Probability Density Function (PDF)
A probability density function is a function that indicates the relative chances of occurrence of different outcomes of a variable. The function integrates to unity over the domain for which it is defined and has the property that the integral over a sub-domain equals the probability that the outcome of the variable lies within that sub-domain. For example, the probability that a temperature anomaly defined in a particular way is greater than zero is obtained from its PDF by integrating the PDF over all possible temperature anomalies greater than zero. Probability density functions that describe two or more variables simultaneously are similarly defined.

[13] This glossary entry builds from the definition in the Stockholm Convention on Persistent Organic Pollutants (Secretariat of the Stockholm Convention, 2001).

Projection
A projection is a potential future evolution of a quantity or set of quantities, often computed with the aid of a model. Unlike predictions, projections are conditional on assumptions concerning, for example, future socioeconomic and technological developments that may or may not be realized. See also Climate prediction and Climate projection.

Proxy
A proxy climate indicator is a record that is interpreted, using physical and biophysical principles, to represent some combination of climate-related variations back in time. Climate-related data derived in this way are referred to as proxy data. Examples of proxies include pollen analysis, tree ring records, speleothems, characteristics of corals, and various data derived from marine sediments and ice cores. Proxy data can be calibrated to provide quantitative climate information.

Public good
A good that is both non-excludable and non-rivalrous in that individuals cannot be effectively excluded from use and where use by one individual does not reduce availability to others.

Radiative forcing
Radiative forcing is the change in the net, downward minus upward, radiative flux (expressed in W m^{-2}) at the tropopause or top of atmosphere due to a change in an external driver of climate change, such as a change in the concentration of carbon dioxide or the output of the Sun. Sometimes internal drivers are still treated as forcings even though they result from the alteration in climate, for example aerosol or greenhouse gas changes in paleoclimates. The traditional radiative forcing is computed with all tropospheric properties held fixed at their unperturbed values, and after allowing for stratospheric temperatures, if perturbed, to readjust to radiative-dynamical equilibrium. Radiative forcing is called instantaneous if no change in stratospheric temperature is accounted for. The radiative forcing once rapid adjustments are accounted for is termed the effective radiative forcing. For the purposes of this report, radiative forcing is further defined as the change relative to the year 1750 and, unless otherwise noted, refers to a global and annual average value. Radiative forcing is not to be confused with cloud radiative forcing, which describes an unrelated measure of the impact of clouds on the radiative flux at the top of the atmosphere.

Reanalysis
Reanalyses are estimates of historical atmospheric temperature and wind or oceanographic temperature and current, and other quantities, created by processing past meteorological or oceanographic data using fixed state-of-the-art weather forecasting or ocean circulation models with data assimilation techniques. Using fixed data assimilation avoids effects from the changing analysis system that occur in operational analyses. Although continuity is improved, global reanalyses still suffer from changing coverage and biases in the observing systems.

Reasons for concern
Elements of a classification framework, first developed in the IPCC Third Assessment Report, which aims to facilitate judgments about what level of climate change may be "dangerous" (in the language of Article 2 of the UNFCCC) by aggregating impacts, risks, and vulnerabilities.

Reference scenario
See Baseline/reference.

Reflexivity
A system attribute where cause and effect form a feedback loop, in which the effect changes the system itself. Self-adapting systems such as societies are inherently reflexive, as are planned changes in complex systems. Reflexive decision making in a social system has the potential to change the underpinning values that led to those decisions. Reflexivity is also an important aspect of adaptive management.

Reforestation
Planting of forests on lands that have previously contained forests but that have been converted to some other use. For a discussion of the term *forest* and related terms such as *afforestation*, *reforestation*, and *deforestation*, see the IPCC Special Report on Land Use, Land-Use Change, and Forestry (IPCC, 2000). See also the Report on Definitions and Methodological Options to Inventory Emissions from Direct Human-induced Degradation of Forests and Devegetation of Other Vegetation Types (IPCC, 2003).

Relative sea level
Sea level measured by a tide gauge with respect to the land upon which it is situated. See also Mean sea level and Sea level change.

Representative Concentration Pathways (RCPs)
Scenarios that include time series of emissions and concentrations of the full suite of greenhouse gases and aerosols and chemically active gases, as well as land use/land cover (Moss et al., 2008). The word *representative* signifies that each RCP provides only one of many possible scenarios that would lead to the specific radiative forcing characteristics. The term *pathway* emphasizes that not only the long-term concentration levels are of interest, but also the trajectory taken over time to reach that outcome (Moss et al., 2010).

RCPs usually refer to the portion of the concentration pathway extending up to 2100, for which Integrated Assessment Models produced corresponding emission scenarios. Extended Concentration Pathways (ECPs) describe extensions of the RCPs from 2100 to 2500 that were calculated using simple rules generated by stakeholder consultations, and do not represent fully consistent scenarios.

Four RCPs produced from Integrated Assessment Models were selected from the published literature and are used in the present IPCC Assessment as a basis for the climate predictions and projections in WGI AR5 Chapters 11 to 14:

RCP2.6 One pathway where radiative forcing peaks at approximately 3 W m^{-2} before 2100 and then declines (the corresponding ECP assuming constant emissions after 2100).

RCP4.5 and RCP6.0 Two intermediate stabilization pathways in which radiative forcing is stabilized at approximately 4.5 W m^{-2} and 6.0 W m^{-2} after 2100 (the corresponding ECPs assuming constant concentrations after 2150).

RCP8.5 One high pathway for which radiative forcing reaches greater than 8.5 W m^{-2} by 2100 and continues to rise for some amount of time (the corresponding ECP assuming constant emissions after 2100 and constant concentrations after 2250).

For further description of future scenarios, see WGI AR5 Box 1.1.

Resilience
The capacity of social, economic, and environmental systems to cope with a hazardous event or trend or disturbance, responding or reorganizing in ways that maintain their essential function, identity, and structure, while also maintaining the capacity for adaptation, learning, and transformation.[14]

Return period
An estimate of the average time interval between occurrences of an event (e.g., flood or extreme rainfall) of (or below/above) a defined size or intensity. See also Return value.

Return value
The highest (or, alternatively, lowest) value of a given variable, on average occurring once in a given period of time (e.g., in 10 years). See also Return period.

Risk
The potential for consequences where something of value is at stake and where the outcome is uncertain, recognizing the diversity of values.[15] Risk is often represented as probability of occurrence of hazardous events or trends multiplied by the impacts if these events or trends occur. Risk results from the interaction of vulnerability, exposure, and hazard. In this report, the term *risk* is used primarily to refer to the risks of climate-change impacts.

Risk assessment
The qualitative and/or quantitative scientific estimation of risks.

Risk management
Plans, actions, or policies to reduce the likelihood and/or consequences of risks or to respond to consequences.

Risk perception
The subjective judgment that people make about the characteristics and severity of a risk.

Risk transfer
The practice of formally or informally shifting the risk of financial consequences for particular negative events from one party to another.

Runoff
That part of precipitation that does not evaporate and is not transpired, but flows through the ground or over the ground surface and returns to bodies of water. See also Hydrological cycle.

Salt-water intrusion/encroachment
Displacement of fresh surface water or groundwater by the advance of salt water due to its greater density. This usually occurs in coastal and estuarine areas due to decreasing land-based influence (e.g., from reduced runoff or groundwater recharge, or from excessive water withdrawals from aquifers) or increasing marine influence (e.g., relative sea level rise).

Scenario
A plausible description of how the future may develop based on a coherent and internally consistent set of assumptions about key driving forces (e.g., rate of technological change, prices) and relationships. Note that scenarios are neither predictions nor forecasts, but are useful to provide a view of the implications of developments and actions. See also Climate scenario, Emission scenario, Representative Concentration Pathways, and SRES scenarios.

Sea level change
Sea level can change, both globally and locally due to (1) changes in the shape of the ocean basins, (2) a change in ocean volume as a result of a change in the mass of water in the ocean, and (3) changes in ocean volume as a result of changes in ocean water density. Global mean sea level change resulting from change in the mass of the ocean is called barystatic. The amount of barystatic sea level change due to the addition or removal of a mass of water is called its sea level equivalent (SLE). Sea level changes, both globally and locally, resulting from changes in water density are called steric. Density changes induced by temperature changes only are called thermosteric, while density changes induced by salinity changes are called halosteric. Barystatic and steric sea level changes do not include the effect of changes in the shape of ocean basins induced by the change in the ocean mass and its distribution. See also Relative sea level and Thermal expansion.

Sea Surface Temperature (SST)
The sea surface temperature is the subsurface bulk temperature in the top few meters of the ocean, measured by ships, buoys, and drifters. From ships, measurements of water samples in buckets were mostly switched in the 1940s to samples from engine intake water. Satellite measurements of skin temperature (uppermost layer; a fraction of a millimeter thick) in the infrared or the top centimeter or so in the microwave are also used, but must be adjusted to be compatible with the bulk temperature.

Semi-arid zone
Areas where vegetation growth is constrained by limited water availability, often with short growing seasons and high interannual variation in primary production. Annual precipitation ranges from 300 to 800 mm, depending on the occurrence of summer and winter rains.

Sensitivity
The degree to which a system or species is affected, either adversely or beneficially, by climate variability or change. The effect may be direct (e.g., a change in crop yield in response to a change in the mean, range,

[14] This definition builds from the definition used in Arctic Council (2013).
[15] This definition builds from the definitions used in Rosa (1998) and Rosa (2003).

or variability of temperature) or indirect (e.g., damages caused by an increase in the frequency of coastal flooding due to sea level rise).

Significant wave height
The average trough-to-crest height of the highest one-third of the wave heights (sea and swell) occurring in a particular time period.

Sink
Any process, activity, or mechanism that removes a greenhouse gas, an aerosol, or a precursor of a greenhouse gas or aerosol from the atmosphere.

Social Cost of Carbon (SCC)
The net present value of climate damages (with harmful damages expressed as a positive number) from one more tonne of carbon in the form of CO_2, conditional on a global emissions trajectory over time.

Social protection
In the context of development aid and climate policy, social protection usually describes public and private initiatives that provide income or consumption transfers to the poor, protect the vulnerable against livelihood risks, and enhance the social status and rights of the marginalized, with the overall objective of reducing the economic and social vulnerability of poor, vulnerable, and marginalized groups (Devereux and Sabates-Wheeler, 2004). In other contexts, social protection may be used synonymously with social policy and can be described as all public and private initiatives that provide access to services, such as health, education, or housing, or income and consumption transfers to people. Social protection policies protect the poor and vulnerable against livelihood risks and enhance the social status and rights of the marginalized, as well as prevent vulnerable people from falling into poverty.

Socioeconomic scenario
A scenario that describes a possible future in terms of population, gross domestic product, and other socioeconomic factors relevant to understanding the implications of climate change.

Southern Annular Mode (SAM)
The leading mode of variability of Southern Hemisphere geopotential height, which is associated with shifts in the latitude of the midlatitude jet. See SAM Index in WGI AR5 Box 2.5.

Species distribution modeling
Simulation of ecological effects of climate change. Species distribution modeling uses statistically or theoretically derived response surfaces to relate observations of species occurrence or known tolerance limits to environmental predictor variables, thereby predicting a species' range as the manifestation of habitat characteristics that limit or support its presence at a particular location. Species distribution models are also referred to as environmental niche models. Bioclimate envelope models can be considered as a subset of species distribution models that predict species occurrence or habitat suitability based on climatic variables only.

SRES scenarios
SRES scenarios are emission scenarios developed by Nakićenović and Swart (2000) and used, among others, as a basis for some of the climate projections shown in Chapters 9 to 11 of IPCC (2001) and Chapters 10 and 11 of IPCC (2007). The following terms are relevant for a better understanding of the structure and use of the set of SRES scenarios:

Scenario family Scenarios that have a similar demographic, societal, economic, and technical change storyline. Four scenario families comprise the SRES scenario set: A1, A2, B1, and B2.

Illustrative scenario A scenario that is illustrative for each of the six scenario groups reflected in the Summary for Policymakers of Nakićenović and Swart (2000). They include four revised marker scenarios for the scenario groups A1B, A2, B1, and B2, and two additional scenarios for the A1FI and A1T groups. All scenario groups are equally sound.

Marker scenario A scenario that was originally posted in draft form on the SRES web site to represent a given scenario family. The choice of markers was based on which of the initial quantifications best reflected the storyline, and the features of specific models. Markers are no more likely than other scenarios, but are considered by the SRES writing team as illustrative of a particular storyline. They are included in revised form in Nakićenović and Swart (2000). These scenarios received the closest scrutiny of the entire writing team and via the SRES open process. Scenarios were also selected to illustrate the other two scenario groups.

Storyline A narrative description of a scenario (or family of scenarios), highlighting the main scenario characteristics, relationships between key driving forces, and the dynamics of their evolution.

Storm surge
The temporary increase, at a particular locality, in the height of the sea due to extreme meteorological conditions (low atmospheric pressure and/or strong winds). The storm surge is defined as being the excess above the level expected from the tidal variation alone at that time and place.

Storm tracks
Originally, a term referring to the tracks of individual cyclonic weather systems, but now often generalized to refer to the main regions where the tracks of extratropical disturbances occur as sequences of low (cyclonic) and high (anticyclonic) pressure systems.

Stratosphere
The highly stratified region of the atmosphere above the troposphere extending from about 10 km (ranging from 9 km at high latitudes to 16 km in the tropics on average) to about 50 km altitude.

Stressors
Events and trends, often not climate-related, that have an important effect on the system exposed and can increase vulnerability to climate-related risk.

Subsistence agriculture
Farming and associated activities that together form a livelihood strategy in which most output is consumed directly but some may be sold at market. Subsistence agriculture can be one of several livelihood activities.

Surface temperature
See Global mean surface temperature, Land surface air temperature, and Sea Surface Temperature.

Sustainability
A dynamic process that guarantees the persistence of natural and human systems in an equitable manner.

Sustainable development
Development that meets the needs of the present without compromising the ability of future generations to meet their own needs (WCED, 1987).

Thermal expansion
In connection with sea level, this refers to the increase in volume (and decrease in density) that results from warming water. A warming of the ocean leads to an expansion of the ocean volume and hence an increase in sea level. See also Sea level change.

Thermocline
The layer of maximum vertical temperature gradient in the ocean, lying between the surface ocean and the abyssal ocean. In subtropical regions, its source waters are typically surface waters at higher latitudes that have subducted and moved equatorward. At high latitudes, it is sometimes absent, replaced by a halocline, which is a layer of maximum vertical salinity gradient.

Thermohaline circulation (THC)
Large-scale circulation in the ocean that transforms low-density upper ocean waters to higher-density intermediate and deep waters and returns those waters back to the upper ocean. The circulation is asymmetric, with conversion to dense waters in restricted regions at high latitudes and the return to the surface involving slow upwelling and diffusive processes over much larger geographic regions. The THC is driven by high densities at or near the surface, caused by cold temperatures and/or high salinities, but despite its suggestive though common name, is also driven by mechanical forces such as wind and tides. Frequently, the name THC has been used synonymously with Meridional Overturning Circulation. See also Meridional Overturning Circulation.

Tipping point
A level of change in system properties beyond which a system reorganizes, often abruptly, and does not return to the initial state even if the drivers of the change are abated.[16]

Traditional knowledge
The knowledge, innovations, and practices of both indigenous and local communities around the world that are deeply grounded in history and experience. Traditional knowledge is dynamic and adapts to cultural and environmental change, and also incorporates other forms of knowledge and viewpoints. Traditional knowledge is generally transmitted orally from generation to generation. It is often used as a synonym for indigenous knowledge, local knowledge, or traditional ecological knowledge.

Transformation
A change in the fundamental attributes of natural and human systems.

Tree line
The upper limit of tree growth in mountains or at high latitudes. It is more elevated or more poleward than the forest line.

Tropical cyclone
A strong, cyclonic-scale disturbance that originates over tropical oceans. Distinguished from weaker systems (often named tropical disturbances or depressions) by exceeding a threshold wind speed. A tropical storm is a tropical cyclone with 1-minute average surface winds between 18 and 32 m s^{-1}. Beyond 32 m s^{-1}, a tropical cyclone is called a hurricane, typhoon, or cyclone, depending on geographic location.

Troposphere
The lowest part of the atmosphere, from the surface to about 10 km in altitude at mid-latitudes (ranging from 9 km at high latitudes to 16 km in the tropics on average), where clouds and weather phenomena occur. In the troposphere, temperatures generally decrease with height. See also Stratosphere.

Tsunami
A wave, or train of waves, produced by a disturbance such as a submarine earthquake displacing the sea floor, a landslide, a volcanic eruption, or an asteroid impact.

Tundra
A treeless biome characteristic of polar and alpine regions.

Uncertainty
A state of incomplete knowledge that can result from a lack of information or from disagreement about what is known or even knowable. It may have many types of sources, from imprecision in the data to ambiguously defined concepts or terminology, or uncertain projections of human behavior. Uncertainty can therefore be represented by quantitative measures (e.g., a probability density function) or by qualitative statements (e.g., reflecting the judgment of a team of experts) (see Moss and Schneider, 2000; Manning et al., 2004; Mastrandrea et al., 2010). See also Confidence and Likelihood.

United Nations Framework Convention on Climate Change (UNFCCC)
The Convention was adopted on 9 May 1992 in New York and signed at the 1992 Earth Summit in Rio de Janeiro by more than 150 countries and the European Community. Its ultimate objective is the "stabilization of greenhouse gas concentrations in the atmosphere at a level that would prevent dangerous anthropogenic interference with the climate system." It contains commitments for all Parties. Under the Convention, Parties included in Annex I (all OECD countries and countries with economies in transition) aim to return greenhouse gas emissions not controlled by the Montreal Protocol to 1990 levels by the year 2000. The convention entered in force in March 1994. In 1997, the UNFCCC adopted the Kyoto Protocol.

[16] The glossary for the Working Group I contribution to the Fifth Assessment Report defines tipping point in the context of climate: "In climate, a hypothesized critical threshold when global or regional climate changes from one stable state to another stable state. The tipping point event may be irreversible."

Uptake
The addition of a substance of concern to a reservoir. The uptake of carbon containing substances, in particular carbon dioxide, is often called (carbon) sequestration.

Upwelling region
A region of an ocean where cold, typically nutrient-rich waters well up from the deep ocean.

Urban heat island
The relative warmth of a city compared with surrounding rural areas, associated with changes in runoff, effects on heat retention, and changes in surface albedo.

Volatile Organic Compounds (VOCs)
Important class of organic chemical air pollutants that are volatile at ambient air conditions. Other terms used to represent VOCs are *hydrocarbons* (HCs), *reactive organic gases* (ROGs), and *non-methane volatile organic compounds* (NMVOCs). NMVOCs are major contributors (together with NO_X and CO) to the formation of photochemical oxidants such as ozone.

Vulnerability[17]
The propensity or predisposition to be adversely affected. Vulnerability encompasses a variety of concepts and elements including sensitivity or susceptibility to harm and lack of capacity to cope and adapt. See also Contextual vulnerability and Outcome vulnerability.

Vulnerability index
A metric characterizing the vulnerability of a system. A climate vulnerability index is typically derived by combining, with or without weighting, several indicators assumed to represent vulnerability.

Water cycle
See Hydrological cycle.

Water-use efficiency
Carbon gain by photosynthesis per unit of water lost by evapotranspiration. It can be expressed on a short-term basis as the ratio of photosynthetic carbon gain per unit transpirational water loss, or on a seasonal basis as the ratio of net primary production or agricultural yield to the amount of water used.

References

AMS, 2000: *AMS Glossary of Meteorology, Second Edition* [Glickman, T.S. (ed.)]. American Meteorological Society (AMS), Boston, MA, USA, http://glossary.ametsoc.org/?s=A&p=1.

Arctic Council, 2013: Glossary of terms. In: *Arctic Resilience Interim Report 2013*. Stockholm Environment Institute (SEI) and Stockholm Resilience Centre, Stockholm, Sweden, p. viii.

CBD, 2000: *COP 5 Decision V/6: Ecosystem Approach.* Fifth Ordinary Meeting of the Conference of the Parties to the Convention on Biological Diversity, 15 - 26 May 2000, Nairobi, Kenya, Secretariat of the Convention on Biological Diversity (CBD), Montreal, QC, Canada, www.cbd.int/decision/cop/?id=7148.

CBD, 2002: *Decision VI/23: Alien Species that Threaten Ecosystems, Habitats or Species.* Sixth Ordinary Meeting of the Conference of the Parties to the Convention on Biological Diversity, 7 - 19 April 2002, The Hague, Netherlands, Secretariat of the Convention on Biological Diversity (CBD), Montreal, QC, Canada, www.cbd.int/decision/cop/?id=7197.

CBD, 2009: *Connecting Biodiversity and Climate Change Mitigation and Adaptation: Report of the Second Ad Hoc Technical Expert Group on Biodiversity and Climate Change.* Technical Series No. 41, Secretariat of the Convention on Biological Diversity (CBD), Montreal, QC, Canada, 126 pp.

Cobo, J.R.M., 1987: *Study of the Problem of Discrimination Against Indigenous Populations. Volume 5: Conclusions, Proposals and Recommendations*. Sub-commission on Prevention of Discrimination and Protection of Minorities, United Nations, New York, NY, USA, 46 pp.

Devereux, S. and R. Sabates-Wheeler, 2004: *Transformative Social Protection.* IDS Working Paper 232, Institute of Development Studies (IDS), University of Sussex, Brighton, UK, 30 pp.

FAO, 2000: *State of Food Insecurity in the World 2000.* Food and Agricultural Organization of the United Nations (FAO), Rome, Italy, 31 pp.

Hegerl, G.C., O. Hoegh-Guldberg, G. Casassa, M.P. Hoerling, R.S. Kovats, C. Parmesan, D.W. Pierce, and P.A. Stott, 2010: Good practice guidance paper on detection and attribution related to anthropogenic climate change. In: *Meeting Report of the Intergovernmental Panel on Climate Change Expert Meeting on Detection and Attribution of Anthropogenic Climate Change* [Stocker, T.F., C.B. Field, D. Qin, V. Barros, G.-K. Plattner, M. Tignor, P.M. Midgley, and K.L. Ebi (eds.)]. IPCC Working Group I Technical Support Unit, University of Bern, Bern, Switzerland, 8 pp.

Heywood, V.H. (ed.), 1995: *The Global Biodiversity Assessment.* United Nations Environment Programme (UNEP), Cambridge University Press, Cambridge, UK, 1152 pp.

IPCC, 1992: *Climate Change 1992: The Supplementary Report to the IPCC Scientific Assessment* [Houghton, J.T., B.A. Callander, and S.K. Varney (eds.)]. Cambridge University Press, Cambridge, UK and New York, NY, USA, 116 pp.

IPCC, 1996: *Climate Change 1995: The Science of Climate Change. Contribution of Working Group I to the Second Assessment Report of the Intergovernmental Panel on Climate Change* [Houghton, J.T., L.G. Meira Filho, B.A. Callander, N. Harris, A. Kattenberg, and K. Maskell (eds.)]. Cambridge University Press, Cambridge, UK and New York, NY, USA, 572 pp.

IPCC, 2000: *Land Use, Land-Use Change, and Forestry. Special Report of the Intergovernmental Panel on Climate Change* [Watson, R.T., I.R. Noble, B. Bolin, N.H. Ravindranath, D.J. Verardo, and D.J. Dokken (eds.)]. Cambridge University Press, Cambridge, UK and New York, NY, USA, 377 pp.

IPCC, 2001: *Climate Change 2001: The Scientific Basis. Contribution of Working Group I to the Third Assessment Report of the Intergovernmental Panel on Climate Change* [Houghton, J.T., Y. Ding, D.J. Griggs, M. Noguer, P.J. van der Linden, X. Dai, K. Maskell, and C.A. Johnson (eds.)]. Cambridge University Press, Cambridge, UK and New York, NY, USA, 881 pp.

IPCC, 2003: *Definitions and Methodological Options to Inventory Emissions from Direct Human-induced Degradation of Forests and Devegetation of Other Vegetation Types* [Penman, J., M. Gytarsky, T. Hiraishi, T. Krug, D. Kruger, R. Pipatti, L. Buendia, K. Miwa, T. Ngara, K. Tanabe, and F. Wagner (eds.)]. The Institute for Global Environmental Strategies (IGES), Hayama, Japan, 32 pp.

IPCC, 2007: *Climate Change 2007: The Physical Science Basis. Contribution of Working Group I to the Fourth Assessment Report of the Intergovernmental Panel on Climate Change* [Solomon, S., D. Qin, M. Manning, M. Marquis, K. Averyt, M.M.B. Tignor, H.L. Miller Jr., and Z. Chen (eds)]. Cambridge University Press, Cambridge, UK and New York, NY, USA, 996 pp.

IPCC, 2011: *Workshop Report of the Intergovernmental Panel on Climate Change Workshop on Impacts of Ocean Acidification on Marine Biology and Ecosystems* [Field, C.B., V. Barros, T.F. Stocker, D. Qin, K.J. Mach, G.-K. Plattner, M.D. Mastrandrea, M. Tignor, and K.L. Ebi (eds.)]. IPCC Working Group II Technical Support Unit, Carnegie Institution, Stanford, CA, USA, 164 pp.

IPCC, 2012a: *Managing the Risks of Extreme Events and Disasters to Advance Climate Change Adaptation. A Special Report of Working Groups I and II of the Intergovernmental Panel on Climate Change* [Field, C.B., V. Barros, T.F. Stocker, D. Qin, D.J. Dokken, K.L. Ebi, M.D. Mastrandrea, K.J. Mach, G.-K.

[17] Reflecting progress in science, this glossary entry differs in breadth and focus from the entry used in the Fourth Assessment Report and other IPCC reports.

Plattner, S.K. Allen, M. Tignor, and P.M. Midgley (eds.)]. Cambridge University Press, Cambridge, UK, and New York, NY, USA, 582 pp.

IPCC, 2012b: *Meeting Report of the Intergovernmental Panel on Climate Change Expert Meeting on Geoengineering* [Edenhofer, O., R. Pichs-Madruga, Y. Sokona, C. Field, V. Barros, T.F. Stocker, Q. Dahe, J. Minx, K. Mach, G.-K. Plattner, S. Schlömer, G. Hansen, and M. Mastrandrea (eds.)]. IPCC Working Group III Technical Support Unit, Potsdam Institute for Climate Impact Research, Potsdam, Germany, 99 pp.

IUCN, 2000: *IUCN Guidelines for the Prevention of Biodiversity Loss Caused by Alien Invasive Species*. Prepared by the Species Survival Commission, Invasive Species Specialist Group, International Union for Conservation of Nature (IUCN), Approved by the 51st Meeting of the IUCN Council, Gland, Switzerland, 24 pp., https://portals.iucn.org/library/efiles/documents/Rep-2000-052.pdf.

Jagers, S.C. and J. Stripple, 2003: Climate governance beyond the state. *Global Governance*, **9**, 385-399.

Kelly, P.M. and W.N. Adger, 2000: Theory and practice in assessing vulnerability to climate change and facilitating adaptation. *Climatic Change*, **47**, 325-352.

Manning, M.R., M. Petit, D. Easterling, J. Murphy, A. Patwardhan, H.-H. Rogner, R. Swart, and G. Yohe (eds.), 2004: *IPCC Workshop on Describing Scientific Uncertainties in Climate Change to Support Analysis of Risk of Options*. Workshop Report, Intergovernmental Panel on Climate Change (IPCC), Geneva, Switzerland, 138 pp.

Mastrandrea, M.D., C.B. Field, T.F. Stocker, O. Edenhofer, K.L. Ebi, D.J. Frame, H. Held, E. Kriegler, K.J. Mach, P.R. Matschoss, G.-K. Plattner, G.W. Yohe, and F.W. Zwiers, 2010: *Guidance Note for Lead Authors of the IPCC Fifth Assessment Report on Consistent Treatment of Uncertainties*. Intergovernmental Panel on Climate Change (IPCC), Published online at: www.ipcc-wg2.gov/meetings/CGCs/index.html#UR.

Matsuyama, K., 2008: Poverty Traps. In: *The New Palgrave Dictionary of Economics, 2nd Edition* [Blume, L. and S. Durlauf (eds.)]. Palgrave Macmillan, New York, NY, USA, www.dictionaryofeconomics.com/article?id=pde2008_P000332.

MEA, 2005: Appendix D: Glossary. In: *Ecosystems and Human Well-being: Current States and Trends. Findings of the Condition and Trends Working Group, Vol. 1* [Hassan, R., R. Scholes, and N. Ash (eds.)]. Millennium Ecosystem Assessment (MEA), Island Press, Washington, DC, USA, pp. 893-900.

Moss, R. and S. Schneider, 2000: Uncertainties in the IPCC TAR: recommendations to lead authors for more consistent assessment and reporting. In: *IPCC Supporting Material: Guidance Papers on Cross Cutting Issues in the Third Assessment Report of the IPCC* [Pachauri, R., T. Taniguchi, and K. Tanaka (eds.)]. Intergovernmental Panel on Climate Change (IPCC), Geneva, Switzerland, pp. 33-51.

Moss, R., M. Babiker, S. Brinkman, E. Calvo, T. Carter, J. Edmonds, I. Elgizouli, S. Emori, L. Erda, K. Hibbard, R. Jones, M. Kainuma, J. Kelleher, J.F. Lamarque, M. Manning, B. Matthews, J. Meehl, L. Meyer, J. Mitchell, N. Nakicenovic, B. O'Neill, R. Pichs, K. Riahi, S. Rose, P. Runci, R. Stouffer, D. van Vuuren, J. Weyant, T. Wilbanks, J.-P. van Ypersele, and M. Zurek, 2008: *Towards New Scenarios for Analysis of Emissions, Climate Change, Impacts and Response Strategies*. IPCC Expert Meeting Report, 19-21 September, 2007, Noordwijkerhout, Netherlands, Intergovernmental Panel on Climate Change (IPCC), Geneva, Switzerland, 132 pp.

Moss, R., J.A. Edmonds, K.A. Hibbard, M.R. Manning, S.K. Rose, D.P. van Vuuren, T.R. Carter, S. Emori, M. Kainuma, T. Kram, G.A. Meehl, J.F.B. Mitchell, N. Nakićenović, K. Riahi, S.J. Smith, R.J. Stouffer, A.M. Thomson, J.P. Weyant, and T.J. Wilbanks, 2010: The next generation of scenarios for climate change research and assessment. *Nature*, **463**, 747-756.

Nakićenović, N. and R. Swart (eds.), 2000: *Special Report on Emissions Scenarios. A Special Report of Working Group III of the Intergovernmental Panel on Climate Change*. Cambridge University Press, Cambridge, UK and New York, NY, USA, 599 pp.

O'Brien, K., S. Eriksen, L.P. Nygaard, and A. Schjolden, 2007: Why different interpretations of vulnerability matter in climate change discourses. *Climate Policy*, **7**, 7-88.

OECD, 2003: *OECD Glossary of Statistical Terms*. Organisation for Economic Co-operation and Development (OECD), Paris, France, http://stats.oecd.org/glossary/detail.asp?ID=1313.

Park, S.E., N.A. Marshall, E. Jakku, A.M. Dowd, S.M. Howden, E. Mendham, and A. Fleming, 2012: Informing adaptation responses to climate change through theories of transformation. *Global Environmental Change*, **22**, 115-126.

Rosa, E.A., 1998: Metatheoretical foundations for post-normal risk. *Journal of Risk Research*, **1(1)**, 15-44.

Rosa, E.A., 2003: The logical structure of the social amplification of risk framework (SARF): metatheoretical foundation and policy implications. In: *The Social Amplification of Risk* [Pidgeon, N., R.E. Kasperson, and P. Slovic (eds.)]. Cambridge University Press, Cambridge, UK, pp. 47-79.

Secretariat of the Stockholm Convention, 2001: *The Stockholm Convention on Persistent Organic Pollutants (as amended in 2009)*. Secretariat of the Stockholm Convention, Châtelaine, Switzerland, 63 pp.

UNCCD, 1994: *Article 1: Use of Terms*. United Nations Convention to Combat Desertification (UNCCD), Paris, France, www.unccd.int/en/about-the-convention/Pages/Text-Part-I.aspx.

UNISDR, 2009: *2009 UNISDR Terminology on Disaster Risk Reduction*. United Nations International Strategy for Disaster Reduction (UNISDR), United Nations, Geneva, Switzerland, 30 pp.

WCED, 1987: *Our Common Future*. World Commission on Environment and Development (WCED), Oxford University Press, Oxford, UK, 300 pp.